BIODIVERSITY ASSESSMENT

REVIEW

A guide to good practice

London : HMSO

ISBN 0117530700

Citation: This work should be referred to as :
Jermy, Long, Sands, Stork, Winser (Eds.) (1995).
Biodiversity assessment: a guide to good practice.
Department of the Environment/HMSO, London.

Printed on recycled paper: Reprise Matt

CONTENTS

Volume One

U.K. Resources as a Contribution to Global Biodiversity Assessment

Editors' Introduction

Acknowledgements and photo-credits

Chapter One Introduction to biodiversity
Nigel Stork

Chapter Two Legal and ethical aspects of biodiversity assessment: The Convention on Biological Diversity
Clive Jermy

Chapter Three Inventories, monitoring and management: International approaches
Harriet Eeley and Nigel Stork

Chapter Four The contribution of UK research institutes to biodiversity assessment
Shane Winser

Chapter Five Biodiversity research and training in UK universities and learned societies
David Rae

Chapter Six Specimen data banks for biodiversity studies
Clive Jermy and Martin Sands

Chapter Seven Other information resources: Libraries, maps and databases
Shane Winser

Appendix 1 Universities and centres of further education in the UK

Appendix 2 Research institutes, museums, botanical gardens, zoos and other useful addresses

Appendix 3 Offices of the British Council throughout the world

Appendix 4 Glossary of acronyms

Introduction

At the 1992 UN Conference on Environment and Development (the 'Rio Earth Summit'), 157 countries, including the United Kingdom, signed the Convention on Biological Diversity. The Convention provides a legal framework for concerted action to conserve and use sustainably the world's flora and fauna and share in a fair and equitable way the benefits arising from its utilisation. It also calls upon countries to develop national strategies for achieving these goals. As in many other developed countries, British biologists worked with government officers and others involved in environmental planning to prepare a comprehensive UK strategy, subsequently published as *Biodiversity: The UK Action Plan* (Department of the Environment, 1994).

The Government is aware that in its own research institutes, the many universities and in organisations such as provincial museums and botanic and zoological gardens, knowledge and information on the world's wider biodiversity which had been built up over many years, is continually being augumented. In 1993 it launched a unique venture: *The Darwin Initiative for the Survival of Species*, to harness Britain's skills and expertise in the fields of biodiversity. It provided substantial funding for British organisations to work with those in countries rich in biodiversity but poor in financial resources. To date 116 projects have been funded (see Box 5.27 in Chapter Five).

The purpose of this manual is to help those who have the responsibility to survey and assess the biodiversity of their own countries. It is not intended to be definitive in the techniques presented, and reference is frequently made to more comprehensive works should the user want to pursue them. It is particularly directed to those field workers who want to put their own work into perspective. In Volume One, Chapter One gives the background to biodiversity studies. Chapter Two discusses ownership in relation to the Convention on Biological Diversity, and its implications for scientists of one nation who are studying or assessing, and possibly using, the biodiversity of another nation. Chapter Three gives an outline of international initiatives so that those beginning such surveys can see that they are not working in isolation and can see their own programme in a wider context.

The second approach is seen in Chapters Four to Seven where an attempt has been made to show the resources available in the UK, in terms of preserved specimen collections, living gene banks, bibliographical and archival resources and technical knowledge which can help those making biodiversity assessments abroad. In Chapter Four research funded directly by government or through Ministries and Research Councils is discussed. We have included a number of leading international organisations that have their headquarters or origins in the UK.

Chapter Five is a review of academic research on biodiversity (which is becoming increasingly applied in its approach) seen in the universities. The full extent of this research is difficult to assess but we hope we have shown something of the range and expertise available. The potential for training and sharing technical skills through specialised courses, and through working with UK-based scientists on joint projects, or for higher degrees, is also discussed. The input to such studies by specialist Societies and other non-government organisations (NGOs) is shown to be significant in the UK and demonstrates the role of those who want to share their knowledge of, and enthusiasm for, particular groups of organisms.

Chapter Six discusses the vast database of knowledge that exists in the natural history collections in national and provincial museums and other depositories built up over a span of 200 or more years. It highlights the importance of living collections in biodiversity assessment and extensive holdings in genebanks of useful plants and domestic animals, as well as the significant conservation role and potential for re-introducing endangered plants and animals seen in botanic gardens and zoos. Chapter Seven gives information on library resources and the large arena of electronic databasing, including the internationally funded World Conservation Monitoring Centre, to demonstrate the potential for information retrieval as well as storage.

Volumes Two and Three are practical manuals and produced in a smaller (A5) format for field use. They review in detail the methods by which plants and fungi (Volume Two) and animals (Volume Three) can be surveyed, collected and preserved as voucher or research specimens. The first chapter in Volume Two discusses making inventories at different levels and for different purposes.

We believe that these books will have a readership within the UK, namely, those scientists, both established and as students, who wish to cooperate with overseas counterparts, and contribute to the concerted action encouraged by the Biodiversity Convention. This includes those two to three hundred 'expeditions' that leave the UK each year, all with the approval or the invitation of the host country, and many with the official approval of the Royal Geographical Society to carry out one or another biodiversity study abroad. We hope it will put their work into perspective, improve standards and establish more fruitful links, whereby students from both countries can learn from each others' knowledge.

In drawing together information for this book, the RGS has relied heavily on the cooperation of the three 'biodiversity' institutes (The Natural History Museum, the Royal Botanic Garden at Edinburgh, and the Royal Botanic Gardens, Kew) who have guided the preparation of this manual through their representation on the Advisory Editorial Board, and whose staff have written, or commented on the text. Many people at universities, research institutes, museums, botanic gardens, zoos, scientific societies and professional organisations freely contributed material for boxed texts, tables etc. The RGS is grateful for this cooperation and hopes that in editing we have not misconstrued the information given to us. Furthermore, in some instances, due to lack of space, texts of early drafts have had to be severely cut. If any errors appear in the final text they will be ours and not those of our contributors.

Acknowledgments

We would like to thank the members who served with us at various times on the Advisory Editorial Board: Ian Haines (ODA), Valerie Richardson (DoE) and Nigel Winser (RGS).

We are grateful to the following for writing chapter texts, boxed items or for reviewing, commenting on or correcting text, or for help in other ways:

L. Alderson, L. Ambridge, J. Ambrose, M. Ambrose, J. Anderson, M. Andrews, M. Angel, N. Arnold, P. Atkinson, R. Atkinson, M. Austin, A. Baker, J-A. Baker, M. Balick, R. Banks, M. Beasley, P. Barnard, A. Barnett, R. Barnett, J. Barrett, M. Barrett, I. Bishop, G. Bloemers, J. Bodrell, B. Bolton, G. Boxshall, R. Bray, M. Brendell, D. Bridson, D. Brooks, A. Broome, L. Brown, H. Buckley, D. Burt, S. Cable, G. Caldicott, P. Cannon, D. Chamberlain, M. Chase, M. Cheek, J. Cherfas, H. Chesney, J. Chimonides, M. Claridge, P. Clark, J. Coll, M. Collins, J. Cooke, B. Coppins, P. Cornelius, J. Cory, Q. Cronk, J. Croxall, R. Cubey, J. Cuff, C. Culver, C. Curds, D. Cutler, J. Davies, P. Davies, P. Davis, S. Davis, J. Day, J. Dickson, J. Dodd, N. Donlon, G. Douglas, M. Dunbar, J. Edmundson, J. Edwards, S. Edwards, H. Eeley, G. Ellis, L. Ellis, P. Ellis, S. Everett, J. Fa, T. Ferrero, M. Fitton, S. FitzGerald, S. Flint, D. Frodin, E. Fryd, C. Furk, T. Furnass, D. Galloway, A. Gammell, N. Garwood, I. Gauld, M. Gee, D. George, M.Gibby, D. Gibson, A. Gill, H. Gillett, M. Gillman, C. Gokce, P. Goriup, M. Gosling, F. Greenaway, N. Gregory, R. Grimble, K. Grose, L. Guarino, P.Hackney, M. Haggis, G. Halliday, P. Hammond, G. Hancock, P. Harding, E. Harris, J. Harrison, J. Hawkes, D. Hawksworth, T. Heilbronn, T. Henson, F. Herbert, D. Hollis, J. Holloway, F. Howie, C. Humphries, R. Huys, T. Irwin, S. James, D. Janzen, P. Jenkins, D. John, E. Jones, K. Joysey, S. Jury, C. Keates, M. Kelly-Borges, I. Kitching, S. Knapp, K. Kumari, S. Lambing, J. Lambshead, C. Langley, R. Lankaster, E. Leadley, G. Legg, R. Lincoln, S. Linington, S. Loat, M. Lock, W. Loder, V. Lutman, M. McBride, G. Mace, C. McCarthy, H. McConville, J. MacDonald, G. MacKinnon, P. Maplestone, J. Marsden, M. Massey, M. Maunder, N. McGough, N. McWilliam, D. Mann, A. Margetts, N. Maxted, N. Merrett, C. Milner, J. Moore, P. Mordan, P. Morgan, S. Mosedale, A. Muir, J. Noyes, L. Nunn, G. Oliver, P. Olney, R. Ormond, M. OíShea, J. Palmer, T. Parmenter, G. Paterson, G. Patisson, A. Paul, D. Pegler, M. Penn, S. Perrin, C. Pettitt, K. Pipe-Wolferstan, R. Prys-Jones, A. Pickering, R. Potter, J. Press, F. Priestly, W. Purvis, S. Pusinelli, P. Raines, J. Ratter, D. Reid, T. Rice, N. Riddiford, M. Riley, T. Riordan, J. Robertson-Vernhes, E. Robin, G. Robinson,

E. Rodgers, J. Rodwell, B. Rosen, R. Rowe, P. Ryan, P. Sanders, B. Schrire, M. Schultz, M. Scoble, S. Seal, M. Shaw, K. Shawe, K. Sherman, D. Shiel, D. Siebert, K. Singh, D. Skibinski, M. Simmonds, R.D. Smith, R.O. Smith, H. Stanley, M. Stevenson, P. Surrey, D. Sutton, S. Sutton, A. Tatham, J. Taylor, R. Teeuw, L. Thompson, D. Timms, D. Turner, M. Turner, J. Upton, R. Vane-Wright, C. Walker, M. Walkey, K. Walter, F. Wanless, R. Warwick, P. Waterman, R. Watling, A. Watt, E. Watt, S. Webster, G. Whalley, M. Wilkinson, T. Wilkinson, M. Williams, P. Williams, B. Wilson, C. Wood, P. Wyse-Jackson and G.Young.

We are grateful to the following for allowing us to reproduce their photographs:

M. Ambrose, G. Argent, E.K. Balls, J. Banks, G. Beccaloni, BirdLife International, S. Blackmore, The Body Shop, M.J.D. Brendell, British Antarctic Survey, British Council Press Office, British Library, CAB International, Cambridge University Dept. of Aerial Photography, P.A. Chapman, M. Cheek, M.F. Claridge, P. Clark, B. Clayton-Jolly, N.M. Collins, Coral Cay Conservation, B. Critchley, J. Davies, K.L. Denham, R. Didham, J. Dransfield, Dunstaffnage Marine Laboratory, Edinburgh Zoo, J. Ellaway, Forestry Commission, R.M. Fuller, N. Garwood, I.D. Gauld, J.D. George, A.C. Gill, L. Gill, M.P Gillman, Glasgow Art Gallery and Museums, G. Halliday, M. Hakansson, R. Hanbury Tenison, J.G. Hawkes, Henry Doubleday Research Association, J. Houghton, B. Ing, Institute of Arable Crop Research, Institute of Freshwater Ecology, Institute of Oceanographic Science, Institute of Terrestrial Ecology, International Institute of Entomology, International Institute of Mycology, A. Jackson, A.C. Jermy, D. John, John Innes Centre, S. Jury, J.G. Kairo, T. Kelly, G. Legg, Linnean Society of London, D. Long, C.H.C. Lyal, N. McWilliams, G. Maguire, M.E.N. Majerus, J.H. Martin, N. Maxwell, National Museum of Wales, Natural History Museum, Natural Resources Institute, National Council for the Conservation of Plants and Gardens, R.F.G. Ormond, M. O'Shea, G.L.J Paterson, G. Patterson, G. Perelló, Plymouth Marine Laboratory, R.D.Pope, H.D.V. Prendergast, W.O. Purvis, Raleigh International, A.C. Rice, R. Ridgeway, G. Robinson, Royal Botanical Garden, Edinburgh, Royal Botanical Gardens, Kew, Royal Geographical Society, Royal Society, Royal Society for the Protection of Birds, M.J.S. Sands, W.A. Sands, Scottish Crop Research Institute, D. Scott, Scottish National Heritage, S. Seal, D.J. Siebert, P.A. Sims, H. Stanley, M. Stevenson, C. Stirton, N.E. Stork, D.A. Sutton, H. Taylor, N. Taylor, D.M. Tererson, I. Tittley, M. Turner, M.W.F. Tweedie, J. Vogel, C. Walker, A. Warren, R.M.Warwick, A. Watt, Wildlife and Wetland Trust, P. Williams, World Conservation Monitoring Centre, P. York, Zoological Society of London.

The following publishers/authors are gratefully acknowledged as the source of the listed boxes: *Volume* 1, IUBS/F. di Castri (Box 1.1); The Royal Society/M. Embley *et al.* (Box 1.4); WCMC/P.M. Hammond (Boxes 1.7, 1.17); F. Fawcett & Q. Wheeler (Box 1.8); Allen Lane (Penguin Press)/ E.O. Wilson (Boxes 1.9, 1.10); Elsevier Trends Journals /F.D.M. Smith *et al.* (Box 1 11); The Geological Society/A.M.C. Sengör *et al.*/C.R. Scotese (Box 1.13); Blackwell Scientific/ Boxwood Press/V.Pearse (Box 1.14); Trustees of The Natural History Museum (Box 1.15), CAB International/ R. Booth/G. du Heaume (Box 1.18); Edward Arnold/J. Phillipson (Box 1.20); Harper Collins/ E.R. Pianka (Box 1.21); World Resources Institute (Boxes 1.24, 1.25); IUBS/F. di Castri et al. (Box 3.4); G. Sherman (Box 3.9); WCMC/R. Smith (Boxes 3.11, 3.12); BirdLife International (Box 3.13). *Volume* 2, Academic Press/K.S. Brown (Box 1.5);Board of Trustees of the Royal Botanic Gardens, Kew/ Sally E. Dawson and RGS/A.C. Jermy (Boxes 4.1, 4.2, 4.4, 4.5). *Volume* 3, Academic Press/C. R. Bibby *et al.* (Box 2.1).

We are grateful to Valerie Richardson and her colleagues at DoE for critical reviewing and proof-reading. We also appreciate the contribution of Chris Pease and Chris Phillips (DoE Information Section), and the skill and patience of Tony Johnson and his team at COI. The unfailing help of Jacquie Ujetz, and that of Deborah Boys, Ben Coles, Fay Hercod, Asmilen bin Ramlee, Alexander Townend and Alex Walters, at the RGS is acknowledged with gratitude.

A.C. *Jermy*, D. *Long*, M.J.S. *Sands*, N.E. *Stork*, S.*Winser*

Royal Geographical Society (with The Institute of British Geographers), London

Names and affiliations of chapter author/editors

Coordinating Editor A. Clive Jermy (Royal Geographical Society, Kensington Gore, London SW7 2AR).

Volume 1

Chapter 1 Nigel E. Stork (Natural History Museum, Cromwell Road, London SW7 5BD).

Chapter 2 A. Clive Jermy (Royal Geographical Society, Kensington Gore, London SW7 2AR).

Chapter 3 Harriet Eeley & Nigel E. Stork (Natural History Museum, Cromwell Road, London SW7 5BD).

Chapter 4 Shane Winser (Royal Geographical Society, Kensington Gore, London SW7 2AR).

Chapter 5 David Rae (Royal Botanic Garden, Edinburgh, Inverleith Terrace, Edinburgh EH3 5LR).

Chapter 6 A. Clive Jermy (Royal Geographical Society, Kensington Gore, London SW7 2AR) & Martin J.S. Sands (Royal Botanic Gardens, Kew, Richmond, Surrey TW9 3AB).

Chapter 7 Shane Winser (Royal Geographical Society, Kensington Gore, London SW7 2AR).

Volume 2

Chapter 1 Nigel E. Stork & Jon Davies (Natural History Museum, Cromwell Road, London SW7 5BD).

Chapter 2 Colin R. Curds & Nigel E. Stork (Natural History Museum, Cromwell Road, London SW7 5BD).

Chapter 3 Paul F. Cannon (International Mycological Institute, Egham, Surrey TW20 9TY) & Chris Walker (Forestry Commission, Northern Research Station, Roslin, Midlothian, EH25 9SY).

Chapter 4 A. Clive Jermy & Jon Davies (Natural History Museum, Cromwell Road, London SW7 5BD).

Volume 3

Chapter 1 Jon Davies & Nigel E. Stork (Natural History Museum, Cromwell Road, London SW7 5BD).

Chapter 2 Nigel E. Stork & Jon Davies (Natural History Museum, Cromwell Road, London SW7 5BD).

INTRODUCTION TO BIODIVERSITY

1.1 What is biodiversity?

1.2 The dimensions of biological diversity: How much do we know?

1.2.1 Phyletic diversity

1.2.2 Genetic diversity

1.2.3 Genetic analysis and species phylogeny

1.2.4 How many species have been described?

1.2.5 How many species are there?

1.2.6 How much do we know about organisms?

1.2.7 The evolution and extinction of life

1.3 The distribution of life

1.3.1 The geological past

1.3.2 The present day distribution of life

1.3.2.1 Marine environments

1.3.2.2 Freshwater environments

1.3.2.3 Terrestrial environments

1.4 The role of systematics in biodiversity assessment

1.4.1 What is systematics?

1.4.2 Taxonomic research

1.4.3 Taxonomic products

1.4.3.1 Descriptions

1.4.3.2 Revisionary studies

1.4.3.3 Keys

1.4.3.4 Catalogues, checklists and inventories

1.4.3.5 Faunas and Floras

1.4.3.6 Handbooks and field guides

1.5 The ecological context

1.5.1 Ecosystem structure

1.5.2 Ecosystem function

1.5.3 Human impacts on ecosystems

1.5.3.1 Habitat destruction and fragmentation

1.5.3.2 Land-use and artificial chemical additions

1.5.3.3 Over-exploitation

1.5.3.4 Global climate change

1.5.3.5 Changes to soil structure

1.5.3.6 Species introductions

1.5.3.7 Freshwater systems

1.6 The human dimension of biodiversity

1.6.1 Population rise and the use of resources

1.6.2 The economics of biodiversity

1.6.3 Biodiversity and agriculture

1.7 Summary – the need for inventorying and monitoring

1.8 References and bibliography

Chapter authorship

The chapter has been written by Nigel Stork *with additional contributions or comments by*: J. Barrett, M. Chase, C. Culver, J. Edwards, H. Eeley, I. Gauld, M. Gibby, C. Humphries, C. Jermy, K. Kumari, I. Kitching, J. Lambshead, N. Merrett, G. Paterson, J. Robertson-Vernhes, B. Rosen, M. Sands, J. Schneller, M. Scoble, D. Siebert, H. Stanley, R. Vane-Wright and J. Vogel.

1.1 What is biodiversity?

The term 'biodiversity', the contraction of biological diversity, came into common usage throughout the world, following the signing of the Biodiversity Convention at the United Nations Conference on Environment and Development, in Rio de Janeiro in 1992. Prior to that 'biodiversity' was already a rapidly developing concept in biology, having been launched by the book edited by E.O. Wilson (Wilson, 1988). But what is biodiversity? The Convention defines biodiversity as *'the variability among living organisms from all sources including terrestrial, marine and other aquatic ecosystems and ecological complexes of which they are a part: this includes diversity within species, between species and of ecosystems'*.

Box 1.1 shows just how complex the situation may be as biodiversity encompasses not just hierarchies of taxonomic and ecological scale but also temporal and other more or less independent scales (e.g. geographical scales and scaling in the body size of organisms). Scientists still know surprisingly little about many aspects of biodiversity and there are many myths associated with it (**Box 1.2**). Biodiversity may represent very different things to different people. To those working in museums and herbaria it perhaps represents a new thrust for efforts to describe the Earth's fauna and flora. To ecologists it may represent a growing concern about the balance of nature and how well ecosystems can function as biological diversity rapidly decreases.

Box 1.1 Temporal and spatial scales affecting biodiversity (Source: di Castri, 1991)

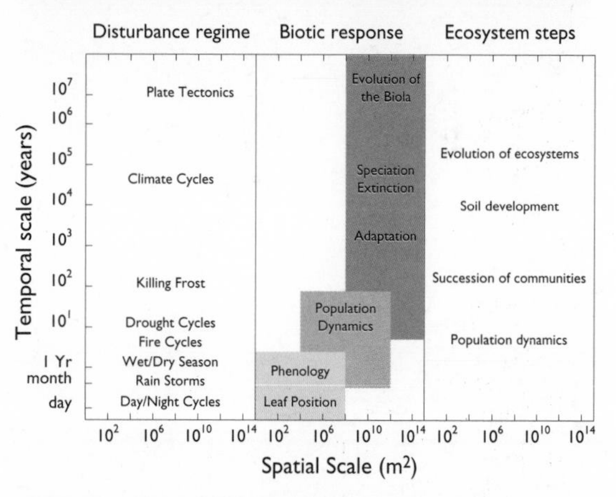

Temporal and spatial scales of disturbance regimes and biotic responses as determinants of a stepwise evolution of terrestrial ecosystem, nesting biological entities (populations and communities through their own dynamics) and abiotic components (soils, through pedogenesis).

To economists and politicians it may represent a new and largely untapped source of urgently needed income for developing nations.

Interesting and crucial scientific issues of why biodiversity is so important – such as the loss of species through habitat change and global climate change (see **Box 1.3**, page 4), and the possible effects of these species losses on ecosystem function (see Section 1.5.2) – are over-ridden by those of a socio-economic nature. The 'ownership' of biodiversity and who should pay for its conservation are emotive subjects. These, and other issues which relate to the sustainable utilisation of biological and non-biological resources as well as the maintenance of well-nurtured populations of humans throughout the world, are extremely complex. They are not dealt with in this Manual and are only discussed here in a general context. For a more detailed discussion of such problems see IUCN/UNEP/WWF (1991, 1993) and Sections 9–11 of the UNEP Global Biodiversity Assessment (UNEP, 1995).

1.2 The dimensions of biological diversity: How much do we know?

1.2.1 Phyletic diversity

In the past much attention has focused on biological variation at the species level and recent attempts to determine the diversity of life at this level are discussed below. However, diversity at the highest and lowest levels – kingdoms and genes – is surprisingly uncertain. Traditionally, two kingdoms, Animalia and Plantae, have been recognised and most organisms have been included in these. However, in the last few decades this view has been questioned by experts with other kingdoms being recognised. Whittaker (1959), for example, recognised five kingdoms: Animalia, Plantae, Fungi, Protista and Monera. More recent work (Woese *et al.*, 1990) would suggest that a new taxonomic level, the domain, is required. The two traditional kingdoms of animals and plants and several others are grouped within the domain Eucarya and the other kingdoms (mostly fungi, bacteria and their relatives) fall within two other

Box 1.2 Biodiversity myth, fact and action

Myth

- We know what biodiversity exists and what purpose it serves.
- Every species can be saved.
- Conserving biodiversity in protected areas necessarily means prohibiting resource use.
- Protected natural areas meet most conservation needs.
- Only tropical countries need to be concerned with conservation of biodiversity.

A. C. Jermy/NHM

Yorkshire Moors, UK. Reduced diversity in tree monoculture and managed heather communities for recreation.

Fact

- Most species have yet to be identified and studied. Generally accepted estimates today range from 5 to 15 million species (discounting earlier estimates of up to 100 million) while only 1.5 million have been actually described. The degree to which most species are threatened is unknown. The role of different species in maintaining natural systems is largely unknown.
- Species appear to have a finite life span. Some loss of biodiversity is inevitable. Conservation focused on individual species is often not feasible.
- Local support for conservation is critical and possible.
- Most biodiversity is found outside protected natural areas (i.e. natural, semi-natural and cultivated areas). Protected natural areas are affected by activities in surrounding areas.

A. C. Jermy/NHM

High density of farming on edge of Tjibodas National Park, Java. Incursions to collect fuel wood are not uncommon.

Action

- Minimise habitat loss and other threats until more is known.
  Improve knowledge about biodiversity, threats to biodiversity and the ecosystem role of biodiversity.
- Focus conservation efforts primarily at the ecosystem and landscape levels to minimise loss of biodiversity.
- Provide new options for sustainable use of biodiversity and other economic choices.
  Involve local people in planning and managing protected areas.
- Integrate conservation and regional development planning.
  Protect the diversity of cultivated and domesticated species and their wild relatives.
- Develop a global framework for conservation and sustainable use.

J. G. Kairo/Mombasa University

Plantation of mangrove *Rhizophora mucronata*, 8 months after planting, Gazi community, Kenya.

(*Source*: UNESCO, 1994)

Box 1.3 Why biodiversity is crucial

- Provides food, construction material, raw material for industry, medicine.
- Provides the basis for all improvements to domesticated plants and animals.
- Maintains ecosystem functions, including evolutionary processes.
- Stores and cycles nutrients essential for life, e.g. carbon, nitrogen and oxygen.
- Absorbs and breaks down pollutants, including organic wastes, pesticides, heavy metals .
- Recharges groundwater, protects catchment basins and buffers extreme water conditions.
- Produces soil and protects it from excessive erosion.

(*Source*: UNESCO, 1994)

domains, Bacteria and Archaea. This latter view is based on studies of genetic distance as measured through recent molecular analyses (see **Box 1.4**). These and other studies indicate that most genetic diversity is in the micro-organisms and that much of the genetic diversity found in vertebrates, invertebrates and plants, are common to organisms such as yeasts.

1.2.2 Genetic diversity

In the *Origin of Species*, Darwin observed that it should be remembered "that systematists are far from being pleased in finding variability in important characters . . ." and yet it is this variation, or rather its genetic component, that is the very foundation of evolution. Without genetic variation, evolution and adaptation cannot occur. This is as true for common species as it is for species on the verge of extinction. Even when a habitat has been preserved, evolution is still going on. It may be longer term adaptation to climatic and other environmental

Box 1.4 Diagram of one view of the division of kingdoms and major groups (adapted from Embley *et al.*, 1995)

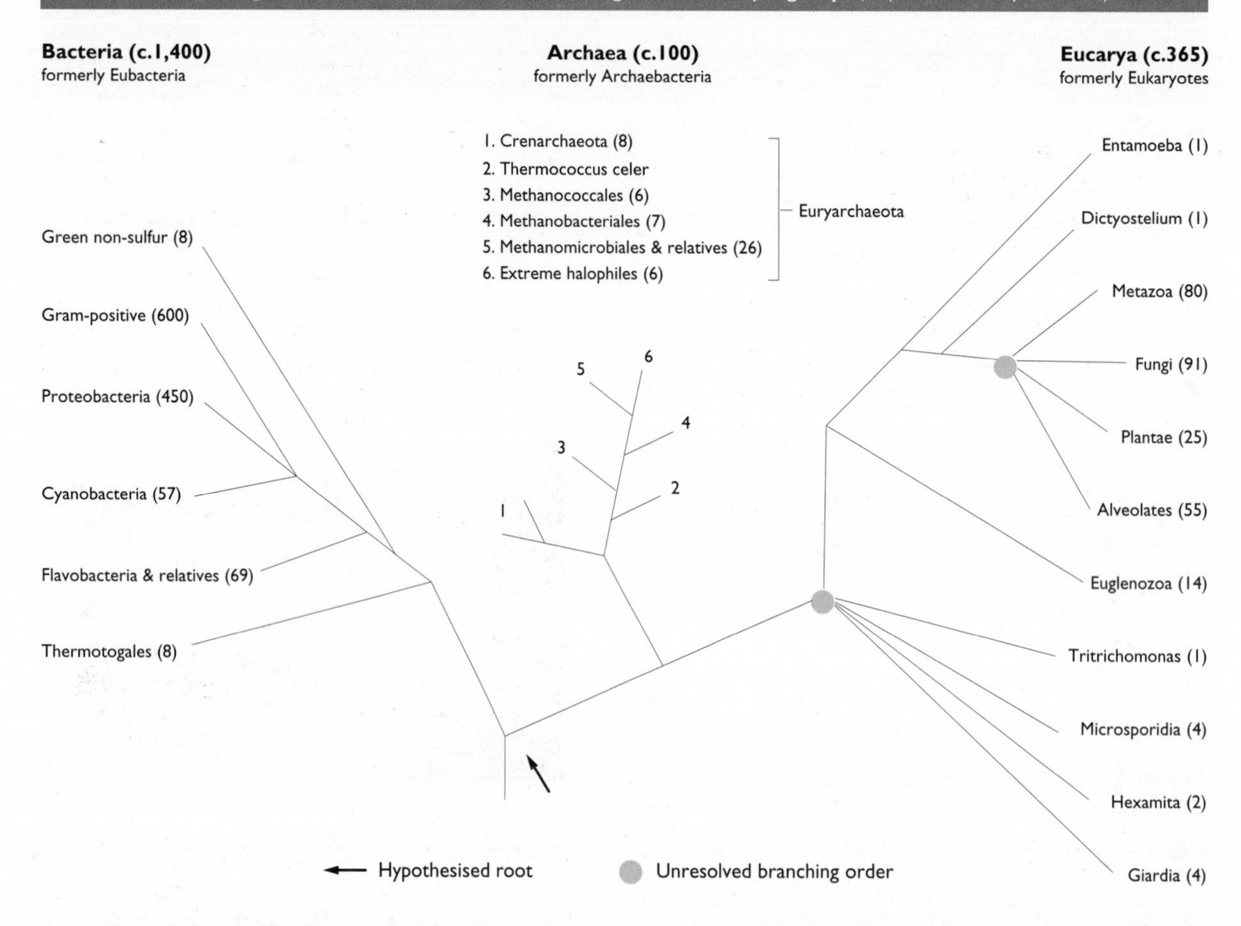

The domains of life based upon small subunit (ssu) ribosomal (r)RNA sequence comparisons. The figure is adapted from Woese *et al.* (1990) and shows only a representative set of lineages from each domain. Figures in brackets indicate the approximate number of ssu rRNA sequences for taxa. The root is currently thought to occur on the bacterial branch.

change, or it may be on a shorter time scale because even in the best preserved habitats the ecological community is a dynamic entity with species interacting with one another. For example, pathogens (and parasites) adapt to their host species and host species evolve in response to pathogens and parasites. The preservation of natural habitats may be seen as the primary objective of any conservation effort in order to maintain the species diversity. However, the longer term success of such efforts will be determined by the genetic diversity within the species as they adapt to the more subtle forces within the ecosystem.

Except in the rarest of localised species, species are divided into populations and below that into subpopulations and even smaller units. How genetic variation is distributed within and between these units can tell us a great deal about how different or similar they are. Indeed, at the systematic level, such differences are often recognised as subspecific taxa, but these differences are often based on rather simple characters that can be easily 'keyed out'. The distribution of genetic variation within and between populations can also tell us a great deal about migration between populations, past processes of colonisation and the effects of habitat disruption. How variation is distributed between individuals within a population, or sub-population can tell us about the breeding structures of the population (inbreeding/ outbreeding/ apomixis), the relative reproductive success of different individuals, patterns of pollen dispersal (and hence pollinator behaviour), patterns of recruitment, the relative importance of sexual reproduction and vegetative propagation (ramets versus genets), the effects of population bottlenecks in the past and the relatedness of individuals.

Whilst actual population size (i.e. how many individuals there are in the population) is important, it is the degree to which individuals vary in reproductive success that affects the maintenance of longterm genetic diversity. If relatively few individuals are contributing the majority of offspring, the effective population (or breeding) size can be substantially less than the actual (census) population. Furthermore, determination of genetic variation within and between populations can identify which are genetically depauperate, and which, relatively speaking, are genetically rich. This information can then be used, for example, to decide the best sources of material for reintroduction or to supplement critically small populations. At an extreme, where decisions may have to be made about which populations should be conserved, and which allowed to go extinct, an assessment of within-population genetic variation can be used to inform those decisions. At the level of enforcement of conservation legislation and conventions, identification of population or taxon, specific markers can aid in tracing the source of plants or products being traded illegally.

The types of characters that can be used for looking at diversity below the species level can be divided onto three broad groups.

- *Morphological characters*

Discrete variation in characters at this level is relatively rare. Often these characters are used as being diagnostic of taxonomic units (variatas or even subspecific rank) but much of this sort of variation may be continuously variable rather than falling into discrete classes, for example, polymorphism for flower colour or fruit shape; ecotypic variation seen in populations of the same species growing in different habitats.

- *'Biochemical' characters*

The characters analysed here are the products (usually enzymes) of individual genes directly, and are therefore less liable to environmental influence. Surveys of genetic variation by these methods are based on the assumption that the genes controlling the different proteins represent a sample of the variation at all genes. Different allelic forms of an enzyme (*allozymes*) controlled by the same gene can be separated using electrophoretic techniques (see **Box 1.5**). However, where variation does exist, these methods will usually underestimate it because not all mutational differences can be detected.

- *Characters based on the properties of the DNA itself.*
  - (*i*) *restriction fragment length polymorphism* (RFLPs);
  - (*ii*) *fingerprinting methods*;
  - (*iii*) *sequence amplification using the polymerase chain reaction* (PCR);
  - (*iv*) DNA *sequencing*.

Because these methods examine the DNA itself, the variation exposed is truly genetic. However, it is known that the different regions of DNA evolve at different rates. It is therefore important to chose DNA markers appropriate to the

Box 1.5 Isoenzyme profiles indicating relationships between hybrids in the fern genus *Asplenium*
(Source: J. Schneller, University of Zurich and J. Vogel, Natural History Museum, London.)

The isozyme profiles for TPI (triose-phosphate isomerase) on an agarose gel, representing two diploid fern taxa, *Asplenium trichomanes* subsp. *trichomanes* (genome formula TT; lanes 21–27) and A. *viride* (formula VV; lanes 17–20), the presumed tetraploid derived from them, A. *adulterinum* (formula TTVV; lanes 1–8), and a triploid backcross with A. *viride*, A. x *poscharskyanum* (formula VVT; lanes 9–16). All plants found in the same field locality in Graubunden, Switzerland. The bands in the tetraploid are additive of the diploids. Note the low dosage of genes from A. *trichomanes* seen in the hybrid.

Asplenium adulterinum

question being asked. For example, the highly variable 'microsatellite' markers are ideal for resolving relationships between individuals and comparing populations which may have only been separated by a few generations, but their high mutation rates make them appropriate for comparison between populations, sub-populations or taxa which may have been separated for a long time.

In terms of cost, the cheapest markers and least demanding on facilities, are simple morphological differences, and the most expensive, and the most demanding of specialist laboratory facilities are the DNA-based technologies. When assessing genetic variation, it is essential that the problem to be addressed is clearly formulated and a technique is chosen that is appropriate to that problem. In formulating the problem, the appropriate statistical analysis must be chosen, and the experimental design and sampling procedure required by the analysis implemented.

Molecular studies including PCR and RFLP (see p. 5) on *Pseudomonas solacearum* at the NRI showed the genetics of this widespread bacterium to be highly complex. Even the strains causing the diseased bananas shown here (above *bugtok*, from the Phillippines; below, *blood disease* from Indonesia) are genetically extremely distinct.

The long-term aim of any conservation effort must be to maintain a self-sustaining dynamic ecological community, with the minimum of human intervention. This objective cannot be attained without due regard to the genetic diversity within the member species of the community.

1.2.3 Genetic analysis and species phylogeny

Floras and faunas are the products of history and should not be viewed simply as a set(s) of taxa that exist independently of each other and

Box 1.6 Molecular evolution and genetic diversity of South American camelids

The **lamini** (South American members of the camel family) are of considerable importance to the Andean economy; yet little is known of the extent and distribution of genetic variation in these animals. The Institute of Zoology of the Zoological Society of London is studying the genetic diversity of these animals in order to elucidate their relationship to their wild ancestors (**vicugna** and **guanaco**) and substantiate, or otherwise, a genetic basis for the current subspecific classification of the wild animals. Such research on these isolated, threatened populations will identify valuable gene pools for conservation. In addition, the Institute is assessing variation among different fibre-producing breeds of domestic animals.

H. Stanley/Zoological Society of London

Domesticated llamas and alpacas in the Parque Nacional Lauca, N. Chile, below Pannacota volcano.

Both **llamas** and **alpacas** were originally domesticated at least 6,000 years ago in the high Andes. Recent DNA analysis at the Institute showed that hybridisation has occurred in the ancestry of the lamini and that defining the origins of the domesticated camelids is difficult using contemporary animals alone. This may now be possible, however, through the analysis of both wild and domesticated camelids using archaeological specimens from Peruvian sites spanning the period from 7,000 BC to the sixteenth century. The identification of camelid genotypes from teeth and bones at different time points, spanning the entire period of domestication, should provide important new information on the evolutionary relationships of these animals. In addition, the discovery of 900–1,000 year old naturally dessicated llamas and alpacas has permitted a study of native Andean breeds prior to European contact. These exceedingly well preserved specimens exhibit a uniformity of fibre colour, distribution and fineness which is absent in contemporary animals, and may represent presently unknown varieties or unhybridised animals.

This research will provide insight into the history of camelid domestication and may have direct implications for breeding and conservation policy for all four South American camelids.

(*Source*: *Helen Stanley, Institute of Zoology, London*)

the land masses they occupy. Their common and unique histories need to be studied in order to develop proper strategies for conservation as well as exploitation. What we view today as a unified and inter-dependent ecosystem may have several distinct subsets that historically had little to do with each other. If these threads of knowledge become better understood then we should be able to know where to look for close relatives and this might in turn shed further light on factors related to rarity and endemism.

In spite of the existence of several taxonomic schemes, controversy over relationships is considerable and many taxa are problematic. We have available today an unprecedented array of modern methologies that could help unravel many of the long-standing problems. This phylogenetic aspect of biodiversity studies has only recently been developed. Work in the Institute of Zoology (see Chapter Five) on seals, and South American camelids described in Box **1.6** is an example. In similar studies in progress at the Royal Botanic Gardens, Kew, it has been found that *Nesiota elliptica* (a rare flowering plant endemic to St. Helena) is closely related to *Phylica*, a genus of ericoid shrubs mainly distributed in Southern Africa. *Nesiota* and *Phylica* are not particularly similar, but, knowing that

Box 1.7 Number of described species for major groups of organisms as proportions of the global total (Source: Hammond, 1992)

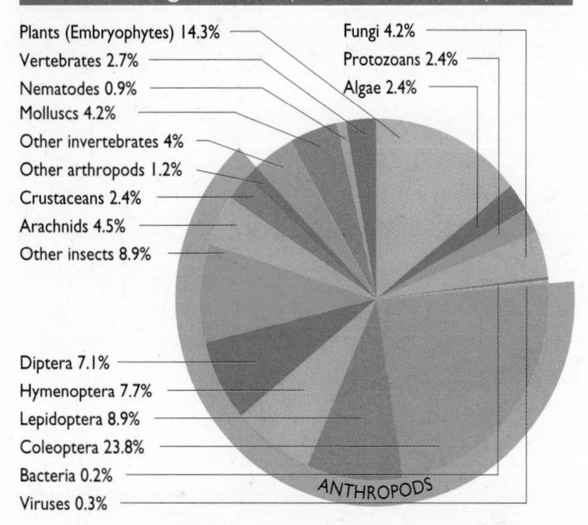

More than two thirds of all described species are insects - indeed there are almost as many named species of beetle as there are of all other insects added together, or all other non-insects (plants and animals)! In contrast, for some groups such as viruses and bacteria, normal species concepts do not apply and the numbers of named species for these groups are surprisingly low.

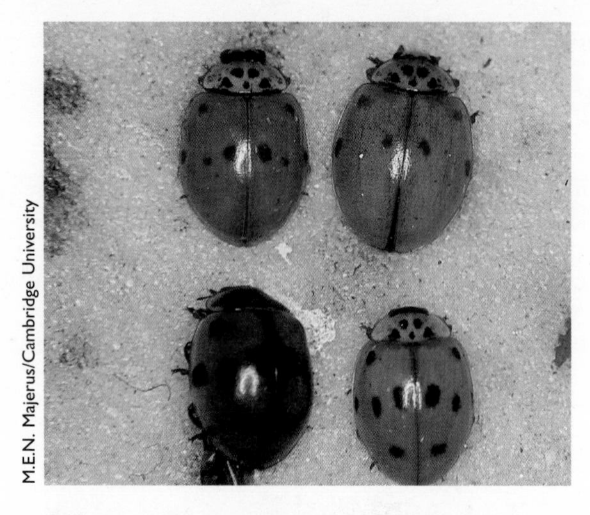

M.E.N. Majerus/Cambridge University

Different colour morphs of ten-spot ladybird (*Adalia decempunctata*).

Box 1.8 The Species Scape key

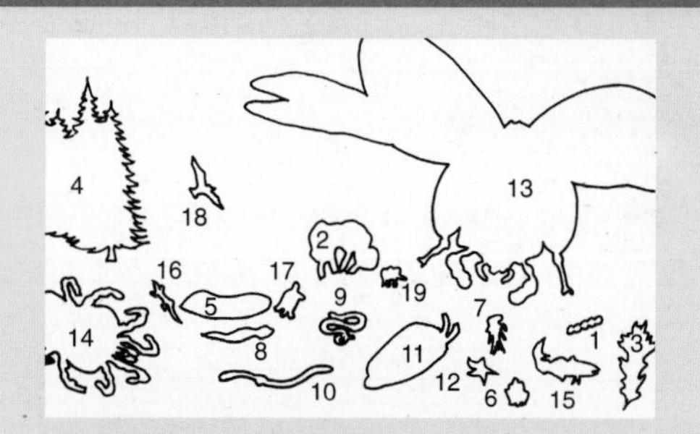

Size of individual organisms represents number of described species in major taxon. Based on Wilson (1992).

Unit Area: ☐ = approximately 1,000 described species.

Taxon

1 Monera (Bacteria, Blue-green Algea)
2 Fungi
3 Algae
4 Plantae (Multicellular Plants)
5 Protozoa
6 Porifera (Sponges)
7 Coelenterata (Jellyfish, Corals, Comb Jellies)
8 Platyhelminthes (Flatworms)
9 Nematoda (Roundworms)
10 Annelida (Earthworms etc.)
11 Mollusca (Molluscs)
12 Echinodermata (Starfish etc.)
13 Insecta
14 Non-insect Arthropoda (Mites, Spiders, Crustaceans etc.)
15 Pices (Fish)
16 Amphibia (Amphibians)
17 Reptilia (Reptiles)
18 Aves (Birds)
19 Mammalia (Mammals)

The illustration opposite is drawn Frances L. Fawcett by and reproduced with the permission of Quentin Wheeler, originally published in Wheeler, Q. D. (1990) Ann. Entomol. Soc. Amer. 83: 1031–1057.

this relationship is likely, we are forced to design follow-up studies that will look at the factors responsible for the high degree of morphological repatterning that took place within the island habitat. Furthermore, it allows us to prioritise species based on an understanding of their relative degree of genetic isolation; those without close extant relatives should be given greater emphasis than those that have closely related and widespread cogeners.

For example, the close relationship of *Phylica* and *Nesiota* makes *Nesiota* less important than *Lactoris fernandeziana* (Lactoridaceae) from the Juan Fernandez Islands (see Chapter Four, **Box 4.5**) or *Medusagyne oppositifolia* (Medusagynaceae) from the Seychelles. The techniques used in these studies include those mentioned above, principally those associated with DNA sequencing and RFLP analysis.

1.2.4 How many species have been described ?

In spite of immense efforts by 19th and 20th century taxonomists in describing the world's fauna and flora we are still uncertain as to the true dimensions of biological diversity. This understanding is hampered by the fact that even the total number of named and described species is unclear, with estimates ranging from 1.4 to 1.8 million species (see **Box 1.7**). Why is there this uncertainty?

Box 1.8 The species scape

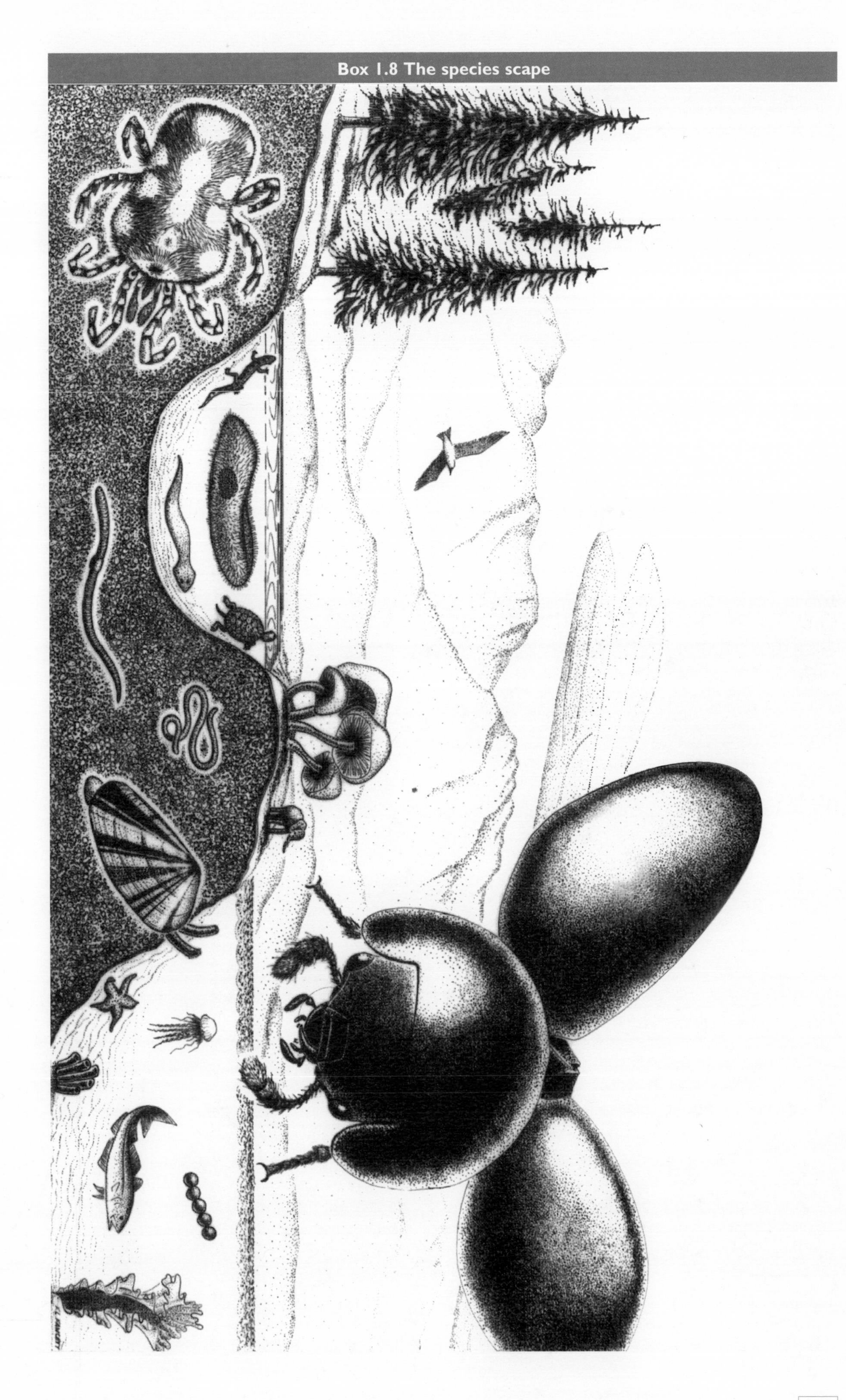

Firstly, it is often difficult to determine whether a series of individuals constitutes one species or several species, or whether a new individual is the same species as others that have been described. Secondly, species may be described more than once. A taxonomist in one part of the world may not realise that the same species has already been described from elsewhere. On some occasions species can be so variable that they are described many times. For example, the common 'ten-spot' ladybird, *Adalia decempunctata* L., has more than 40 synonyms (i.e. redundant names). This species has many colour morphs and, at various times during the last two hundred years, different taxonomists have given names to the colour morphs without realising that they were all one species. Similarly, in the mid-nineteenth century, habitat forms of flowering plants in Europe were often described as distinct species. The level of synonymy in some groups of organisms may be extremely high. For example, Gaston and Mound (1993) estimate 80% and 35% synonymy for Papilionidae and Aphididae, respectively. Finally there is no complete catalogue of names for all organisms and for many groups it is often very difficult to know what has or has not been named and described.

1.2.5 How many species are there?

In 1833 the British natural historian, John Obadiah Westwood, estimated that there might be some 20,000 species of insects worldwide. Today we recognise that there are about this number of insect species in Britain alone! Estimates of how many species there are on Earth have continued to rise and still it seems we do not know the answer to within a factor of ten. Groups such as birds, large mammals and some woody plants are well-known and we can be fairly confident about estimates of their global numbers of species, provided species concepts do not change radically in the near future (but see 1.4.2). However, the scientific rationale for almost all estimates of global species numbers for the remaining taxa is surprisingly thin. Although estimates for global numbers of species (both undescribed and described) vary from as low as 2 million to more than 100 million, much evidence appears to support estimates on the lower end of this scale: 5–15 million species (Hammond, 1992, 1994a, b; May, 1992; Stork, 1988, 1993). However, in this context there remain queries over several key groups of organisms such as nematodes, fungi and insects.

Much of the recent literature on global species estimates has focused on insects and in particular on tropical forest insects. Erwin (1982), for example, estimated that there might be 30 million species of tropical arthropods in the world. This estimate has been criticised on the basis of the assumptions made, such as the relative proportion of ground to canopy species, the relative proportion of beetle species to other groups of insects and, perhaps most important of all, the number of species specific to a given species of tree. Re-analyses with more soundly based assumptions indicate that lower estimates of around 5–10 million species may be more reasonable (Gaston, 1991; Hammond, 1990; Stork, 1993).

There has also been some debate, but perhaps less critical appraisal, as to the global dimensions for two other groups: fungi and nematodes. Hawksworth (1991, 1993) has suggested that, although only some 70,000 species of fungi have been described, there may be more than 1.5 million species yet to be discovered. Lambshead (1993) has predicted even greater numbers of species of nematodes. His argument is supported by estimates of 10 million or more species of marine nematodes and other invertebrates (Grassle, 1991; Grassle & Maciolek, 1992; but see also May, 1992).

Thus the answer to the question, 'how many species are there?' remains unresolved and this problem is likely to continue to puzzle scientists. However, even conservative estimates would indicate that at most only 20% of all species have been described.

1.2.6 How much do we know about organisms?

It may seem that we know a great deal about the biology, distribution and threatened or non-threatened status of plants and animals. In practice, this is far from the truth. For well-known faunas and floras, such as those of Britain and other areas of Europe, virtually all species have been described (but, surprisingly, not all). Even so, distribution maps for these species are often extremely poor and the data used are often based on records more than

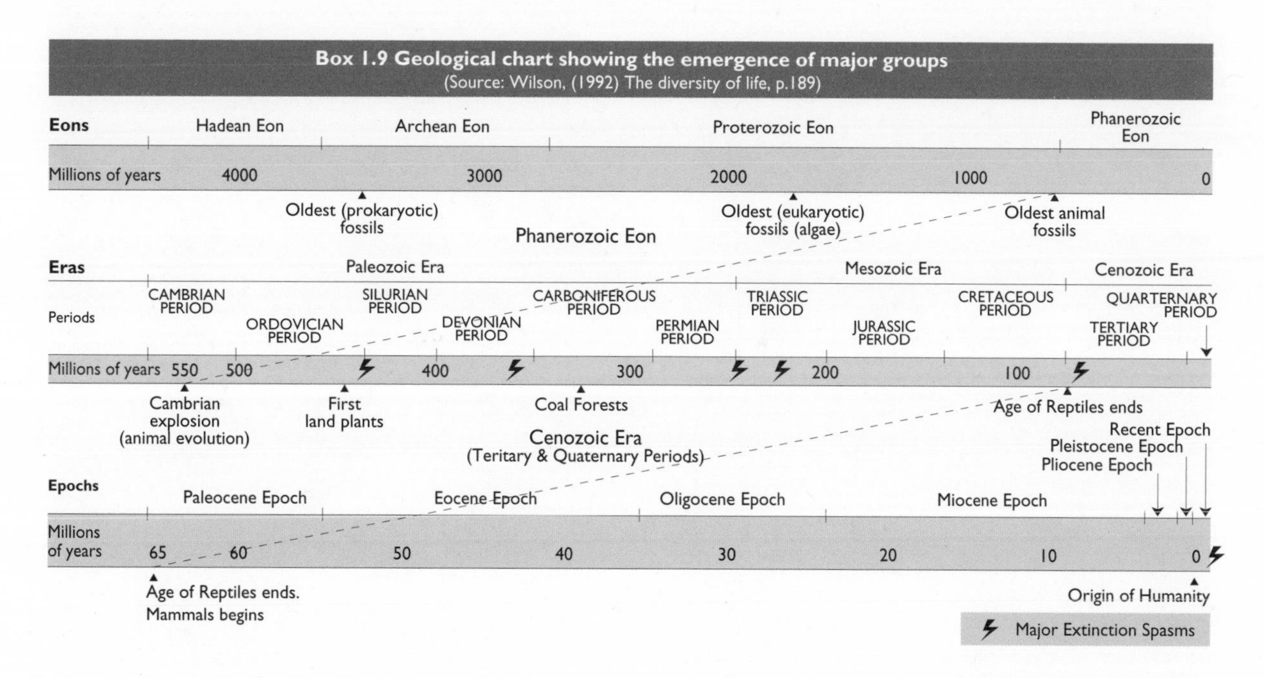

Box 1.9 Geological chart showing the emergence of major groups
(Source: Wilson, (1992) The diversity of life, p.189)

50 years old. For other parts of the world, particularly tropical regions, knowledge of the biota is largely non-existent. Rarely are there even species lists for some of the more well-known groups, let alone taxonomic keys and field guides to identify these and other, less well-known plants and animals.

Much of the information on the distribution and biology of species is housed in the museums, herbaria and libraries of developed countries. In a few instances, some of this information has been databased in card indexes. There is now a desperate need to compile electronic databases for this and other information associated with specimens in the collections and to make this information readily available to those that need it (see also Chapter Six). Similarly, for the vast majority of species their individual ecologies and their conservation status are unknown. For this reason IUCN's Red Data Books on the threatened status of organisms are generally limited to groups of large vertebrates and higher plants.

1.2.7 The evolution and extinction of life

Simply speaking, evolution means 'change through time'. Darwin described it as "descent with modification" and Dobzhanski (1951), more specifically, as "changes in the genetic composition of a population". In genetic terms, evolution is an alteration in the frequency with which different genes are represented in a population and results primarily from the processes of natural selection and random drift. Natural selection operates through differential survival and reproductive success of individuals in a population, which determines their contribution to the genetic composition of the next generation. Natural selection acts on individual phenotypes best suited to the environment. With time, the genes controlling such well-adjusted characteristics increase in frequency within the population and the species is said to become 'adapted' to its environment. Genetic drift results from the random assortment of genes in each generation which means that, by chance, some will increase in frequency and others decrease. In general, genetic drift has a greater effect among small populations.

Evolution, however, does not necessarily result in an increase in the number of species. With sufficient change over time, different species may be recognised within a lineage (*chrono-species* or *successional species*). Leaving aside the problem of the species concept (which is discussed in section 1.4.2), speciation involves vertical evolution with the division of a lineage into two or more discrete and recognisable units. Speciation may occur *allopatrically*, among geographically isolated populations, *sympatrically* within a population or *parapatrically*, among adjoining populations. It is speciation that is responsible for the diversity of organisms on Earth (Wilson, 1992).

To our knowledge there has been life on Earth for at least 3.5 billion of the 4.6 billion years that Earth has existed (Awramik *et al.*, 1983). For

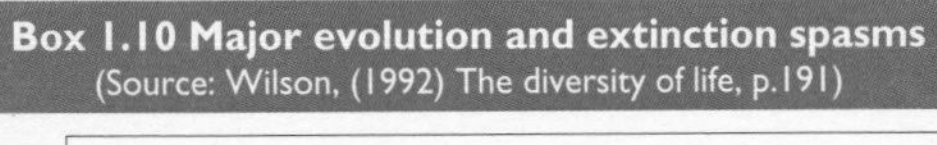

Box 1.10 Major evolution and extinction spasms
(Source: Wilson, (1992) The diversity of life, p.191)

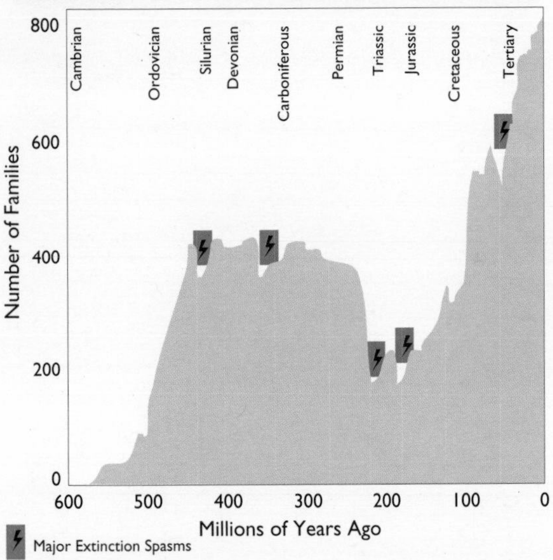

Box 1.11 Trends in recorded animal species extinctions (Source: Smith *et al.*, (1993), TREE 8:376)

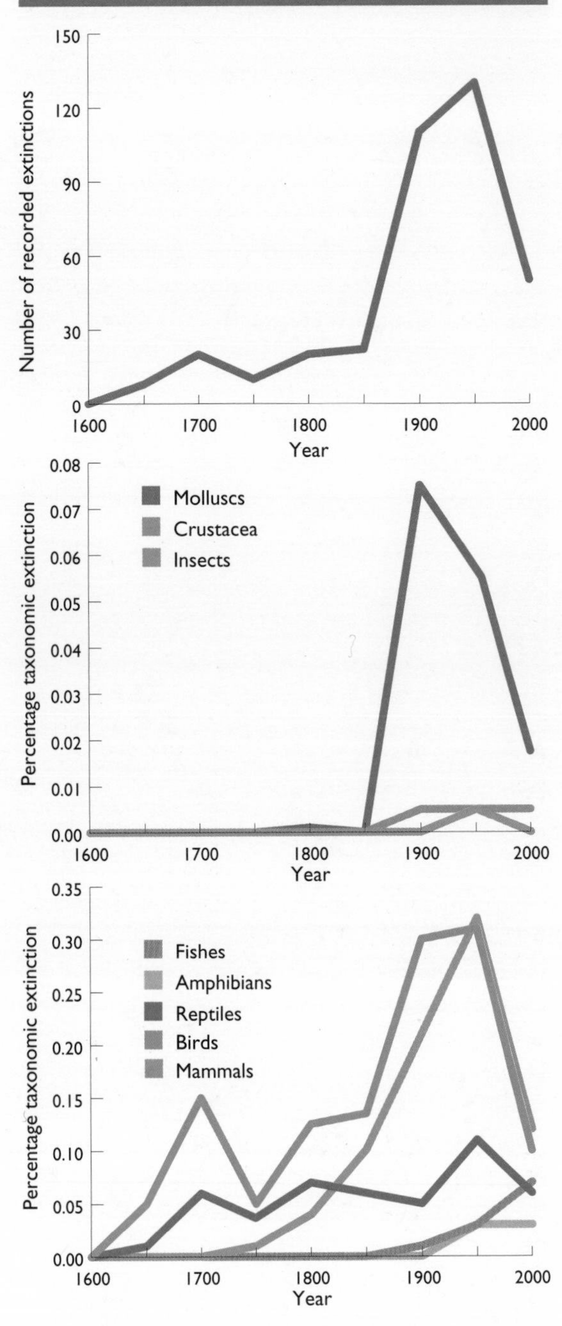

the first 2 billion years such life existed in the form of bacteria and related organisms. Multicellular plants and animals have evolved in just the last 1.4 billion years (Walter *et al.*, 1990) (see **Box 1.9**). Our own species, *Homo sapiens*, has existed for only about 50,000 years and yet in that time humans have come to dominate almost all other forms of life, dramatically altering the distribution of plants, animals and whole ecosystems on all continents. As a result, we are responsible for the present extinction spasm – a spasm possibly greater than any ever seen before in the geological record (May *et al.*, 1995).

The fossil record of some animals and plants is sufficiently detailed to allow us to trace the evolution and demise of whole groups and of individual species (Box **1.10**). Numerous studies have shown that there have been periods of rapid evolution and even more dramatic periods of extinction. Four of the five big episodes of extinction in the last 500 million years of the fossil record saw the removal of approximately 65–85% of those animal species in the ocean that are preserved as fossils, and the fifth resulted in the loss of 95% or more. In spite of these huge losses it is now estimated that, through subsequent rapid evolution, the present day diversity of organisms at both the species level and higher taxonomic levels, is greater than at any other time. May *et al.* (1995) suggest that present day diversity may represent roughly 1% of all the species that have ever existed.

The extinction of species, just like the evolution of species, is a natural process and we should, therefore, expect to see the extinction of existing species at the same time as the evolution of new ones. The World Conservation Monitoring Centre's volume on Global Biodiversity (WCMC, 1992) provides lists of all plants and animals that are recognised as having become extinct in the last few hundred years. In total this amounts to just 654 plant and 535 animal species (Smith *et al.*, 1993; see **Box 1.11**) and yet, since the early 1970s and 1980s, leading environmentalists and scientists have been predicting losses of species amounting to between 2–10% of all species per

WWF International

The recently discovered Vu Quang Ox (*Pseudoryx nohetinheuensis*).

W. O. Purvis/NHM

A recently described lichen (*Psilolechia leprosa*) from Britain.

decade for the period from 1980 to 2030 (see **Box 1.12**). Why is there this big discrepancy between our records of extinction and predicted extinction rates?

Firstly, as Diamond (1987), Mawdsley and Stork (1995) and others have pointed out, determining when a species has become extinct is immensely difficult. As a species becomes rarer and therefore more prone to extinction, keeping track of the number of individuals becomes increasingly more difficult. Those faced with the challenge of estimating the abundance of even very large organisms, such as the Sumatran rhinoceros in a Bornean rain forest, have an almost impossible task. This problem is made more obvious by the fact that scientists are still finding new species of large animals and plants. For example, since 1992 three new species of primate were discovered in Brazil, a new species of bovid (the Vu Quang Ox) was found in Vietnam and a new marsupial in New Guinea. It is not surprising therefore, that the fate of many thousands of threatened species of insects, other invertebrates and fungi is almost completely overlooked. The death of the last passenger pigeon, 'Martha', in 1914 is well known to many conservation biologists and yet the 'co-extinction' of two species of lice that were host-specific to this species of bird went by almost completely unnoticed (Stork & Lyal, 1993)!

Secondly, if only 10–20% of all species on Earth have been named and described, and the distribution and natural population fluctuations are known for a small proportion of these, then no information will be available on the extinction status of the remainder. Some estimated extinction rates (see **Box 1.12**) would indicate that most species are more likely to become extinct than to be named by taxonomists!

Thirdly, it seems that there is a genetic or population threshold below which the survival of a species diminishes rapidly. For some species this 'minimum viable population' may be 10 individuals and for others 100s or 1,000s. Such species, which Janzen (1986) has termed the 'living dead', although not presently extinct,

Natural History Museum Photo Library

Passenger pigeon (*Ectopistes migratorius*) which became extinct in 1914.

Nat. Hist. Museum Photo Library

Tasmanian pouched wolf (*Thylacinus cynocephalus*), hunted out of existence last century.

Box 1.12 Estimated rates of extinction (adapted from Mawdsley and Stork, 1995)			
Estimate	**% global loss per decade**	**Method of estimation**	**Reference**
One million species between 1975 and 2000	4	Extrapolation of past exponentially increasing trend	Myers (1979)
15–20% of species between 1980 and 2000	8–11	Estimated species-area curve; forest loss based on Global 2000 projections	Lovejoy (1980)
50% of species by 2000 or soon after 100% by 2010–2025	20–30	Various assumptions	Ehrlich and Ehrlich (1981)
9% extinction by 2000	7–8	Estimates based on Lovejoy's calculations using Lanly's (1982) estimates of forest loss	Lugo (1988)
12% of plant species in neotropics, 15% of bird species in Amazon basin	–	Species-area curve (z=0.25)	Simberloff (1986)
2000 plant species per year in tropics and subtropics	8	Loss of half the species in area likely to be deforested by 2015	Raven (1987)
25% of species between 1985 and 2015	9	As above	Raven (1988a,b)
At least 7% of plant species	7	Half of species lost over next decade in 10 'hot-spots' covering 3.5.% of forest area	Myers (1988)
0.2–0.3% per year	2–6	Half of rain forest species assumed lost in tropical rain forests to be local endemics and becoming extinct with forest loss	Wilson (1988, 1989, 1992)
2–13% loss between 1990 and 2015	1–5	Species-area curve (0.15 < z < 0.35); range includes current rate of forest loss and 50% increase	Reid (1992)
Red Data Books: selected species: 50% extinct in 50–100 years (palms), 300–400 years (birds & mammals)	1–10	Extrapolating current recorded extinction rates and by the dynamics of threatened status	Smith et al. (1993)
Red data Books: selected vertebrate species	0.6–5	Fitting of exponential extinction functions based on IUCN categories of threat	Mace (1995)

are in fact doomed to extinction in the future. A critical factor in the long-term survival of a single species or group of species is the maintenance of the intricate web of interacting species that is important in some way or other for each other's survival. For example, the brazil nut tree, *Bertholletia excelsa*, relies on euglossine bees for pollination and seed setting (Prance, 1983), while the bees rely on the availability of other resources in the forest to complete their life cycle. Loss of these resources through forest fragmentation or disturbance could lead to the loss of the bees. The subsequent loss of the brazil nut tree, however, might not occur for many years. This is just one example from the continuum of co-survival of species, which ranges from those that are entirely dependent on the existence of one other species (e.g., Passenger pigeon lice) to those that are only in part dependent on one or a number of species. In this way the survival or extinction of species or groups of species is linked to the survival of whole habitats or ecosystems.

1.3 The distribution of life

1.3.1 The geological past

Past geological events have led to the peculiar distribution of land and water on Earth and, in large part, to the diversity and distribution of life forms that we see today. At present, more than 70% of the surface of the Earth is covered by water (Press & Siever, 1986) and of the 30% represented by land more than 70% is in the northern hemisphere. However, the amount of the Earth's surface above sea level has varied

Box 1.13 Fragmentation and spreading of the Earth's continental landmasses (adapted from Sengaor *et al.*, 1988, Geol. Soc. Spec. Publ. 37, and Scotese, 1984)

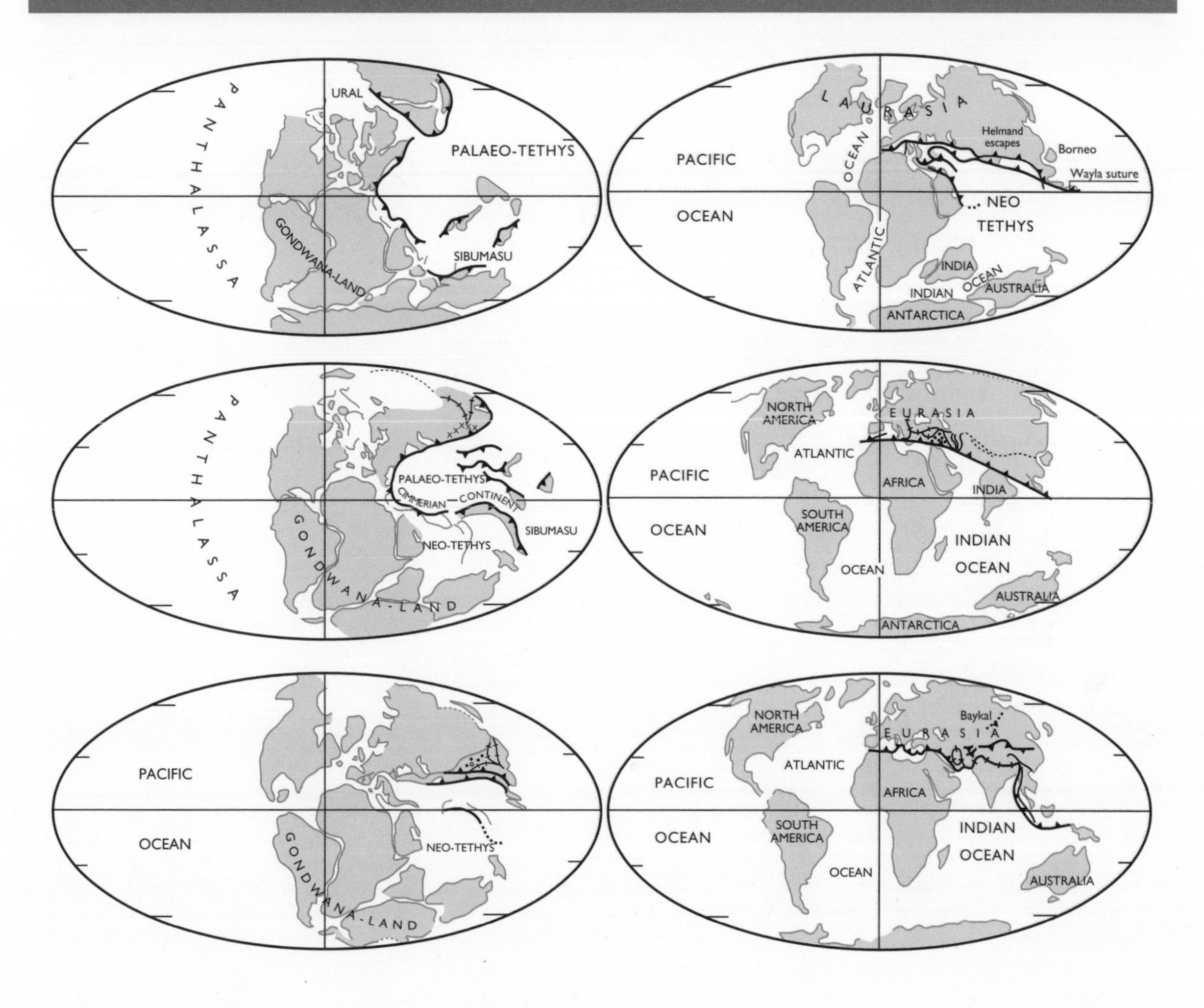

Key to the maps of the Tethyian Domain: A: Late Permian B: Early Tirasic C: Late Jurassic D: Late Cretaceous E: Middle Eocene F: Late Miocene

throughout geological time, depending on the quantity of water held in the polar ice sheets and cloud cover. Just over 10,000 years ago the sea level was sufficiently low for the North Sea to have disappeared and Britain was connected by land to France, the Netherlands and Scandinavia. Patterns of rainfall and solar energy also vary considerably across the Earth's surface. For example, one of the largest deserts on Earth is on the continent of Antarctica – a region noted for ice and snow!

The distribution of land-masses has varied considerably throughout the Earth's history. The framework of the terrestrial land-mass consists of a series of 'tectonic plates' of great age. These solid plates are continually moving, if extremely slowly, over the surface of the Earth, driven by convection currents in the molten mantle beneath. As the currents are in constant motion so are the plates on the surface, resulting in a continual, slowly changing land-mass distribution. It is this movement of continents situated on the plates – their joining and moving apart over millions of years – that has had a profound affect on the distribution of terrestrial animals, plants and microorganisms we see today. At the same time, the location of the oceans and the changing movement of water currents in relation to the distribution of land-masses have greatly affected the distribution and diversity of marine organisms.

About 300 million years ago, during the explosive evolution of terrestrial plants, the land-mass was comprised of five continents: Laurasia, Kazakhstania, 'Siberia', 'China' and Gondwanaland. By 250 million years ago these had all joined to form one super continent,

Pangaea. Over the last 200 million years this supercontinent has fragmented to form today's pattern of continents. While Pangaea existed as a single land-mass many groups of organisms evolved and radiated across this continent. With the break up of Pangaea (**Box 1.13** overleaf), populations became isolated on separate land-masses and adapted to their new and changing environments. Thus, from common ancestors on Pangaea whole groups of higher organisms evolved. New species developed, not only between continents but within them, through the process of adaptive radiation.

1.3.2 The present day distribution of life

1.3.2.1 Marine environments

Marine environments cover more than 70% of the Earth's surface, and of this 70% is more than 3,000m deep. In general, marine ecosystems have been less well studied than terrestrial ones. Only about 160,000 species (or 10% of all described species on Earth) are from marine environments (Barnes, 1989). However, the diversity of marine systems is demonstrated by examination of the higher taxonomic levels; 28 animal phyla are found in the seas, of which thirteen are endemic, compared to 11 terrestrial phyla, of which only one is exclusively terrestrial (see **Box 1.14**). One of the exclusively marine phyla, Loricifera, has only recently been discovered and it is possible that other higher taxa remain to be found in these environments.

Box 1.14 The distribution of animal phyla by habitat*
(Source: Based on V. Pearse, 1987, Blackwell Scientific, Palo Alto, Calif.)

Phylum	Marine	Symbiotic	Freshwater	Terrestrial
Acanthocephala		E		
Annelida	b+p	+	+	+
Arthropoda	b+p	+	+	+
Brachiopoda	B			
Bryozoa	+		+	
Chaetognatha	b+p			
Chordata	b+p	+	+	+
Cnidaria	b+p	+	+	
Ctenophora	P			
Cycliophora	E			
Dicyemida		E		
Echinodermata	b+p			
Echiura	B			
Gastrotricha	+		+	
Gnathostomulida	B			
Hemichordata	B			
Kamptozoa		+	+	
Kinorhyncha	B			
Loricifera	B			
Mollusca	b+p	+	+	+
Nematoda	b+p	+	+	+
Nematomorpha		E		
Nemertea	b+p	+	+	+
Onychophora				E
Orthonectida		E		
Phoronida	B			
Placozoa	B			
Platyhelminthes	b+p	+	+	+
Pogonophora	B			
Porifera	b	+	+	
Priapula	B			
Rotifera	b+p	+	+	+
Sipuncula	b			+
Tardigrada	b		+	+
Total	28	15	14	11

* Non-marine endemic phyla are indicated by E.
Marine phyla with pelagic representatives are indicated by p (P if endemic) and with benthic representatives by b (B if endemic).

Large Marine Ecosystems (LMEs) are regions in the order of 200,000 km² of ocean space characterised by distinct bathymetry, hydrography and productivity, and trophically dependent populations. Globally, as much as 95% of the total biomass yield of the oceans may be from these LMEs and within 200 miles of the coast. Upwelling, the upward movement of nutrient rich cold water resulting from the circulation of ocean currents, is an important factor in the productivity of many LMEs. This nutrient-rich water supports high densities of planktonic organisms which, in turn, support high densities of fish and mammal species. For example, productivity in the Banda Sea is controlled by upwellings associated with the southeast and northwest monsoons. Primary productivity, planktonic biomass, species composition and pelagic fish resources all increase during upwelling periods and the total productivity of the system may be two to three times higher than at other times (Zijlstra & Baars, 1990). The outflow of nutrients from terrestrial systems also contributes to the high productivity of coastal waters.

Several patterns have been identified in the distribution of species diversity among pelagic communities. Firstly, diversity appears to decrease with increasing depth of water. For example, 1,367 species of marine fishes have been identified in the shallow water of the Caribbean, in comparison with 500 demersal (sea-bottom dwelling) species and 600 pelagic (upper open sea dwelling) species in the whole North Atlantic basin (Merrett, 1994). Secondly, studies of fish and marine invertebrates

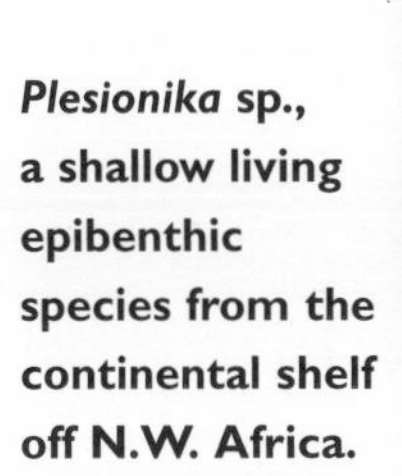

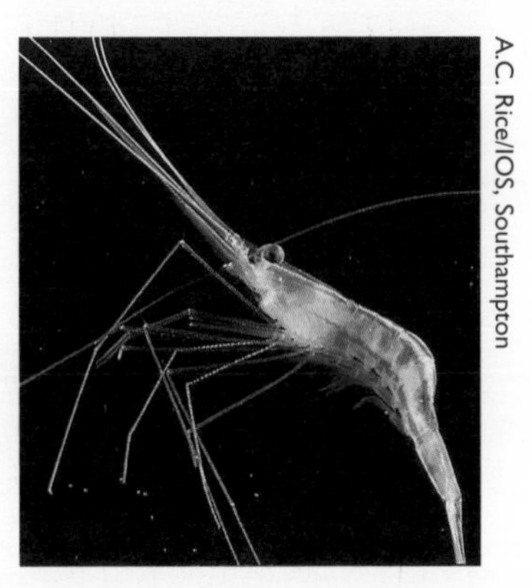

A.C. Rice/IOS, Southampton

Plesionika sp., a shallow living epibenthic species from the continental shelf off N.W. Africa.

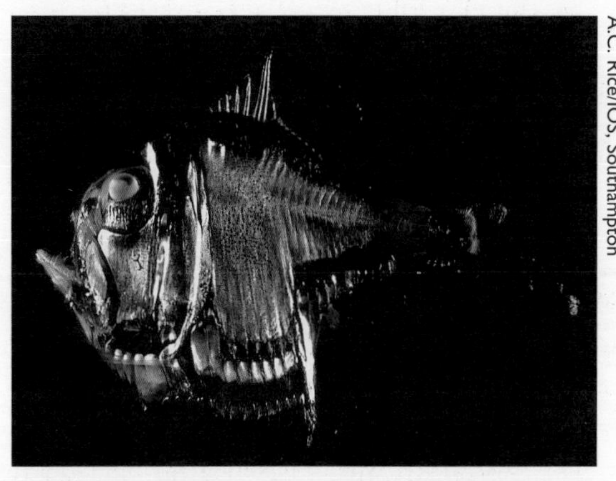

A.C. Rice/IOS, Southampton

Hatchetfish (*Argyropelecus* sp.) a common mesopelgic fish from depth of 400-500m in the N.E. Atlantic showing the typical mirror-like scales and ventral photophores of fishes inhabiting such depths.

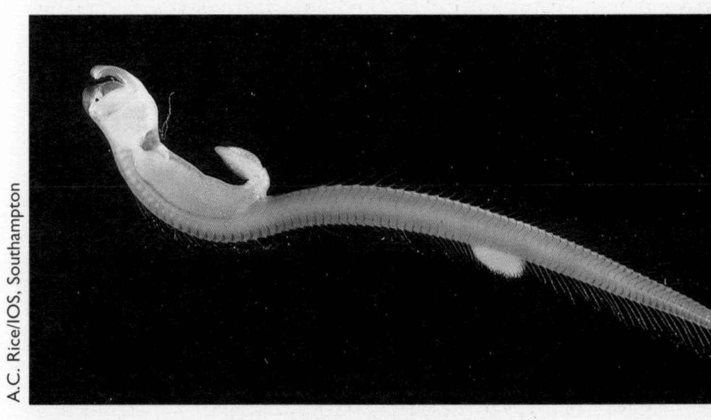

A.C. Rice/IOS, Southampton

***Monognathus*, an extremely rare bottom-dwelling fish from depth of 5,000m, showing great reduction in musculature and sensory organs typical of species inhabiting zones where there is so little food available.**

indicate that species diversity is higher in tropical regions than at higher latitudes, a trend which is apparently inversely related to oceanic productivity which is highest in temperate and polar regions (MacPherson & Duarte, 1994; Rex *et al.*, 1993). Thirdly, in deep water, vertical changes in diversity of both pelagic and demersal fish have also been identified with the highest species-richness occurring around 1,000m depth (Angel, 1993). In general, however, the low diversity of pelagic communities may be due to relatively large species ranges and the lack of barriers within the marine system. Many organisms, from whales to diatoms, are involved in large-scale vertical or horizontal migrations, either seasonally or at some point in their life history. Furthermore, the distribution of pelagic communities tends to correspond to patterns of large scale circulation. Zones of highest species-richness occur at the boundaries between different types of oceanic water where these different faunas mix. These boundaries, however, are not fixed and may shift seasonally by hundreds of kilometres (Angel, 1993).

The deep-sea ocean floor is the largest environment after the marine pelagic and is usually considered to be the area covered by salt water from 200m to deeper than 6,000m (Gage & Tyler, 1991). This environment is noteworthy for an absence of primary production apart from chemotrophs in hydrothermal vents. Until the 1960s it was considered to be stable and unchanging, but with relatively low biological diversity. Recent work has revolutionised this viewpoint. The pioneering work of Hessler and Sanders (1967) demonstrated an astonishingly high local species diversity despite the paucity of life. Recent research (see Lambshead, 1993, for references) has confirmed these results. Bathyal depths (200–2,000m) appear to possess the highest biological diversity of all marine sediments (Boucher & Lambshead, 1995; Rex, 1983). A large biodiversity study (Grassle, 1991; Grassle & Maciolek, 1992) of a 176 km bathyal transect in the northwest Atlantic Ocean showed that there may be up to 10^7 macrofauna (larger metazoans) species in the deep sea – a similar figure to that estimated for terrestrial systems. Grassle and Maciolek suggest that the number of species of meiofauna (smaller metazoans) might be comparable to the arthropod fauna of some of the richest terrestrial ecosystems.

Hydrothermal vents have been described as seafloor 'oases' providing areas of high productivity amid the abyssal plains (Cone, 1991). They are found in association with areas

J.D. George/NHM

Regal angelfish (*Pygoplites diacanthus*), Red Sea.

P. Harris/Coral Cay Conservation

Mangrove forest on Phillippines. This unique habitat of tidal forest has a specialised fauna of fish, shell-fish, crabs and trees which are mud-binders and important in coastal protection.

of tectonic activity, particularly in association with mid-oceanic ridges, at various depths around the world. These communities are unique in that they are supported by chemosynthetic, rather than photosynthetic, primary production. Bacteria which derive energy from compounds such as hydrogen sulphide support many of the vent-dwelling species, while some rely on symbiotic sulphur bacteria for energy. Such hydrothermal vent communities are generally shortlived in any one area, surviving only several years or decades. The whole ecosystem, however, may be over 200 million years old (Grassle, 1985). Although the total biomass of these areas may be high, species diversity is thought to be low (Grassle, 1986). However, endemism in these communities is high, with more than 160 new species having been described from hydrothermal vents (WCMC, 1992).

Coral reefs occur in shallow water 30° north and south of the equator and some of these are among both the most highly productive and the most biologically diverse ecosystems. There are an estimated 60,000km^2 of reefs to a depth of 30m (60% of which are located in the Indian Ocean (WCMC, 1992)). Relatively constant high productivity, efficient recycling and high nutrient retention all contribute to the rich biodiversity of these ecosystems. The multidimensional nature and high diversity of corals provides a variety of different habitats for many other organisms including algae, seaweeds, starfish, sea urchins, fish, crustacea and molluscs.

30% of ocean production occurs at the margins, in shallow water, bays and estuaries (Cherfas, 1990). Such ecosystems provide a link between marine, freshwater and terrestrial environments and exchange both water and species. For example, the outflow of freshwater from major rivers, such as the Amazon, may spread many miles into ocean environments. One particularly important tropical and sub-tropical ecosystem occurring at the ocean edge is mangrove swamp. Although of relatively low diversity in comparison with other systems, mangrove swamps house a variety of endemic specialised trees, shrubs and other organisms which are adapted to shallow, warm, salty tidal conditions. For example, at least 60 species of plants from 22 different genera are exclusively associated with mangroves (Saenger *et al.*, 1983).

1.3.2.2 Freshwater environments

Freshwater systems, habitable by non-microscopic organisms, occupy a very small part of the Earth's surface. Only 2.5% of all water on the Earth is non-marine and most of this is unavailable to life; 69% of all freshwater exists as ice, principally in the polar regions, and another 30% is present underground. Just 0.3% of freshwater is freely available in rivers, streams, lakes and freshwater wetlands, taking up only about 1% of the Earth's surface. Although occupying only a tiny percentage of surface, freshwater ecosystems support a rich and varied biota. For some groups the number of freshwater species is seemingly out of proportion to the amount of habitat available. For example, a little over 40% of all known fishes are freshwater fishes (Nelson, 1984) and, since the global freshwater and marine fish faunas are documented to similar degrees, this percentage is unlikely to change significantly with future discoveries in unexplored areas.

Box 1.15 Map of East Africa and body forms of cichlids (Source: Greenwood, 1981)

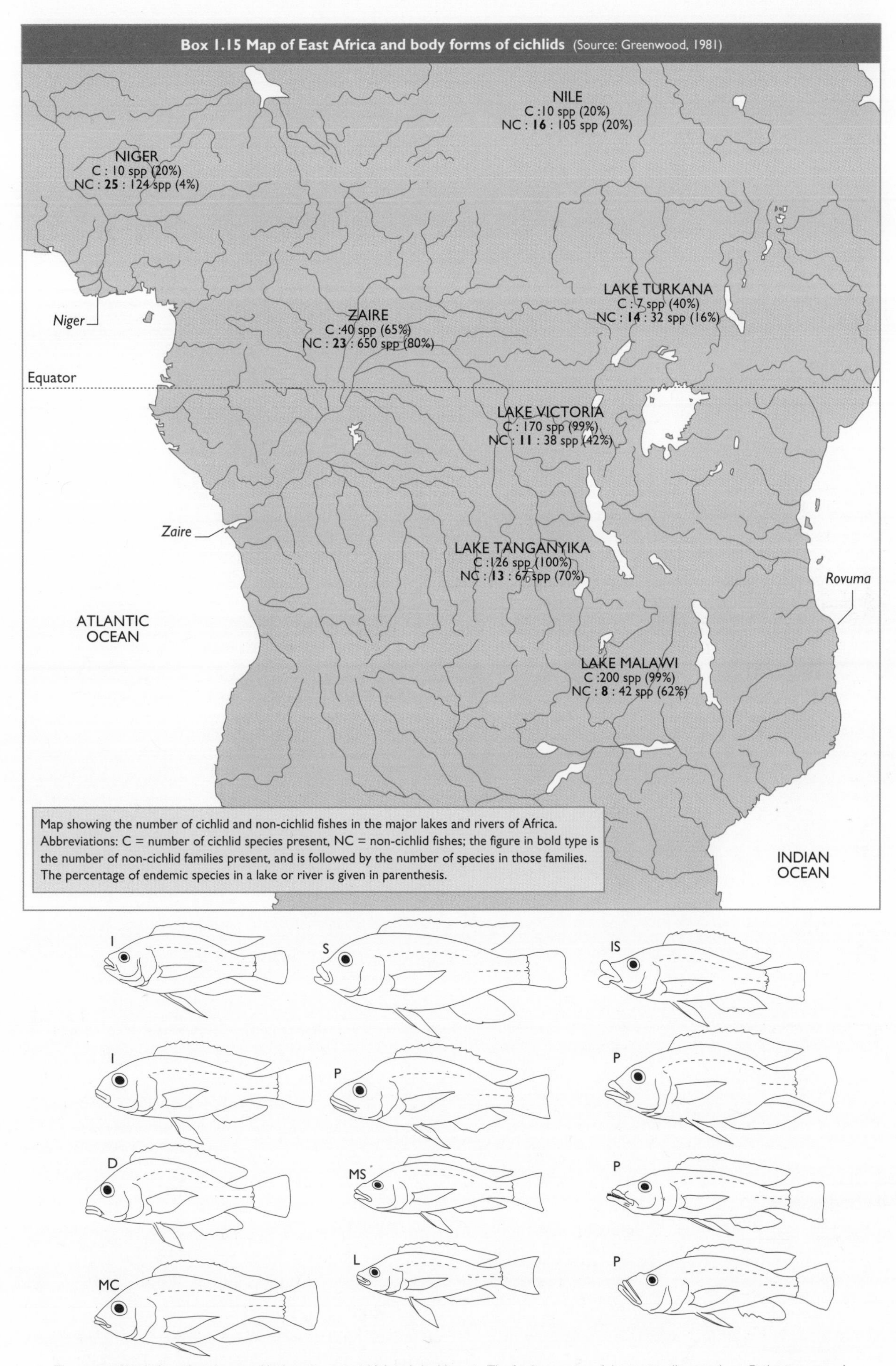

Map showing the number of cichlid and non-cichlid fishes in the major lakes and rivers of Africa. Abbreviations: C = number of cichlid species present, NC = non-cichlid fishes; the figure in bold type is the number of non-cichlid families present, and is followed by the number of species in those families. The percentage of endemic species in a lake or river is given in parenthesis.

The range of body form found among *Haplocymis group* cichlids in Lake Victoria. The feeding groups of the species illustrated are: D-detritus eater, I-insectivore, IS-specialised insectivore (removes larvae and pupae from burrows), L-paedophage, MC-mollusc eater (pharyngeal crusher), MS-mollusc eater (oral sheller), P-piscivore, S-scale eater. The drawings are not to scale.

Box 1.16 Characteristics of world vegetation zones (Source: J.M. Lock, 1994, from Loch and Eltringham, 1985)

Zone	Rainfall	Temperature	Climax vegetation	Area
Equatorial Zone	Rain falls in every month with drier seasons <3months. Ann. rainfall 1–2000mm (rarely >4000mm).	Diurnal temperature variations larger than seasonal; latter only 2–3 degrees. No frost.	Rain forest; vascular epiphytes, lianas, and buttresses are frequent in lowland; bryophytes, vascular epiphytes, bamboos and dwarf palms common in montane (cloud) forest.	N South America, W and W Central Africa, and S E Asia.
Seasonal Tropical Zone	600–1200 mm, with a cooler, dry season of 4–8 months.	Diurnal temperature fluctuations > annual.	Woodland, ± deciduous, <20m tall without lianas,vasc epiphytes rare. Grass spp common ('savanna'): Thick bark and underground woody stem protect from frequent fires.	Africa, South America, India, N Indo-China, and N & C Australia.
Subtropical Arid Zone	Low; usually lacks surface water.	Considerable seasonal variation; diurnal variations often very large.	Flora is often of ephemerals which germinate after rain & complete life cycle in few weeks. Succulent species often prominent.	N of equator in Africa, Arabia, NW India, C Australia. Also on W sides of continents in SW Africa & W coast S America.
Mediterranean type	200–1200mm, mainly in winter.	Summers hot and dry. Winters cooler, frost rare at sea level.	Original climax vegt dry evergreen sclerophyllous forest. Human use and fire have produced evergreen thicket (chaparral in California, fynbos in S Africa, and maquis, garrigue or phrygana, in European Mediterranean region) with spring-flowering annuals and plants growing from bulbs or other underground organs. Highly species-rich.	Europe, around Mediterranean Sea, SW North America (California), in parts of South America (Chile), Cape Province of S Africa, and in S Australia.
Continental Temperate Zone	Low and in winter months.	Winters cold, with several months with mean minima below 0°C. Summers are hot & dry. Marked variation in daylength through the year.	Grasslands predominate: prairies (N America), steppes (Eurasia), & pampas (Argentina). All are very similar in their vegetation which is dominated by perennial grasses.	Centres of continents in N America and Eurasia, with a related region but warmer in S America.
Oceanic Temperate Zone	>1500mm and <4000mm; considerable snowfall.	Seasonal variation but equitable; few months when mean minima below zero.	Temperate forest: in N Hemisphere with gymnosperms (Coniferae); in S Hemisphere, *Nothofagus* & *Podocarpus*. Ferns and bryophytes are often a prominent component. A variant of dwarf shrubland in which *Sphagnum* is abundant, & beneath which peat accumulation is rapid, producing blanket bog, is found in westernmost parts of the British Isles.	Seaboard of Europe, Western N. America, Chile and Tasmania & New Zealand.
Warm Temperate Zone	Occurs throughout the year.	Marked seasonal variation in both temperature and daylength but no cold months with mean minima below zero.	Mostly deciduous forest, with few evergreen trees.	Eastern Asia, E North America, S Europe, and in New Zealand, Tasmania and Chile.
Cool Temperate Zone	800–1500mm, either evenly distributed or with a summer maximum.	Here winter temperatures are lower. Frosts in 4–5 months, some having a mean minimum temperature below zero.	Deciduous woody spp with few gymnosperms less diverse than above zone.	In both hemisheres, lying appropriately north or south of the Warm Temperate Zone. Elsewhere on temperate mountains at appropriate altitudes.
Boreal Zone	150 – 600 mm, but the low temperature mean evaporation is also low.	Mid-winter temps 50 to –60°C; deep snow cover allows both plants and animals to survive. Mean max in the warmest months: 20°C, with a growing season of 150 days.	Boreal forest (northern coniferous forest – taiga: *Pinus*, *Picea* and *Abies* spp and few dwarf shrubs. Mosses and lichens often abundant on the ground.	Confined to the Northern Hemisphere, much of it north of the Arctic Circle.
Tundra	Ppt low most of it falling as snow, with much of the zone receiving less than 200 mm annually. Evaporation is low, and water usually abundant, at least in summer.	Annual fluctuations of daylength are even more extreme than in the boreal zone, as virtually all the zone lies north of the Arctic Circle. No month has a mean maximum temp, above 10°C. The growing season 50 – 100 days, when daylight is ±continuous, and productivity considerable. Sites where snow regularly blown away will have a much harsher winter microclimate than those where snow accumulates.	Trees are absent and dwarf shrubs are the only woody plants. Mosses are abundant in the wetter sites, and lichens in the driest. The most common higher plants are perennial herbs.	Occupies vast areas of N Hemisphere to the north of boreal forest. Absent from the Southern Hemisphere; (corresponding latitudes covered by snow and ice), although a few of the subantarctic islands bear a low treeless vegetation very similar in physiognomy to true tundra of the north.

Polar Regions

The two polar regions could hardly be more different. The North Polar region is a frozen ocean, the islands within this frozen sea are warm enough in summer to support plant growth (the Spitzbergen group, which lie between 76° and 81°N, have 160 species of flowering plants). In contrast, the Antarctic is a real continent, albeit almost entirely covered by ice, much of which lies between 3000 and 4000 metres asl. Although the Antarctic Continent extends to 65°S, only two species of flowering plant occur there.

Factors explaining the surprising diversity of freshwater life are very little understood. Among freshwater fishes, species richness is strongly correlated to latitude, with many more freshwater fishes being found in tropical and subtropical regions than in temperate ones. Comparisons of species richness within broad latitudinal bands reveal complex relationships among a number of variables but strong positive correlations with river basin size and the volume of water discharged have been observed (Hugueny, 1989). Most freshwater fishes have relatively small geographic ranges of single major river basins, a consequence of which is a high level of endemism. Large lakes sometimes present unusual patterns of speciation, the most spectacular examples being the great lakes of the Rift Valley of East Africa. Cichlid fishes in these lakes are extraordinarily speciose and demonstrate incredible morphological, ethological and ecological diversity (**Box 1.15**).

1.3.2.3 Terrestrial environments

In the following sections we examine the diversity of life in a few of the most species-rich and biologically significant ecosystems. One of the most useful systems for analysing the

C. Stirton/RBG, Kew

South African fynbos. A 'mediterranean' type of vegetation on the Cape Peninsula, in which 68% of its plant species are endemic.

distribution of terrestrial life is vegetation mapping. **Box 1.16** describes the characteristics of the world vegetation zones. The major biomes of the world are the result of climate, soil type and resulting vegetation and are shown in **Box 1.17**.

The nature and distribution of life forms in terrestrial environments are determined by climatic variables, such as the amount and periodicity of rainfall, temperature and

Box 1.17 Distribution of the biogeographical biomes of the world

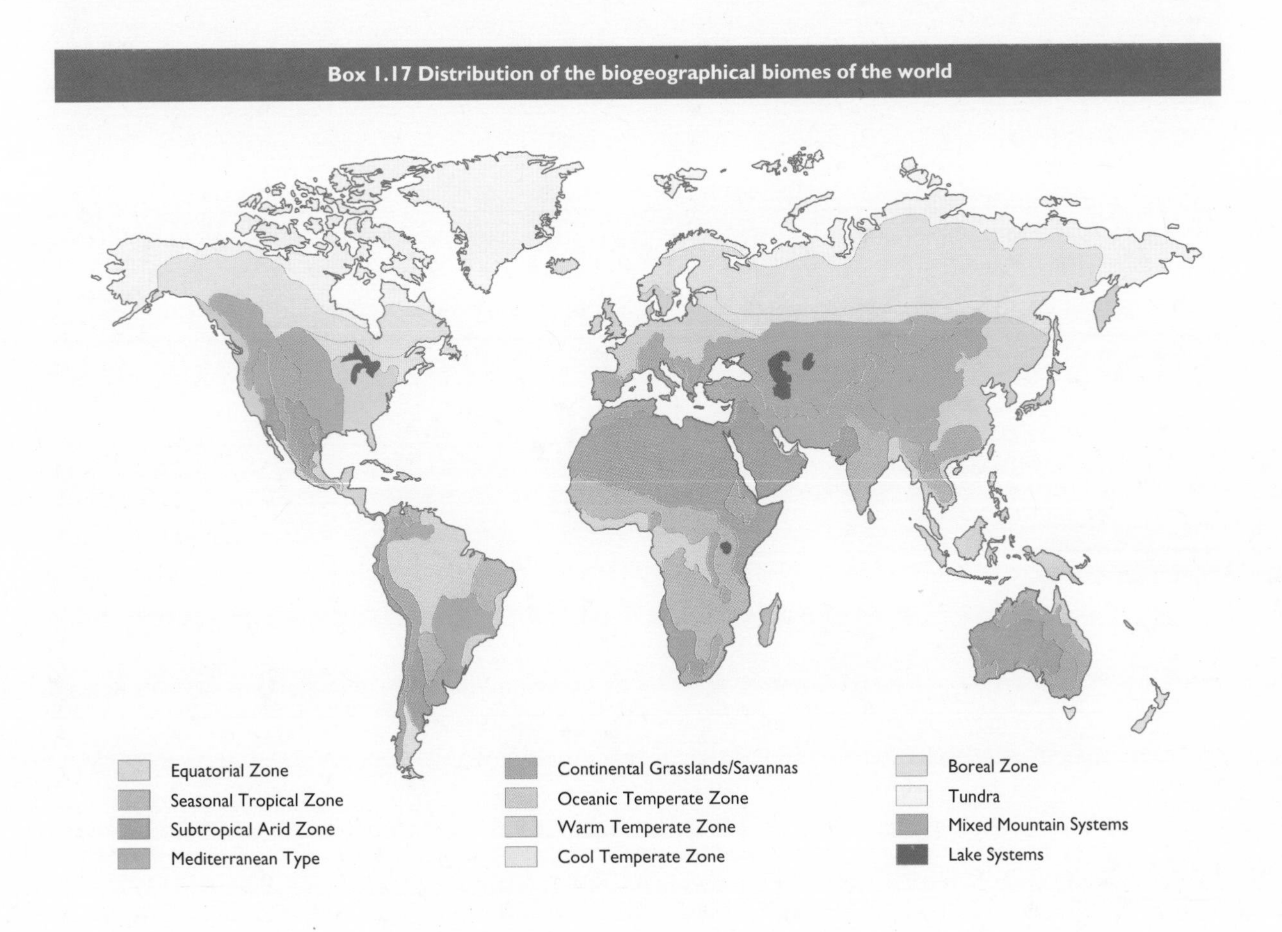

R. Ridgeway/Ulu Barito Project.

Scientists assessing mineral turn-over in tropical rainforest, Ulu Barito, Kalimantan.

S. Blackmore/NHM

Low altitude forest with epiphytes, Honduras.

sunshine and the distribution, topography and geological nature of the land. Factors such as these determine whether a region is desert, alpine meadow, tropical forest, marsh or grassland. These factors and the historical origins of the region in turn determine the diversity and forms of life that are to be found.

Of the world's open forest and shrubland, 75% and 42%, respectively, lie within tropical boundaries. At least two thirds of all plant species are tropical and thus, 6–7% of the Earth's surface may contain 50% to 90% of all species of plants and animals. The high species richness of tropical forests is illustrated by La Selva forest of Costa Rica, 13.7 km^2 of which harbours 1,500 species of plants, more than the total in the 243,500 km^2 of Great Britain (WCMC, 1992). La Selva forest also contains 388 species of birds, 63 bats, 42 fish, 122 reptile and 143 butterfly species (Myers, 1988). The high diversity of tropical rain forest systems was illustrated by Gentry (1988) who identified 300 tree species in single hectare plots in Peru, an enormous diversity when compared with 700 tree species in the whole of North America. Forty three species of ants were recorded from a single tropical tree in Peru: a number approximately equal to the entire ant fauna of the British Isles (Ehrlich & Wilson, 1991). Tropical forests may cover only a small proportion of the Earth's surface but they are vital for the global cycling of energy, water and nutrients. Most terrestrial life is found in temperate and tropical forests and grasslands. Some other vegetation types, such as the fynbos of South Africa, are also extremely species-rich. This system supports more plant species per square metre than any other place on Earth, with more than 8,500 species in total, 68% of which being endemic to this area (Armstrong, 1995).

Two strata in forests, the canopy and the soil, are particularly noteworthy, both for their important roles in the functioning of animal and plant communities, and for their high species richness. The canopy of trees has been called by some the 'last biotic frontier' because of the immense diversity of insects, plants and fungi found there. Forest canopies came to the attention of biologists largely through the work of entomologists using knockdown insecticides to collect insects from the tops of trees. Erwin and Scott (1980), for example, found more than 1,200 species of beetle in insecticide samples from the canopy of 19 individuals of one tree species in Panama. At that time many of these insects were thought to be specific to this tree but it is now believed that many were effectively 'tourists' to the tree concerned (Stork, 1991). Insecticide fogging, is therefore, the forest equivalent to the marine plankton net in that it can sample a large proportion of the available species in an area.

P.V. York/NHM

Soil organisms such as this mite *Macrocheles*, a species of soils with high level of humus.

Perhaps less attention has been paid to the diversity of life in soils. Indeed, we are just beginning to understand the diversity of soil organism assemblages (Stork & Eggleton, 1992) and their importance in ecosystem functioning (see Sections 1.5.2. and 1.5.3). Relatively obscure groups such as fungi, springtails, mites and nematodes are all rich in species in the soil and are extremely important in ensuring that organic material is broken down and the resulting nutrients made available for the growth of plants. Earthworms in temperate regions and termites in tropical regions are critical for the production, turnover and enrichment of the soil. They also help to aerate the soil and increase the through flow of water, hence reducing water run-off and soil erosion.

1.4 The role of systematics in biodiversity assessment

1.4.1 What is systematics?

The science of biology can be divided into two basic types (Nelson, 1970). Physiological biology aims to elucidate fundamental processes and the organisms studied are those best suited to the study of a particular process. In contrast, comparative biology is concerned with the diversity of biological processes and an explanation of that diversity. Systematics is the branch of comparative biology responsible for recognising, comparing, classifying and naming the millions of different sorts of organisms that exist (Wiley, 1981). Taxonomy is the theory and practice of describing the diversity of organisms and the arrangement of these organisms into classifications. Classifications are constructed to facilitate the retrieval of information that the investigator considers to be relevant. For example, classifications might attempt to reflect the phylogenetic ('family-tree') relationships among organisms.

1.4.2 Taxonomic research

The most fundamental unit for taxonomic research is the individual organism. Individuals can be grouped together into many different kinds of larger units on the basis of observed features they hold in common (characters or attributes). These larger units or groups are of two different types (although, in reality, combinations of the two are often used). Some groups of organisms appear superficially to have the same shared characters but are from different 'family-trees', and hence are artificial in a phylogenetic sense. In contrast, natural groups are those that exist in nature independently of our ability to perceive them (Wiley, 1981). They are composed of organisms with homologous characters in common and thus can be used to make far wider predictions and generalisations than can artificial groups (Vane-Wright, in WCMC, 1992).

Widely accepted as the most basic of natural units is the species. However, this consensus belies a lack of agreement as to exactly what a species is. A major problem stems from variation observed among individual organisms, and the species question is largely one of how biologists attempt to classify individual organisms (all of which differ to a greater or lesser extent when compared with one another) into discrete groups or *taxa* (singular: *taxon*). There is a range of definitions that largely reflects the various theories of the origin of diversity. When biological classification was first developed, organisms were considered each to have a fundamental design and the task of the taxonomist was to discover the essential features of these 'types'. Even after the emergence of the Darwin's theory of organic evolution, this concept did not change.

It was only with the emergence of a reliable theory of inheritance, and the development of the disciplines of genetics and population biology, that biologists began to develop rational explanations for the origin of diversity and then apply this knowledge to the species concept. The initial step forward was the recognition of geographical variation, first as 'varieties', then as subspecies. This led to the concept of the species as a group of populations that reflected both common ancestry and adaptation to local conditions. In turn, this view was developed into the biological species concept, which defined the species as *"groups of interbreeding natural populations that are isolated from other such groups"* (Mayr, 1969). This species concept is perhaps the most widely accepted today but only applies to sexually-reproducing species. A surprisingly large number of organisms, particularly protists and plants, reproduce asexually and in these cases the biological species concept is inappropriate.

In contrast to this concept, which places emphasis on the reproductive process, the phylogenetic species concept emphasises the phenomenon of differentiation. From this viewpoint, species are defined as *"the smallest aggregation of populations diagnosable by a unique combination of character states in comparable individuals"* (Nixon & Wheeler, 1990). Species are, therefore, clusters on a phylogenetic tree, between which reproductive isolation may or may not occur.

The choice of species concept can exert considerable influence upon the number of species recognised. In birds, Cracraft (1989) provided some striking illustrations of the disparate numbers of species that can result from different species concepts. For example, using the biological species concept, 40–42 species and 110 subspecies of birds-of-paradise had been recognised. When Cracraft (1992) applied the phylogenetic species concept to these birds, the number of species rose to 90. Furthermore, he abandoned the use of subspecies because he found many of the biologically defined 'subspecies' were less closely related to each other than they were to other 'species'.

After species recognition, taxonomy involves classification. Darwin's theory of evolution was based upon descent with modification. This, in turn, provided a firm foundation upon which the classification of organisms could be securely based: the hierarchical pattern of phylogenetic relationships. Because these relationships are part of the unique history of the Earth, they can never be known with certainty. However, methods have been developed by which they can be estimated. Of these, cladistic analysis is now widely acknowledged as the best (Farris, 1983). It rests upon three basic assumptions (Hennig, 1966; Farris *et al.*, 1970; Vane-Wright, in WCMC, 1992): features shared by organisms (termed *homologies* or *apomorphies*) form a hierarchic pattern; this pattern can be expressed as a branching diagram (*cladogram*); and each branching point symbolises the features held in common by all the species arising from that node. Cladograms are the most efficient method for representing information about organisms (Farris, 1980) and are hence the most predictive of unknown properties of those organisms. A good overview of cladistic analysis, with extensive references, is provided by Forey *et al.* (1992).

Once a cladogram of taxa has been established, the next stage is to recognise and name formally the species and higher taxa. Names are assigned to these taxa according to a system based upon that first developed by the Swedish naturalist, Linnaeus, in the mid-eighteenth century. Species are grouped into genera, and these in turn are grouped into families, orders, classes, phyla and kingdoms. The ultimate goal of this nomenclature is to produce a universal system of unambiguous names for all recognised taxa. Animals, plants and bacteria each have a separate set of rules or codes, which are applied voluntarily by taxonomists and are designed to promote stability and consistency in taxonomic nomenclature, and thus to biological science in general. Nevertheless, the aim is not to produce a fixed, 'official list' of names. However irritating changes in scientific names may be, to insist on fixation would be far more damaging because it would prevent us from adjusting the nomenclature as our understanding of the inter-relationships among organisms steadily improves.

1.4.3 Taxonomic products

The results of taxonomic research are made available in a wide range of different types of products in a variety of formats. Although, still largely published on paper, the increasing availability of computerised techniques is enabling greater flexibility in both data recording and publication of results. Increasingly, taxonomic products are being published as databases and on CD-ROM, or are being made available on computer networks accessible in the public domain.

Box 1.18 An example of a desciption of a new genus and species of coccinellid beetle from india
(Source: Booth, R.G. (1991). Bulletin of Entomological Research, 81: 147–150. [Copyright CABI])

Aponephus, genus novum
Type species: *Aponephus lentiformis*, **sp. nova**

Diagnostic characters. Belonging to Scymnini and closely related to *Nephus* Mulsant by: pubescent eyes; normal, T-shaped proternum, lacking carinae; trimerous tarsi; incomplete, postcoxal plates on basal, abdominal sternite; but differing by rather broad, strongly descending, elytral epipleura; very broad, plate-like fermora; and strongly flattened tibiae.

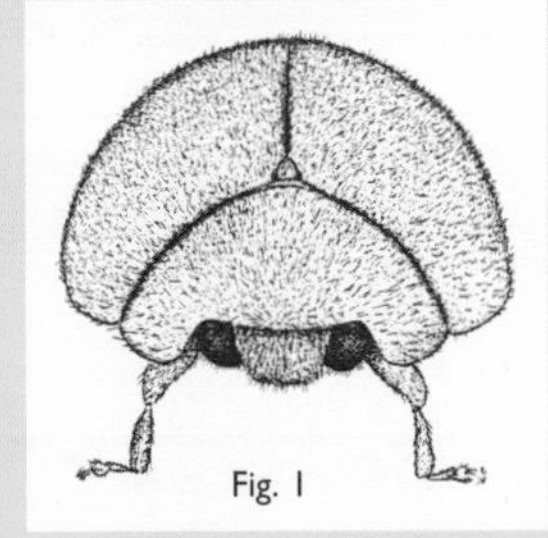
Fig. 1

Description. Body (fig. 1) short oval; outline of head, sides of pronotum and elytra continuous; dorsum moderately and evenly convex, pubescent; venter flat. Antennae (fig. 3) 9-segmented, very short, scarecly longer than half width of head between eyes; basal segment largest, broad medially, not visibly subdivided; second segment more or less quadrate; third to seventh segments short, becoming progressively more transverse; eight quadrate and terminal segment weakly transverse; pubescent and with pale setae, but lacking any long, black setae (as found characteristically in *Pseudoscymnus* Chapin). Head (fig. 2) transverse; eyes of moderate size, moderately finely faceted, with erect pubescence and small emargination behind antennal insertions; clypeus emarginate around antennal insertions, weakly concave anteriorly. Mouthparts typically scymnine; mandibles (fig. 4) bifid apically; maxillary palpi (fig. 5) short, apical segment slightly longer than broad, more or less parrallel sided, obliquely truncate apically; labial palpi (fig. 6) 3-segmented; mentum (fig. 6) of moderate width basally.

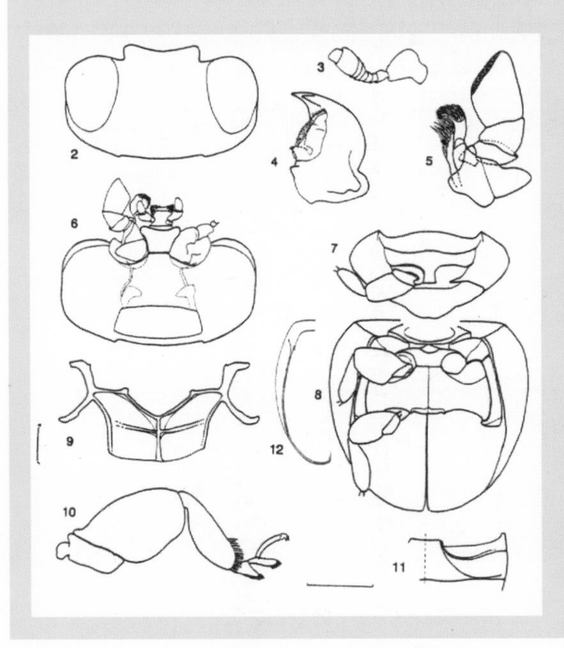

Lateral margins of pronotum strongly descending (fig. 1). Scutellum small (fig. 1). Prostemum (fig. 7) T-shaped, prosternal process convex anteriorly, lacking carinae. Mesocoxae (fig. 8) moderately well separated. Elytral epipleura (fig. 8) rather broad, strongly descending externally, with weak impressions but not obviously foveate to receive femoral apices; internal marginal bead continued anteriorly to base of apipleura (fig. 8), not partially traversing epipleura just behind humera angles (as fig. 12). Legs (fig. 10) with femora very broad, strongly flattened; trochanters very broad; tibiae strongly flattened with apical fringe of setae but lacking tibial spurs; tarsi trimerous. Metendosternite (fig. 9).
Abdomen with six visible sternites in both sexes; postcoxal plates (fig. 11) on first visible sternite almost reaching hind margin of sternite and approaching lateral margin. Apophysis of ninth abdominal segment in male (fig. 18) elongate; hemisternites of female (fig. 19) elongate; sperm duct (fig. 19) without sclerotized infundibulum.

Eltymology. The prefix Apo- is used to suggest that this genus is separate from *Nephus*. Gender is masculine.

***Aponephus lentiformis*, sp. nov.**

Description. 1.9-2.5 mm long, uniformly brownish yellow. Head with frons finely punctured; punctures separated by about one diameter; interstices very finely microsculptured. Pronotum shiny, without microsculpture; punctures fine, separated by about 1.5-3 diameters except adjacent to lateral margins where punctation coaser. Elytral punctation shallow, lacking subsutural rows of larger punctures. Elytral pubescence directed apicad, variable in length, shorter hairs decumbent, longer hairs semi-erect.
Sixth visible sternite notched medially in male (fig. 17), entire in female. Male genitalia with parameres broad, median lobe rather short and asymetric apically (figs 15, 16). sipho (figs 13,14). Female genitalia with spermatheca simple, sperm duct short (fig. 19).

Distribution. Currently known from southern India. Material examined. Holotype M: INDIA, Tamil Nadu, Vriddhachalam, July 1986, on mango with *Rastrococcilis iceryoides*, CIE A18380 (The Natural History Museum, London).
Paratypes: 37. INDIA, 2M, 1F, 16 not sexed, as holotype; 1M, 1F, 10 not sexed, as holtype but collected with Rastrococcus, CIE A18130; 6 not sexed, CIbC India Station, Bangalore, 1986, laboratory reared on *Rostrococcus iceryoides*. Paratypes are deposited in The Natural History Museum, London, the United Sates National Museum, Washington, and in the Zoological Survey of India, Calcutta.

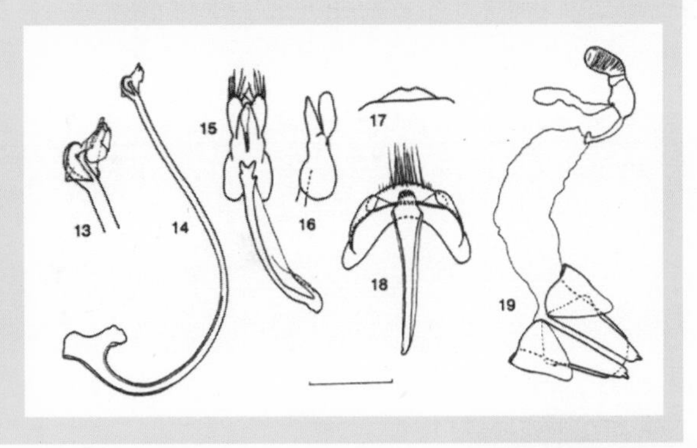

The following are brief descriptions of the types of taxonomic publications that are valuable in biodiversity assessment.

1.4.3.1 Descriptions

The description of new taxa is a fundamental process in taxonomy, but the isolated description of one new species or genus is the least desirable taxonomic product, except, perhaps, in the most well-known groups or in the wake of a more substantial work. Isolated descriptions often result from a superficial knowledge of the group and a larger percentage of synonyms (redundant names) result from this type of work than from larger, more comprehensive treatments.

1.4.3.2 Revisionary studies

Revisionary studies are the foundations of taxonomic research. They range from short synopses of several pages to large monographs comprising several volumes. Reviews are utilitarian and serve to collate current knowledge of a particular group. At best, they provide invaluable summaries of often scattered primary data sources (e.g. Bolton, 1992; Cribb *et al.*, 1992). Revisions and monographs are more comprehensive treatments, incorporating new data and interpretations, which are thoroughly evaluated and integrated with previous knowledge. Generally, revisions will include a taxonomic history of the group, descriptions and/or diagnoses of the included taxa with data on their distributions and biologies, identification keys, a classification (which may or may not be phylogenetic) and all relevant literature citations (e.g. Mattile, 1990; Bremer & Humphries, 1993).

1.4.3.3 Keys

Keys are intended to facilitate the identification of a specimen by presenting appropriate diagnostic characters in a series of alternative choices. Most keys still use the principle of dichotomous couplets, in which a choice between two alternatives has to be made. Traditionally, keys have been published as part of revisions, although they may be published separately (e.g. Bolton, 1994) or as a series (e.g. the Royal Entomological Society *Handbooks for the Identification of British Insects*). Recently, innovations in computer software have made multiple-entry keys more possible. These allow the user to answer questions in any order and, being less affected by missing data, are more efficient (Pankhurst, 1991). Computerised keys use a number of formats, such as DELTA (e.g. Watson & Dallwitz, 1988) or CABIKEY (e.g. Sands, 1992). An innovative and ambitious project, undertaken by the Expert Centre for Taxonomic Identification (ETI) in Amsterdam, combines multiple entry with HYPERTEXT software, producing high-quality, interactive identification and information systems distributed on CD-ROM.

1.4.3.4 Catalogues, checklists and inventories

A catalogue is essentially a directory of taxa, usually to species level. The taxa are generally arranged alphabetically for simplicity. The amount of additional information in a catalogue varies widely, but usually includes references to original descriptions, types, synonymies and geographical ranges (e.g. Crosskey, 1980). A checklist is generally less detailed than a catalogue, consisting simply of

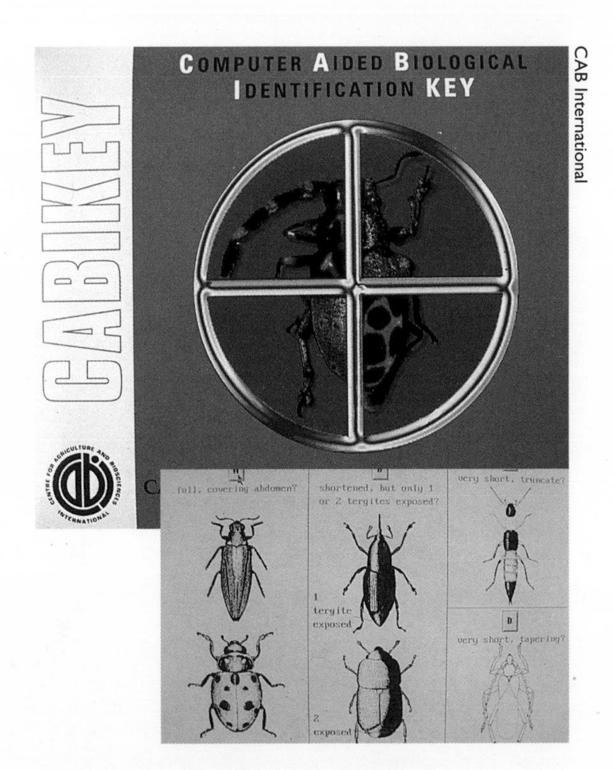

CAB International

a list of species, with minimal other information, in a skeletal classification, sometimes with a general indication of geographical range (e.g. Viette, 1990). Inventories are checklists of particular, usually restricted, geographical locations and provide simple guides to what taxa have been found there (e.g. Conservation Data Centre, 1989). As with keys, catalogues are being increasingly computerised, allowing for simpler and more efficient information storage and retrieval.

1.4.3.5 Faunas and Floras

Faunistic and floristic works differ from all other taxonomic products discussed so far in that their scope is dictated by the limits of a geographical region rather than a particular taxon. Their composition varies greatly, including the styles associated with revisions, catalogues and checklists. Floras and faunas generally include a history of taxonomic work in the area, the species occurring there (including descriptions of any new species), keys to those species, and an analysis of their distributional patterns. Many floristic and faunistic works take the form of a series of revisions of the taxa of a particular area. *Flora Malesiana* is one such undertaking and *Flora Mesoamericana* is another described in **Box 1.19**.

1.4.3.6 Handbooks and field guides

Handbooks and field guides are principally aimed at a non-specialist audience and are

Box 1.19 Flora Mesoamericana

The Flora Mesoamericana Project is an international collaborative initiative, involving tropical botanists, including taxonomists, from around the globe, under the organisation of the Missouri Botanical Gardens, the Universidad Autónoma Nacional de México (UNAM) and The Natural History Museum, London. It comprises an account of all the vascular plants of Mesoamerica, covering tropical Middle America from the Tehuantepec Isthmus, in southern Mexico to the border of Panama and Colombia and including the five southernmost states of Mexico (Campeche, Chiapas, Quitana Roo, Tabasco and Yucatán), as well as Belize, Costa Rica, El Salvador, Guatemala, Honduras, Nicaragua and Panama. The diversity of plant families in this region is particularly high as an overlap occurs here between the floras of the northern and southern Americas, in conjunction with an evolution of unique endemic taxa.

Flora Mesoamericana is the first project of its kind since Biologia Centrali-Americana (published by W. B. Hemsley between 1879 and 1888). When completed, the project will provide a readily available and accessible survey of nearly 18,000 native and cultivated Mesoamerican plant species. This data will be a vital reference for anyone with an interest in the Mesoamerican flora (professionals and amateurs alike) and will be published in both English and Spanish as well as being made available in either electronic or printed format. The first volume of this project includes 28 families of monocotyledons covering 326 genera and 1,891 species and is now available on the World Wide Web; a second, covering pteridophyta, was published in book form in 1995. A further seven volumes will be produced in the series.

(*Source*: *Sandra Knapp, Natural History Museum, London*)

therefore less technical in their content. A handbook is a comprehensive work on a group of organisms, with particular emphasis on biology and distribution (e.g. DeVries, 1987). Field guides are designed to enable specimens to be identified in the field, most frequently by direct comparison with colour illustrations, rather than through the use of keys. Emphasis is placed upon easily observed, diagnostic characters. For larger groups, only a selection of the species most likely to be encountered may be included. Series such as the Observers Books, Collins Field Guides and The Audubon Society Field Guides have inspired generations of young biologists in the developed world. The use of such guides has not been common in developing countries. Many field guides and handbooks that are available are too expensive or are not written in the local language. In developing nations, it is imperative that field guides and handbooks be published that are appropriate to local needs, in order to stimulate interest in biodiversity.

1.5 The ecological context

Ecology may be defined as 'the study of the relations between organisms and the totality of the physical and biological factors affecting them or influenced by them' (Pianka, 1983), or more simply as 'the study of patterns in nature' (Elton, 1966). Ecologists investigate the biology of organisms, looking for consistent patterns in their behaviour, structure and organisation. Although a relatively new field in comparison with systematics, ecology has already provided considerable insights into the organisation of taxa. Three aspects of community ecology are of importance in the context of this Manual: the role of species in the functioning of ecosystems, the importance of ecosystems to man and the effect of man on ecosystems. These are addressed below.

1.5.1 Ecosystem structure

An ecosystem comprises all the individuals, populations and species in an area or habitat, and the interactions among them and with the physical or abiotic environment. It thus comprises both biotic and abiotic components. Together the biotic components of an ecosystem constitute an ecological community. The interactions between community members can be shown as a food web (see **Box 1.20** and **Box 1.21**). Primary producers (or *autotrophs*) comprise the first trophic level. These species (principally green plants but also algae and cyanobacteria in some ecosystems) are responsible for trapping solar energy and converting it to chemical energy and tissue biomass, which may then be utilised by the rest of the ecosystem. All other members of the community are dependent, either directly or indirectly, on the primary producers for energy. Because some energy is used at each trophic level (for example in growth and maintenance of tissues) and because the transfer of energy between levels is never completely efficient, less energy is available to the higher trophic levels. It is estimated that only 10 to 20% of the energy at one level is available to that above (Pianka, 1983). The flow of energy through a system determines many properties of that community, including the number of trophic levels that can be supported and the total biomass at each level.

Box 1.20 Diagram of the different trophic levels in an underwater meadow ecosystem at Silver Springs, Florida, illustrating the decreasing amount of energy availiable to higher trophic levels
(Data redrawn from: E. Pianka, 1983 and J. Phillipson, 1966)

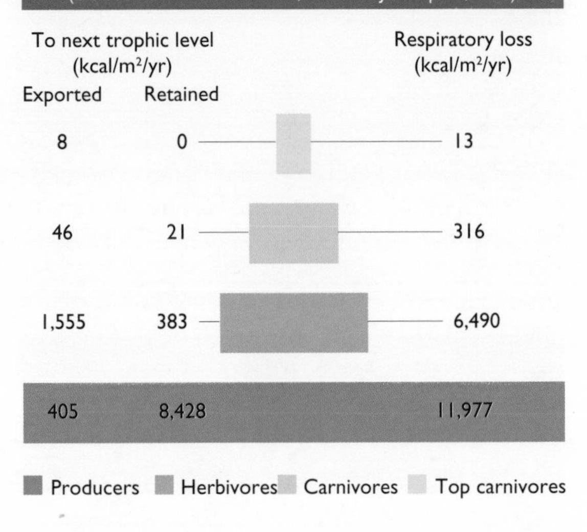

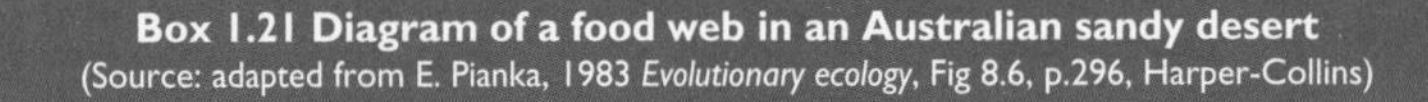

Box 1.21 Diagram of a food web in an Australian sandy desert
(Source: adapted from E. Pianka, 1983 *Evolutionary ecology*, Fig 8.6, p.296, Harper-Collins)

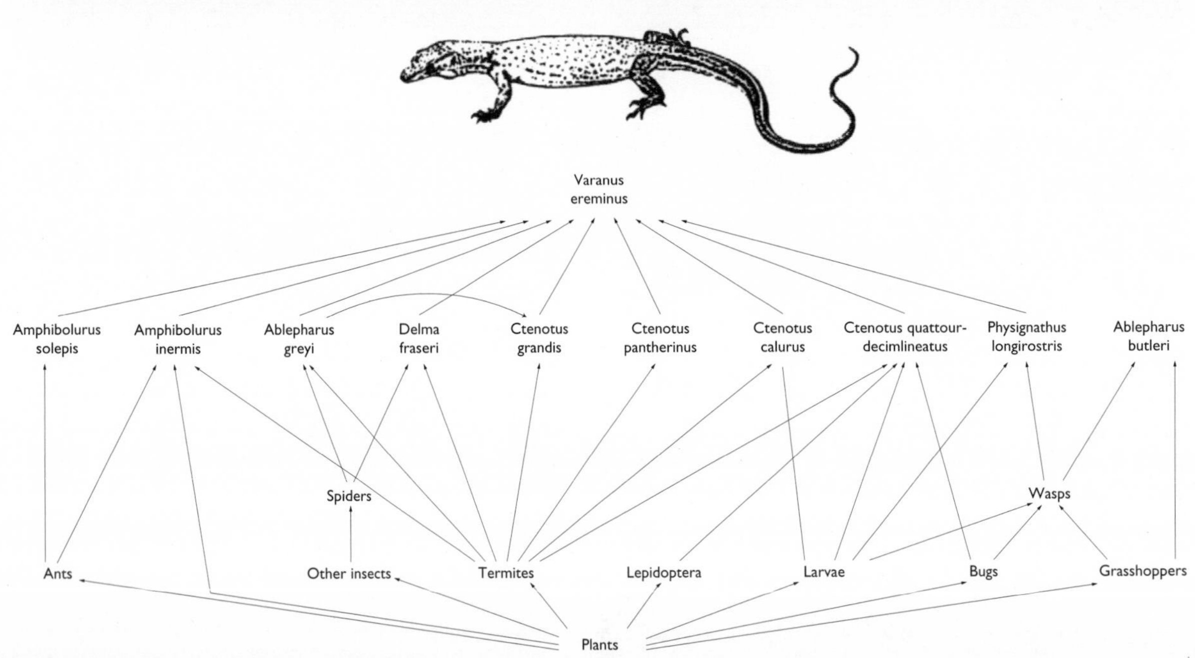

A more detailed food web would separate all food types into species and would indicate the actual rate of flow of energy up each link in the web.

Consumers (or *heterotrophs*) are animals and microorganisms (and occasionally plants) that feed on primary producers and each other. The second trophic level is made up of primary consumers (herbivores which feed directly on primary producers), while secondary consumers (carnivores which feed on herbivores) comprise the third trophic level and tertiary consumers (carnivores which feed on other carnivores) comprise the fourth. Omnivorous animals may be given partial representation in several trophic levels, in proportion to the composition of their diet. Decomposers (also classed as heterotrophs) are the species which feed on, and break down, dead plant and animal material making the component nutrients available to the system again. The role of decomposers is fundamental, especially in terrestrial ecosystems, as these species are responsible for the cycling of nutrients such as calcium, carbon, nitrogen and phosphorous through the ecosystem and for recycling all the primary productivity not utilised by consumers. In some communities 99% of the net annual primary production may be consumed by decomposers. Understanding the flow of energy through an ecosystem is important for conservation and any sustainable exploitation of that system.

1.5.2 Ecosystem function

Ecosystem function refers to the sum total of processes operating at the ecosystem level, such as the cycling of matter, energy and nutrients. The species in a community influence its productivity, nutrient cycling and fluxes of carbon, water and energy (Ehrlich & Wilson, 1991; Lawton, 1994a). Ultimately, species may be responsible for such factors as the maintenance of atmospheric composition, the dispersal and breakdown of waste material, the amelioration of weather patterns and the hydrological cycle, the development of fertile soils and even the protection of many coastal areas.

Biogeochemical cycling is the movement of substances including carbon, nitrogen, phosphorous and calcium through an ecosystem as individuals of different trophic levels are consumed by others at higher levels. These nutrients are returned eventually to the abiotic 'nutrient pool' where they are again available to primary producers. Some of the important roles played by different species in biochemical cycling are outlined below.

By their photosynthetic activity, plants play a fundamental role in the carbon cycle, introducing carbon into the food web. Microorganisms are also crucial. It is estimated that algae and cyanobacteria are responsible for 40% of the carbon fixed by photosynthesis on Earth (WCMC, 1992). At the other end of the process, wood-decaying fungi release approximately 85 billion tonnes of carbon into the atmosphere each year as carbon dioxide (WCMC, 1992). Termites also play an important role in global carbon cycling (and hence, potentially, in global climate change) through their production of methane (Jones, 1990; Lawton *et al.*, 1996). The Earth's nitrogen cycle is dependent on bacteria for nitrogen fixation and the release of nitrogen by denitrification. The microbial community thus controls the amount of nitrogen available to an ecosystem, determining ecosystem productivity in areas where nitrogen is limiting. By absorbing water from soils or other surrounding media, plants have a fundamental effect on the water cycle.

There is an ongoing debate as to whether all species in an ecosystem are important or whether some are 'functionally redundant' (Ehrlich & Mooney, 1983; Lawton, 1994a; Lawton & Brown, 1993). That is, if a species is removed from an ecosystem, can other species fulfil the same role? Two factors influence the importance of a species in ecosystem functioning: the number of ecologically similar species in the community and the extent to which a species has qualitative or quantitative effects on the ecosystem (Mooney *et al.*, 1995).

There are several types of species which have a crucial role in an ecosystem. A 'keystone' species is one which, if it disappears from an ecosystem, results (directly or indirectly) in the virtual or actual disappearance of other species (Soulé & Kohm, 1989). Examples of groups which include keystone species are: predators, parasites and pathogens, which help to maintain population levels of prey and host species; large herbivores and termites, which control ecological succession; species that create and maintain landscape features such as waterholes and wallows in arid areas; pollinators, seed dispersers and other obligate mutualists; and plants that provide a resource in times of scarcity. For example, in one Peruvian forest, twelve species from only two plant groups (palms and figs) maintain almost all vertebrate frugivores for three months of the year when other tree species are not in fruit.

Some keystone species are essential for the formation of the biotope in which they live and as such are 'ecosystem engineers'. For example, soil invertebrates such as earthworms, termites, springtails and nematodes are responsible for the production and maintenance of high quality soils and thus help to determine plant productivity (Jones *et al.*, 1994; Lawton 1994a; Stork & Eggleton, 1992).

Just as coal miners in the UK used to take caged birds down mines to provide an early warning of dangerous coal gas levels, it is possible to use some groups, such as lichens, birds and butterflies, as 'ecological canaries' to give an early warning of ecosystem change so that remedial action may be taken. In contrast, many charismatic 'flagship' species, which are important in raising the public awareness of conservation issues, may not be crucial in the functioning of an ecosystem. However, they do provide a focus for conservation and their protection may provide safety for whole communities.

There are at least three competing theories as to how an ecosystem might respond to loss of species diversity (Lawton *et al.*, 1996). The 'redundant species' hypothesis suggests that there is a minimum set of species required for an ecosystem to function and that adding or losing others does not affect processes (Walker, 1992; Lawton & Brown, 1993). The 'rivet' hypothesis (Ehrlich & Ehrlich, 1981) would suggest that all species are essential and that, like a machine, as rivets (species) are lost, functioning is impaired. The 'idiosyncratic response' hypothesis suggests that functions change when diversity changes but in an unpredictable way (see Lawton, 1994a; Vitousek & Hooper, 1993). The first experimental tests of these hypotheses in controlled laboratory conditions using a controlled environmental chamber (*Ecotron*) for the purpose, suggest that most processes (decomposition rates, nutrient uptake, etc.) varied idiosyncratically with species richness but that both uptake of carbon dioxide and plant productivity declined as species richness declined, as predicted by the rivet hypothesis (Naeem *et al.*, 1994, 1995).

A greater number of links in a food web provides more opportunities for checks and

N. Garwood/NHM

Natural tree fall due to severe but localised wind storm, producing damage similar to that seen from hurricanes. Barro Colorado Island.

balances should any environmental change occur. Because all species respond differently to change and as changes are unlikely to eliminate all members of a functional group, species diversity is important for ameliorating large changes in ecosystem processes. For example, there is evidence that more diverse grassland plant communities are more resistant to and recover more quickly from major drought (Tilman & Downing, 1994). Natural disturbance is an important component in the evolution of biological diversity at all levels and of the normal functioning of ecosystems (Bazzaz, 1983; Reiners, 1983). For example, natural tree falls in forests provide light and space for many species to grow and may promote diversity. Often natural disturbance is related to climatic patterns but these patterns may be difficult to identify. For example, it is only in the past few decades that we have come to understand the complex and far reaching El Niño climatic conditions (see **Box 1.22**).

The strength and duration of a disturbance determines its impact on an ecosystem. Understanding patterns of natural disturbance may be useful for management policies and the maintenance of natural biodiversity levels. Humans have increased the frequency and severity of natural disturbance such that, overall, the impact of anthropogenic disturbance is now greater than that of most natural disturbance (Likens, 1991; Reiners, 1983; Woodwell, 1983).

1.5.3 Human impacts on ecosystems

The current growth of human populations and the resulting increase in demand for food and other products has already had a huge effect on

Box 1.22 El Niño climatic cycles

The El Niño event is caused by an irregular alternation of climatic conditions in the tropical Pacific region that can affect climatic patterns across the globe. The 1982–1983 El Niño was the strongest this century, causing droughts, floods, erosion and storms, and having a severe effect on both bird and marine life. In Ecuador and Peru 100 inches of rain fell in six months and coastal deserts were transformed into grasslands. The insect, bird and amphibian populations soared as did the incidence of malaria. In other areas, coral reefs were exposed and destroyed, sea temperatures rose and typhoons changed paths. Monsoon rains fell in the central Pacific causing winter storms and flooding in California, while Indonesia and Australia suffered droughts and fires. In total, the world economy lost an estimated $8 billion. Understanding these patterns and monitoring the necessary elements is therefore essential.

The El Niño event generally occurs at irregular intervals, from two years to a decade. Under normal climatic conditions, easterly winds blow along the equator and south-easterly winds blow along the coasts of Equador and Peru, raising the thermocline in the eastern Pacific and causing an upwelling of nutrient-rich, cool waters off the South American coast. This supports high levels of phytoplankton production, which in turn supports large fisheries. During El Niño years the easterly winds weaken and retreat eastward. The Pacific Ocean thermocline becomes flattened so that the cooler water lies deeper in the eastern Pacific. The coastal upwelling is unable to tap the nutrient rich cooler waters, decreasing nutrient availability to the fisheries. The sea level flattens across the Pacific, dropping in west and rising in east, and warm equatorial surface water flows eastward, further raising sea levels. The heavy rain, which in normal years falls over the western Pacific, moves east causing droughts and fires in the west and heavy rain in the east.

In recent years, El Niño events have become both more frequent and stronger, for the first time on record there have been four events in the past four years. It is now imperative to understand what exactly precipitates El Niño and whether this apparent change in the events is part of a natural pattern or whether it is related to a real climatic change, possibly related to global warming.

(*Source: Reports to the Nation on Our Changing Planet*, 1994; *Wuethrich*, 1995)

global biodiversity. More than 1,000 species are recorded as having become extinct since 1600 (WCMC, 1992) although this is probably the tip of the iceberg because the extinction of many species goes unrecorded (Stork & Lyal, 1993) and many others are doomed to extinction (Heywood *et al.*, 1994). It is essential for us to understand the implications of these losses for ecosystem function. Like 'keystone' species, human activity affecting one ecosystem may have repercussions for other ecosystems, altering the whole character of the landscape or region. The ecosystems most likely to be exploited are those with the highest production and the impacts of such exploitation are likely to be far reaching. It has now been estimated that a third of marine primary production is sequestered by human fisheries (Pauly in Lawton, 1994b) and, similarly, that 40% of terrestrial net primary production is consumed by humans (Vitousek, 1994; Vitousek *et al.*, 1986). It is also sobering to realise that 20% of the world's population is responsible for using 75% of the Earth's resources (Lawton, 1994b).

At the regional level, ecosystem diversity is important in the same way that species diversity is important at the ecosystem level. The diversity of ecosystems may ameliorate larger scale changes. Thus any decrease in the variety of ecosystems is likely to have important consequences. Changes in ecosystems world-wide have already fundamentally altered hydrological cycles, atmospheric chemistry, terrestrial carbon storage, soil erosion rates and 'ultimately' human welfare (Ehrlich & Mooney, 1983; Ehrlich & Wilson, 1991; McNeely, 1990; Vitousek, 1994).

Humans change the character of ecosystems in many ways, most importantly by habitat

Degraded mangrove, Kenya.

Tutoh river on the edge of Mulu National Park, Sarawak, showing soil erosion and copious orange silt in river.

destruction (e.g. through flooding, draining, logging, paving and ploughing); changing agricultural and land-use practices (including monoculture and the addition of artificial pesticides and fertilisers); pollution; over-exploitation of resources (including over-fishing and over-grazing) and by the introduction of exotic faunas and floras (including pests and pathogens). All may result in reductions of biodiversity. A few examples are described below.

1.5.3.1 Habitat destruction and fragmentation

Over the past three centuries, land-cover change has vastly reduced areas of forests, savannahs and grasslands. As smaller areas generally support fewer organisms, habitat fragmentation tends to result in the loss of species and may also affect patterns of dispersal and migration, further reducing population levels. Through its effect on rainfall interception and the evapotranspiration of water, forest removal has major consequences for local, regional and, ultimately, global hydrological cycles and temperature regimes, causing them to become less stable. The loss of terrestrial habitats also affects aquatic systems. Shade reduction increases water temperatures, while deforestation leads to increased soil erosion, silting of rivers and lakes and an increased likelihood of flooding.

1.5.3.2 Land-use and artificial chemical additions

As well as causing habitat destruction and fragmentation, modern, intensive agricultural practices such as the addition of pesticides and fertilisers, have had major consequences for natural ecosystems. On land, the addition of nutrients to agricultural systems causes an increase in the number of microorganisms and a subsequent reduction in the overall oxygen available to the system. The increasing area of croplands acts as a net source of carbon and nutrients, accelerating global change. Some agroforestry systems may also affect local meteorological patterns, as decreased diversity results in more regular canopy features. The removal of trees alters the hydrological cycle and soil water availability, while the replacement of perennial plants with annuals leaves soil periodically exposed, encouraging erosion. The decreased complexity of the rooting system, reduction in soil organic matter and change in the soil structure, also increases susceptibility to pests and pathogens. Modern agroecosystems usually result in a reduction in diversity of the associated plants, microbes and invertebrates, especially under larger-scale intensive agriculture (Edwards & Lofty, 1982; Swift & Anderson, 1993).

***Microcytis aeruginosa*, an algal bloom in Loe Pool, Cornwall, UK.**

Chemical additions to any watershed also affect aquatic systems. The addition of sulphate and nitrate causes acidification of streams and lakes. Increases in phosphate levels promote algal growth, resulting in oxygen deficits for the rest of the system and the killing of many fish species. Such eutrophication of waters, with increased primary production, decreased decomposition and reduced oxygen levels has become common. The ability of organisms to process nutrients and remove harmful materials may also be reduced. These effects on freshwater ecosystems have a direct impact on humans because sources of fresh water provide a focus for human settlement and are needed for drinking, irrigation, transportation, industry and fishing (Schindler & Bayley, 1990). The addition of nutrients from sewage, industry and soil run-off is having similar effects on coastal marine ecosystems, particularly mangroves and coral reefs.

1.5.3.3 Over-exploitation

The over-exploitation of many species has led to reduced population levels and, in some cases, extinction. Over-fishing is known to have caused severe population reductions in many fish species, including cod, herring, anchovies and salmon, in several regions of the world. One example of the consequences of over-exploitation is provided by the sea otters of the north-eastern Pacific Ocean kelp forests. Sea-otters are a 'keystone' species which feed on sea urchins, keeping their population level down. The crash in otter numbers has led to a huge increase in the population of urchins and the subsequent overgrazing by them of kelp forests (Estes & Palmisano, 1974; Estes *et al.*, 1978). This has altered primary production, nutrient cycling and decomposition in the ecosystem and affected many other species, including plankton, invertebrates, fish and marine mammals, which depend on the kelp. The removal of the kelp forests has also destroyed the protection they provide to coastal areas.

1.5.3.4 Global climate change

Increasing atmospheric concentrations of carbon dioxide, nitrous oxide, methane and chloroflurohydrocarbons are expected to lead globally to rising temperatures and increased precipitation (McNeely, 1990). Many species could be threatened as natural ecosystems are disrupted by global warming (Peters & Lovejoy, 1992). In addition, rapid climate change could disrupt agriculture (Ehrlich, 1988). Crop yields could become more difficult to maintain in some areas in the face of climatic changes, soil erosion, loss of dependable water supplies, decline of pollinators and assaults by pests. Added to this, increasing incidence of respiratory diseases, epidemics, natural disasters and famine may all have an effect on human populations.

1.5.3.5 Changes to soil structure

Human activities affect soil structure and permeability by compacting it, removing burrowing species and soil microfauna and by altering the litter type and quantity. Removing vegetation from slopes causes erosion, affecting local nutrient and mineral concentrations. Changes in the soil composition and structure also affect plant productivity above ground (Swift & Anderson, 1993).

1.5.3.6 Species introductions

The introduction of exotic species has been responsible for great perturbations in many ecosystems. The arrival of predators, competitors, pests and pathogens has caused decreases in populations of native species in many areas. Native or endemic species often occupy narrow ranges, have small population sizes and lack defences, all of which make them vulnerable to species introductions. The arrival of alien species is generally more of a problem on islands than in continental areas, especially where those islands are remote. This is illustrated by the dramatic changes that have occurred on the Hawaiian islands since human colonisation (see **Box 1.23**). Although introductions may increase local diversity, most colonisers are cosmopolitan and are not endangered, while many endemic species are potentially threatened. Ultimately, many local ecosystem types may be lost world-wide, leading to a more homogenous global biota (Cronk & Fuller, 1995; Lodge, 1993; Vitousek, 1988).

1.5.3.7 Freshwater systems

Man affects freshwater biodiversity both directly, through such activities as harvesting, diversion and habitat alteration by damming and channelling, and indirectly, through activities such as the introduction of exotic

Box 1.23 Species introductions on the Hawaiian islands

The Hawaiian islands are the most isolated archipelago in the world and their age as well as their climatic and topographical diversity have contributed to the evolution of a large number of endemic species. There have been two main waves of colonisation of the Hawaiian islands by man. The first, Polynesian colonisation took place around 1,500 to 1,000 years ago, while the European colonisation took place much more recently, with the first missionaries arriving in 1820. Since their colonisation, many exotic species have been introduced and have become established, even in protected areas. This, along with severe habitat destruction, has precipitated major changes in the islands' flora and fauna.

Prior to colonisation, 94–98% of vascular plant species were endemic to the islands, now more than 4600 species of introduced vascular plants inhabit the islands, at least 700 of which have populations reproducing in the wild. Concurrently, 200 endemic species have become extinct and 800 are presently endangered. Similar, dramatic changes can be seen in the avian fauna. Of the 125–145 species once thought to have inhabited the islands, it is estimated that 21 species of passerines and 8–12 species of non-passerines became extinct due to the Polynesian colonisation and 19 species have become extinct since European colonisation. The Polynesian extinctions, along with the introduction of alien species at that time, reduced non-passerine endemism from 93% to 67%. This figure is further reduced to 62% by more recent extinctions. In contrast, since 1780, 50 species of alien passerines have become established. Such modern species introductions have reduced non-passerine endemism to 16% and currently only one third of passerine birds are endemic. Furthermore, 12 species are currently endangered and 12 are so rare that they are unlikely to survive. More Hawaiian bird species are currently listed as endangered or threatened than for the entire continental USA.

Spectacular changes have also occurred in other taxa. From an original native mammalian fauna comprising a single species of bat, there are now 18 alien species of mammal present on the islands and where there were no native species of amphibians or reptiles, 20 have been introduced. Among insect species also, Simberloff (1986) reports that by 1948 there were 1,476 introduced species, in comparison to 3,638 native insects, representing 29% of the entire Hawaiian entomofauna. Lowland areas in the Hawaiian islands are now entirely dominated by introduced species, and the majority of these species introductions have occurred since the mid 1800s.

(*Source*: *Moulton & Pimm*, 1986; *Pimm et al.*, 1994; *Vitousek*, 1988)

species which compete with and prey on native species, discharge of pollutants and thermal waste and land use on catchment areas. Effects include reduced numbers of freshwater organisms, lower reproductive success, altered freshwater community structures and, in the extreme, extinction of freshwater life forms and the loss of whole systems. For example, extraction of water for municipal and agricultural purposes is nearly 100% in some systems, with disastrous results for freshwater life and local peoples that depend on the maintenance of such systems for their livelihood. Water flow from the U.S.A. into Mexico through the Colorado River is maintained in normal years only by an international treaty limiting water use in parts of the river basin in the U.S.A. So much water is diverted from the streams and rivers supplying the internal lake basin of the Aral Sea in the C.I.S. that it is drying up, with the calamitous loss of a once productive, large fishery.

1.6 The human dimension of biodiversity

1.6.1 Population rise and the use of resources

The Earth's human population continues to rise exponentially. In 1950 the global population measured 2.5 billion. By 1975 it had reached four billion and in the following 12 years it increased to five billion. Projections for the future present a major cause for concern as the population is expected to rise to 6, 7 and 8 billion in successive 10 to 11 year intervals (WRI, 1994). By the year 2050 it is estimated that the world population will have reached ten billion. This huge, projected increase in the human population will undoubtedly have a major impact on all the world's ecosystems (see **Box 1.24**).

More people will mean a greater need for resources, including more extensive settlements, increased industrial and agricultural production and a rise in transportation and infrastructure. This demand will, in turn, place increasing pressures on natural areas, resulting in serious impacts on biodiversity levels and the loss of more aesthetic global qualities (Boserup, 1965). Both expansion of the population base and declining death rates due to improved global health conditions have contributed to this rapid growth in population. Population growth has a two-fold effect on biodiversity: a direct effect represented by an increase in resource consumption and an indirect effect shown in the breakdown of natural resource management, which in turn fuels poverty, migration and increases in city-dwelling populations.

The Brundtland Report (WCED, 1987) highlights the necessity for a five to ten-fold increase in economic activity around the globe over the next half century to meet the demands of this growing population. However, if the projected rises in population and economic growth are realised, much of the biodiversity on Earth (at least for groups such as birds, mammals and flowering plants) will be threatened with extinction or severe genetic erosion.

The most notable human impacts on the environment remain pollution and habitat alteration (see Section 1.5.3), with habitat loss in particular being considered the biggest single current threat to biodiversity. Another cause for concern has been the huge increase in the consumption of energy. Between 1850 and 1990 there was more than a 100-fold increase in the use of commercial energy including oil, coal, gas, hydro-electric and nuclear power (an example of the continuing trends can be seen in **Box 1.25**), and the use of biomass energy (that is, fuel-wood, dung and crop wastes) has approximately tripled. Without improved levels of population management or greater efficiency in energy usage, future population levels will exceed the sustainable capacity of the biosphere.

Box 1.24 Population increase predictions (Source: WRI,1994)

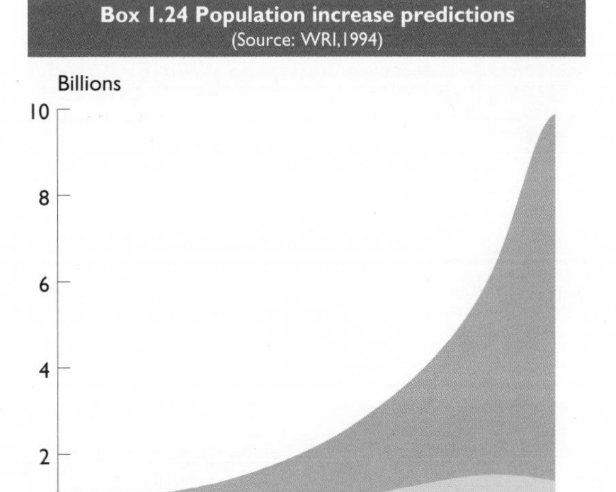

Box 1.25 Increase in commercial energy consumption, 1971-91 (Source: WRI,1994)

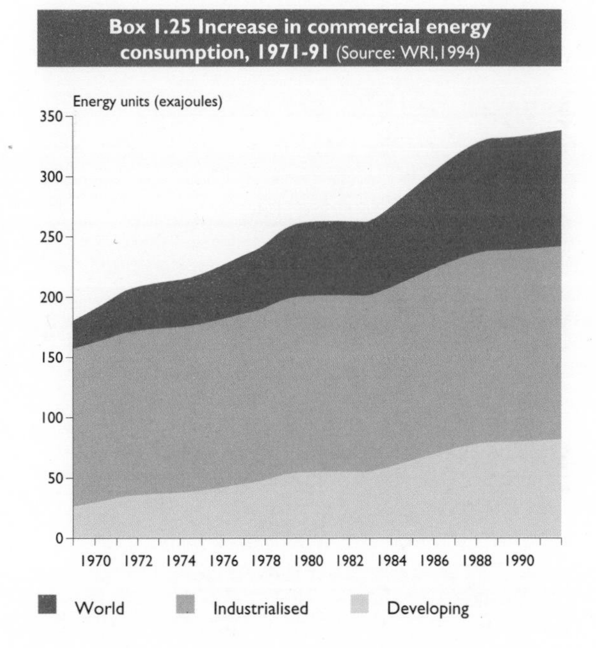

Further reading on human impacts on biodiversity can be found in Beazley (1993), World Commission on Environment and Development (1987), WRI (1994), Tolba and El-Kholy (1992); McNeely *et al.* (1995).

1.6.2 The economics of biodiversity

For a long time ecosystems in their natural state were thought to be unproductive areas and that the only way they could be made productive was to convert them in some way; for example, the

A. C. Jermy/NHM

Traditional farming, Papua New Guinea.

massive clearance of large areas of tropical forest to provide grazing for cattle. However, it has now been shown that the exploitation of natural forest species and products may be a more economic and sustainable use of such areas (e.g. Prance, 1990). Thus, the valuation of natural ecosystems is a vital step in communicating to resource managers and planners the economic importance of biodiversity for national development objectives (Ehrenfeld, 1988). Unfortunately, present methods of valuation often fail to reflect the real socio-economic and environmental value of ecosystems and natural resources. The aim then is to find approaches to valuation that suit both economists and non-economists, which are culturally acceptable and which make biological diversity accessible. With respect to this, one of the most important current issues is the fair and equitable sharing of benefits resulting from the use of biodiversity, an issue which forms one of the major tenets of the Biodiversity Convention. See also Pearce & Moran (1994), McNeeley (1988, 1993) and IUCN/UNEP/WWF (1991).

1.6.3 Biodiversity and agriculture

Traditional methods of agriculture have played an important role in conserving genetic diversity by ensuring the perpetuation of a range of species, sub-species and races. For example, Dover and Talbot (1987) have shown that traditional farming methods in Java involve the use of 607 different species. However, farming methods have undergone an increasingly rapid change from older and more traditional methods. In many areas nomadic herding, subsistence farming and shifting cultivation have given way to intensive animal husbandry, herding and the expansion of cereal cultivation. Diverse forest and grassland ecosystems have been transformed, with the increasing use of high-yield, monoculture tree and crop plantations. Such methods of farming and forestry have resulted in a depletion of genetic variability, especially where tree plantations are based on exogenous species, and a simplification of other ecosystem components, most importantly the soil flora and fauna.

The trend away from nomadic herding and shifting cultivation has also promoted permanent village settlements and then towns and cities, which must be supported by more intensive farming in the surrounding countryside. Furthermore, concentrations of permanently sited populations have led to the evolution and spread of new strains of epidemic diseases, for example plague and smallpox (Cohen & Armelagos, 1984; Eaton *et al.*, 1988). Increased productivity and permanent, concentrated populations have also led to an increase in trade. As a result, transportation, initially by animals such as horse and camel, and later by road, ship, train and air has become more and more necessary. These demands for transportation have in turn had a major impact on land use.

The increasing use of modern genetic and reproduction methods, the acceleration of animal selection and the need to improve the economic performance of the livestock sector have resulted in a reduction of genetic variability in livestock and poultry species. This is demonstrated by the number of extinct and endangered breeds and the depletion of genetic variability within breeds.

1.7 Summary – the need for inventorying and monitoring

The need to inventory and monitor the world's biota has been recognised by numerous international and national bodies. At present, most species inventories focus on a small number of taxonomic groups (for example trees, birds, fish, vertebrates and butterflies), occupying discrete geopolitical or ecopolitical areas. Using such a limited range of taxa in surveys for conservation, environmental impact assessments and monitoring can, however, produce incorrect or misleading results. As Prendergast *et al.* (1993) have shown areas of endemism or species richness do not necessarily coincide for different taxa and the pattern of distribution of birds, for example, is not the same as for lichens or butterflies. It is, therefore, essential that we rapidly increase our knowledge of regional and global diversity.

Action 4 of the Global Biodiversity Strategy recommends an early warning network to "monitor potential threats to biodiversity and mobilise action against them" (WRI/IUCN/UNEP, 1992) and a similar global network, monitoring ecological processes and warning of 'undesirable changes', is proposed by a worldwide strategy for marine biodiversity (Norse, 1993). Throughout the late 1980s UNEP and IUCN, with help from numerous experts including FAO and WWF, were involved in preparations for a global convention on biodiversity. Formal negotiations began in February 1991 by the Intergovernmental Negotiating Committee for a Convention on Biological Diversity. These culminated in the signing of the Convention on Biological Diversity by over 150 states at the Conference on Environment and Development in Rio de Janeiro in June 1992 (see **Box 1.26** for a summary of the principles adopted). Article 7 of this Convention requires all parties to identify and monitor components of biodiversity which are important for conservation and sustainable use (Glowka *et al.*, 1994). Similarly, Chapter 15 of Agenda 21 calls for the "systematic sampling of the components of biological diversity identified by country studies" (Johnson, 1993). Inventorying and monitoring efforts are central to the *Guidelines for Country Studies on Biological Diversity* (UNEP, 1993). Such studies provide information for the development of National Biodiversity Strategies. This aspect and the wider topic of inventorying and monitoring, is discussed further in Volume Two, Chapter One.

Box 1.26 Principles embodied in the Convention on Biological Diversity

- Recognises the importance of biological diversity for evolution and maintaining life sustaining systems of the biosphere.
- Conservation of biological diversity is the common concern of humankind.
- States have sovereign rights over their own biological diversity.
- States are responsible for conserving their biological diversity and for using their biological resources in a sustainable manner.

(*Source*: Glowka *et al.*, 1994)

From Section 1.2.1 (p. 2) it is clear that we have only a rough estimate of the true extent of global biological diversity. We know little about the geographical distribution of most species and we are even unclear about the total number of species that have been named and described. New empirical information is urgently needed for the assessment of a wide range of groups and particularly for the most threatened species.

Increased inventory effort will go some way towards providing a clear view of the magnitude of diversity on Earth and its rate of loss. Inventorying is also important for the assessment and sustainable utilisation of natural resources. Furthermore, inventorying is a fundamental component of biodiversity prospecting, the discovery of naturally occurring compounds (primarily from insects, plants and microorganisms) for use in biotechnology and the pharmaceutical and medical industries. With the fair and equitable sharing of benefits arising from the use of biodiversity, such bioprospecting may in future play an important role in the economies of many developing countries. Increased information about the distribution of species is also necessary for the effective conservation of biological diversity, while monitoring both the distribution and population health of species is important for providing an early warning of gradual and rapid changes in biodiversity through environmental changes. Practical approaches to the international co-ordination of such inventory and monitoring efforts have been presented by several organisations and these are discussed in Chapter Five of this volume.

The present manual should serve to improve our knowledge of the fauna and flora of the Earth and to monitor changes in the populations and distributions of species. It will also make a serious contribution to assist Parties fulfil their obligations to the Convention on Biological Diversity.

1.8 References and bibliography

Angel, M.V. (1993). Biodiversity of the pelagic ocean. *Conservation Biology*, 7: 760–772.

Avise, J.C. (1994). *Molecular markers, natural history and evolution*. Chapman & Hall, New York.

Awramik, S.M., Schopf, J.W. and Walter, M.R. (1983). Filamentous fossil bacteria from the Archean of Western Australia. *Precambrian Research*, 20: 357–374.

Armstrong, S. (1995). Rare plants protect Cape's water supplies. *New Scientist*, 1964: 8.

Barnes, R.D. (1989). Diversity of organisms: how much do we know? *Amer. Zool.*, 29: 1075–84.

Bazzaz, F.A. (1983). Characteristics of populations in relation to disturbance in natural and man-modified ecosystems. In: Mooney, H.A. and Godron, M. (eds.), *Disturbance and ecosystems*. (Ecological Studies No. 44), pp. 259–275. Springer-Verlag, Berlin.

Boserup, E. (1965). *The conditions of agricultural growth: the economics of agrarian change under population pressure*. Aldine, Chicago.

Boucher, G. and Lambshead, P.J.D. (1995). Ecological biodiversity of marine nematode in samples from temperate, tropical and deep-sea regions. *Conservation Biology*, 9(6):1–12.

Bolton, B. (1992). A review of the ant genus *Recurvidris*, a new name for *Trigonogaster*. *Psyche*, 99: 35–48.

Bolton, B. (1994). *Identification guide to the ant genera of the World*. Harvard University Press, Cambridge, Massachusetts & London.

Bremer, K. and Humphries, C.J. (1993). A monograph of the Compositae: Anthemidae. *Bull. Brit. Mus. (Nat. Hist.). Botany*, 23: 71–177.

Castri, F. di (1991). Ecosystem evolution and global change. In: Solbrig, O.T. and Nicolis, G. (eds), *Perspectives on biological complexity*, pp. 189–217. IUBS Monograph Series No. 6, Paris.

Cherfas, J. (1990). The fringe of the ocean: under siege from land. *Science*, 248: 163–165.

Cohen, M. and Armelagos, G. (1984). *Paleopathology at the origins of agriculture*. Academic Press, Orlando.

Cone, J. (1991). *Fire under the sea: The discovery of the most extraordinary environment on earth*. William Morrow, New York.

Conservation Data Centre. (1989). *Birds of Khao Yai National Park check-list*. Mahidol University, Bangkok.

Cracraft, J. (1989). Speciation and its ontology: the empirical consequences of alternative species concepts for understanding patterns and processes of differentiation. In: Otte, D. and Endler, J.A. (eds), *Speciation and its consequences*. pp. 28–59. Sinauer, Sunderland, Massachusetts.

Cracraft, J. (1992). The species of the birds-of-paradise (Paradisaeidae): applying the phylogenetic species concept to a complex pattern of diversification. *Cladistics*, 8: 1–43.

Cribb, T.H., Bray, R.A. and Barker, S.C. (1992). A review of the family Transversotrematidae (Trematoda: Digenea) with the description of a new genus *Crusziella*. *Invertebrate Taxonomy*, 6: 909–935.

Cronk, Q.C.B. and Fuller, J.L. (1995). *Plant Invaders*. Chapman and Hall, London.

Crosskey, R.W. (1980). *Catalogue of the Diptera of the Afrotropical Region*. British Museum (Natural History), London.

DeVries, P.J. (1987). *The butterflies of Costa Rica and their natural history Papilionidae, Pieridae, Nymphalidae*. Princeton University Press, Princeton & Guildford.

Diamond, J.M. (1987). Extant unless proven extinct? Or, extinct unless proven extant? *Conservation Biology*, 1: 77–79.

Dobzhanski, T. (1951). *Genetics and the Origin of Species*. (edn 2).Columbia University Press, New York.

Dover, N. and Talbot, L.M. (1987). *To feed the Earth: agro-ecology for sustainable development*. World Resources Institute, Washington, D.C.

Eaton, S.B., Shostak M. and Konner, M. (1988). *The Palaeolithic prescription*. Harper & Row, New York.

Edwards, C.A. and Lofty, J.R. (1982). The effect of direct drilling and minimal cultivation on earthworm populations. *J. Appl. Ecol.*, 19: 723–734.

Ehrlich, P.R. (1988). The loss of diversity: causes and consequences. In: Wilson E.O. (ed.), *Biodiversity*. National Academy Press, Washington, D.C.

Ehrlich, P.R. and Ehrlich.A. (1981). *Extinction. The causes of the disappearance of species*. Random House, New York.

Ehrlich, P.R. and Mooney, H.A. (1983). Extinction, substitution and ecosystem services. *BioScience*, 33: 248–254.

Ehrlich, P.R. and Wilson, E.O. (1991). Biodiversity studies: science and policy. *Science* 253: 758–762.

Ehrenfeld, D. (1988). Why put a value on biodiversity? In: Wilson, E.O. and Peter, F.M. (eds), *Biodiversity*, pp. 212–216. National Academy Press, Washington D.C..

Elton, C.S. (1966). The pattern of animal communities. Methuen, New York.

Embley, T.M., Hirt, R.P. and Williams, D.M. (1995). Biodiversity at the molecular level: the domains, kingdoms and phyla of life. In: Hawksworth, D.L. (ed.), *Biodiversity measurement and estimation*, pp.21–33. The Royal Society & Chapman and Hall, London.

Erwin, T.L. (1982). Tropical forests: their richness in Coleoptera and other arthropod species. *Coleopterists' Bulletin*, 36: 74–75.

Erwin, T.L. and Scott, J.C. (1980). Seasonal and size patterns, trophic structure and. richness of Coleoptera in the tropical arboreal ecosystem: the fauna of the tree *Luehea seemannii* Triana & Planch in the canal Zone of Panama. *Coleopterist's Bulletin*, 34: 305–322.

Estes, J.A. and Palmisano, J.F. (1974). Sea otters: their role in structuring nearshore communities. *Science*, 185: 1058–1060.

Estes, J.A., Smith, N.S. and Palmisano, J.F. (1978). Sea otter predation and community organisation in the western Aleutian Islands, Alaska. *Ecology*, 59: 822–33

Evans, M.E.G. (1975). *The life of beetles*. George Allen & Unwin, London.

Farris, J.S. (1980). The information content of the phylogenetic system. *Syst. Zool.*, 28: 483–519.

Farris, J.S. (1983). The logical basis of phylogenetic analysis. *Advances in Cladistics*, 2: 7–36.

Farris, J.S., Kluge, A.G. and Eckhardt, M.J. (1970). A numerical approach to phylogenetic analysis. *Syst. Zool.*, 19: 172–189.

Forey, P.L., Humphries, C.J., Kitching, I.J., Scotland, R.W., Siebert, D.J. and Williams, D.M. (1992). *Cladistics: a practical course in systematics*. (Systematics Association Publication No. 10). Oxford University Press, Oxford.

Gage, J.D and Tyler, P.A. (1991). *Deep sea biology*. Cambridge University Press, Cambridge.

Gaston, K.J. (1991). The magnitude of global insect species richness. *Conservation Biology*, 5: 283–296.

Gaston, K.J. and Mound, L.A. (1993). Taxonomy, hypothesis testing and the biodiversity crisis. *Proc. Roy. Soc.*, Ser. B, 251: 139–152.

Gentry, A.H. (1988). Tree species richness of upper Amazonian forests. Proc. *Nat. Acad. Sci.*, 85: 156–159.

Glowka, L., Burhenne-Guilmin, F. and Synge, H. (1994). A *Guide to the Convention on Biological Diversity*. IUCN, Gland and Cambridge.

Grassle, J.F. (1985). Hydrothermal vent animals: distribution and biology. *Science*, 229.

Grassle, J.F. (1986). The ecology of deep-sea hydrothermal vent communities. *Advances in Marine Biology* 23. Academic Press, London.

Grassle, J.F. (1991). Deep sea benthic biodiversity. *BioScience*, 41: 464–469.

Grassle, J.F. and Maciolek, N.J. (1992). Deep-sea species richness: regional and local diversity estimates from quantitative bottom samples. *Amateur Naturalist*, 139: 313–341.

Greenwood, P.H. (1981). Species-flocks and explosive evolution. pp. 61–74. In: Forey, P.L. (ed.), *The evolving biosphere*. (*Chance, change and challenge*). British Museum (Natural History), London.

Hammond, P.M. (1990). Insect abundance and diversity in the Dumoga-Bone National Park, N. Sulawesi, with special reference to the beetle fauna of lowland rain forest in the Toraut region. In: Knight, W.J. and Holloway, J.D. (eds), *Insects and the rain forests of South East Asia (Wallacea)*, pp. 197–254. Royal Entomological Society of London, London.

Hammond, P.M. (1992). Species inventory. In: WCMC, Global *Biodiversity, status of the earths living resources*. Chapman and Hall, London.
Hammond, P.M. (1994a). Described and estimated species numbers: an objective assessment of current knowledge. In: Allsopp, D., Hawksworth, D.L. and Colwell, R.R. (eds), *Microbial biodiversity and ecosystem function*. pp. 11–25. CAB International, Wallingford.
Hammond, P.M. (1994b). Practical approaches to the estimation of the extent of biodiversity in speciose groups. In: Hawksworth, D.L. (ed.), *Biodiversity measurement and estimation*, pp. 119–136. The Royal Society and Chapman & Hall, London.
Hawksworth, D.L. (1991). The fungal dimension of biodiversity: magnitude, significance and conservation. *Mycol. Res.*, 95: 641–655.
Hawksworth, D.L. (1993). The tropical fungal biota: census, pertinence, prophylaxis and prognosis. In: Isaac, S., Frankland, J.S., Whalley, A.J.S. and Watling, R. (eds), *Aspects of tropical mycology*, pp. 265–293. Cambridge University Press, Cambridge.
Hennig, W. (1966). *Phylogenetic systematics*. University of Illinois Press, Urbana.
Hessler, R.R. and Saunders, H.L. (1967). Faunal diversity in the deep-sea. *Deep-Sea Research*, 14: 65–78.
Heywood, V.H., Mace, G.M., May, R.M. and Stuart, S.N. (1994). Uncertainties in extinction rates. *Nature*, 368: 105.
Hillis, D.M. and C. Moritz. (1990). Molecular systematics. Sinauer, Sunderland, Mass.
Hugueny, B. (1989). West African rivers as biogeographic islands: species richness of fish communities. *Oecologia*, 79: 236–243.
IUCN/UNEP/WWF. (1991). *Caring for the Earth: A Strategy for Sustainable Living*. IUCN/UNEP/WWF, Gland, Switzerland.
IUCN/UNEP/WWF. (1993). *Caring for the Earth: A Strategy for Survival*. Reed International Books,.
Janzen, D.H. (1986). The future of tropical biology. *Ann. Rev. Ecol. Sys.*, 17: 305–23.
Johnson, S.P. (ed.). (1993). *The Earth Summit. The United Nations Conference on Environment and Development* (UNCED). Graham & Trotman/Martinus Nijhoff, London.
Jones, C.G.,. Lawton, J.H. and Shachak, M. (1994). Organisms as ecosystem engineers. *Oikos*, 69: 373–386.
Jones, J.A. (1990). Termites, soil fertility and carbon cycling in dry tropical Africa: a hypothesis. *J. Trop. Ecol.*, 6: 291–305.
Lambshead, P.J.D. (1993). Recent developments in marine benthic biodiversity research. *Océanis*, 19: 5–24.
Lawton, J.H. (1994a). What do species do in ecosystems? *Oikos* 71: 367–374.
Lawton, J.H. (1994b). What will you give up? John Lawton's 'View from the Park' 12. *Oikos* 71: 353–354.
Lawton, J.H. and Brown, V.K. (1993). Redundancy in ecosystems. In: Schulz, E-D. and. Mooney, H,A. (eds), *Biodiversity and ecosystem function*. Ecological Studies No. 99, pp.255–270. Springer-Verlag, Berlin.
Lawton, J.H., Bignell, D.E., Bloemers, G.F., Eggleton, P. and Hodda, M.E. (1996). Carbon flux and diversity of Nematoda and termites in Cameroon forest soils. *Biodiversity and Conservation*, 5: 261–273.
Likens, G.E. (1991). Human accelerated environmental change. *BioScience*; 41: 130
Lock, J.M. and Eltringham, S.K. (1985). In Friday, A. and Ingram, D.S. (eds), *The Cambridge Encyclopaedia of Life Sciences*. Cambridge University Press, Cambridge
Lodge, D.M. (1993). Biological invasions: lessons for ecology. TREE, 8: 133–137.
Lovejoy, T.E. (1980). A projection of species extinctions. In: Council on Environment Quality (CEQ), *The Global 2000 Report to the President*, Volume 2, pp 328–331. CEQ, Washington D.C.
Lugo, A.E. (1988). Estimating the reductions in the diversity of tropical forest species. In: Wilson, E.O. and Peter, F.M. (eds), *Biodiversity*, pp. 58–70. National Academy Press, Washington D.C.
McNeely, J.A. (1988). *The economics of biological diversity*. IUCN, Gland, Switzerland.
McNeely, J.A. (1990). Climate change and biological diversity: policy implications. In: Boer, M.M., and de Groot, R.S. (eds), *Landscape-ecological impact of climatic change*, pp. 406–429. IOS Press, Amsterdam.
McNeely, J.A. (1993). Economic incentives for conserving biodiversity – lessons for Africa. *Ambio*, 22: 144–150.
McNeely, J.A., Gadgil, M., Levèque, C., Padoch, C. and Redford, K. (eds). (1995). Human influences on biodiversity. UNEP, *Global Biodiversity Assessment*, pp. 711–823, Cambridge University Press, Cambridge.

MacPherson, E. and Duarte, C. M. (1994). Patterns in species richness, size and latitudinal range of East Atlantic fishes. *Ecography*, 17: 242–248.

Mace, G.M. (1995). Classification of threatened species and its role in conservation planning. In Lawton, J.H. and May, R.M. (eds), *Extinction rates*, pp. 197–213. Oxford University Press, Oxford.

Mattile, L. (1990). Recherches sur la systématique et l'évolution des Keroplatidae (Diptera, Mycetophiloidea). *Mém. Mus. nat. Hist. nat.* (A) *Zool.*, 148: 1–682.

Mawdsley, N.A. and Stork, N.E. (1995). Species extinctions in insects: ecological and biogeographical considerations. In: Harrington, R. and Stork, N.E. (eds), *Insects in a changing environment*, pp. 321–369. Academic Press, London.

May, R.M. (1992a). How many species inhabit the earth?. *Scientific American* October: 18–24.

May, R.M. (1992b). Bottoms up for the oceans. *Nature* 357: 278–9.

May, R.M., Lawton, J.H. and Stork, N.E. (1995). Assessing extinction rates. In: Lawton, J.H. and May, R.M. (eds), *Extinction rates*. Oxford University Press, Oxford.

Mayr, E. (1969). *Principles of systematic zoology*. McGraw-Hill, New York.

Memmott, J., Godfray, H.C.J. and Gauld, I.D. (1994). The structure of a typical host-parasitoid community. *J. Animal Ecol.*, 63: 521–540.

Merrett, N. (1994). Reproduction in the North Atlantic oceanic ichthyofauna and the relationship between fecundity and species' sizes. *Environ. Biol. Fishes*, 41: 207–245.

Mooney, H.A., Lubchenco, J., Dirzo, R. and Sala, O.E. (1995). Biodiversity and ecosystem function: Ecosystem analysis. In UNEP, *Global Biodiversity Assessment*, pp. 327–452. Cambridge University Press, Cambridge.

Moulton, M.P. and Pimm, S.L. (1986). Species introductions to Hawaii. In: Mooney H.A. and Drake, J.A. (eds), *Ecology of biological invasions of North America and Hawaii*, pp. 231–249. (Ecological Studies No. 58), Springer-Verlag, Berlin.

Myers, N. (1979). *The Sinking Ark. A New Look at the problem of disappearing species*. Pergamon, New York.

Myers, N. (1988). Threatened biotas: "hot spots" in tropical forests. *The Environmentalist*, 8: 187–208.

Naeem, S., Thompson, L.J., Lawler, S.P., Lawton, J.H. and Woodfin, R.M. (1994). Declining biodiversity can alter the performance of ecosystems. *Nature*, 368: 734–737.

Naeem, S., Thompson, L.J., Lawler, S.P., Lawton, J.H. and Woodfin, R.M. (1995). Empirical evidence that declining species diversity may alter the performance of terrestrial ecosystems. *Phil. Trans. R. Soc. Lond. Series B.*, 347: 249–262.

Nelson, G.J. (1970). Outline of a theory of comparative biology. *Syst. Zool.*, 19: 373–384.

Nelson, J.S. (1984). *Fishes of the World*. (edn.2) John Wiley, New York.

Nixon, K.C. and Wheeler, Q.D. (1990). An amplification of the phylogenetic species concept. *Cladistics*, 6: 211–223.

Norse, E.A. (1993). *Global marine biodiversity: a strategy for building conservation into decision making*. Island Press, Washington DC.

Pankhurst, R.J. (1991). *Practical taxonomic computing*. Cambridge University Press, Cambridge.

Pearce, D.W. and Moran, D. (1994). *The economic value of biodiversity*. Earthscan, London.

Pearse, V. (ed.) (1987) *Living invertebrates*. Blackwell Scientific/Boxwood Press, Palo Alto, California.

Peters, R.L. and Lovejoy, T.E. (eds). (1992). Global warming and biological diversity. Yale University Press, USA.

Phillipson, J. (1966). *Ecological energetics*. Arnold, London.

Pianka, E.R. (1983). *Evolutionary ecology*. (edn.3) Harper & Row, New York

Pimm, S.L., Moulton, M.P. and Justice , L.J. (1994). Bird extinctions in the central Pacific. *Phil. Trans. R. Soc. Lond.* B.

Prance, G.J. (1983). Pesquisas botanicas e a conservaçao da floresta Amazonica. In: *Anais do XXXIV Congresso Nacional de Botanico* Vol 1; pp. 63–71. Puerto Alegre, Rio Grande do Sul, Brazil.

Prance, G.T. (1990). Fruits of the rainforest. *New Scientist*, 13 January: 42–45.

Prendergast, J.R., Quinn, R.M., Lawton, J.H., Eversham, B.C. and Gibbons, D.W. (1993). Rare species, the coincidence of diversity hotspots and conservation strategies. *Nature*, 365: 335–337.

Press, F. and Siever, R. (1986). *Earth*. (edn. 4) W.H Freeman, New York.

Raven, P.H. (1987). The scope of the plant conservation problem world-wide. *In*: Bramwell, D., Hamann, O., Heywood, V. and Synge, H. (eds.), *Botanic Gardens and the World Conservation Strategy*, pp. 19–29. Academic Press, London.

Raven, P.H. (1988a). Biological resources and global stability. *In*: Kawano, S., Connell, J.H. and Hidaka, H. (eds). *Evolution and coadaptation in biotic communities*, pp. 3–27. University of Tokyo Press, Tokyo.

Raven, P.H. (1988b). Our diminishing tropical forests. *In*: Wilson, E.O. and Peter, F.M. (eds), *Biodiversity*; pp. 119–122. National Academy Press, Washington D.C..

Reid, W.V. (1992). How many species will there be?. *In*: Whitmore, T.C. and Sayer, J.A. (eds), *Tropical deforestation and species extinction*, pp. 55–73. Chapman and Hall, London.

Reiners W.A. (1983). Disturbance and basic properties of ecosystem energetics. In: Mooney, H.A. and Godron, M. (eds), *Disturbance and ecosystems*, pp 83–98. (Ecological Studies 44). Springer-Verlag, New York.

Reports to the Nation on our Changing Planet, (1994). El Niño and climate prediction. Spring 1994, No. 3.

Rex. M.A. (1983). Geographic patterns of species diversity in deep-sea benthos. In:. Rowe, G.T. (ed.) *Deep-sea biology*. Volume 8, *The Sea*. New York: John Wiley, New York.

Rex, M.A.,. Stuart, C.T., Hessler, R.R., Allen, J.A., Sanders, H.L. and Wilson, G.D.F. (1993). Global-scale latitudinal. patterns of species diversity in the deep-sea benthos. *Nature*, 365: 636–639.

Seanger, P., Hegerl, E.J. and Davie, J.D.S. (eds). (1983). *Global status of mangrove ecosystems*. (Commission on Ecology Papers Number 3), IUCN, Gland, Switzerland.

Sands, W.A. (1992). The termite genus *Amitermes* in Africa and the Middle East. *Bull. Nat. Res. Inst.*, 51: i–iv, 1–140.

Schindler, D.W. and S.E. Bayley. (1990). Freshwaters in cycle. In: Mungall, C. and McLaren, D.J. (eds), *Planet under stress*, pp. 149–167. Oxford University Press, Oxford.

Simberloff, D. (1986). Are we on the yerge of a mass extinction in tropical rain forests? *In*: Elliot, D.K. (ed.), *Dynamics of extinction*, pp. 165–180. Wiley, New York.

Scotese, C.R. (1984). An introduction to this volume: Paleozoic paleomagnetism and the assembly of Pangea. *Geodyn.* Ser., 12: 1–10.

Sengör, A.M.C., Altmer D., Cin, A., Ustaömer, T. and Hsü, K.J. (1988). Origin and assembly of the Tethyside orogenic collage at the expense of Gondwana Land. In Audley-Charles, M.G. and Hallam, A. (eds), *Gondwana and Tethys* (Geological Soc. Spec. Publ. 37), pp. 119–81. University Press, Oxford.

Simberloff, D. (1986). Introduced insects: A biogeographic and systematic perspective. In: Mooney, H.A. and Drake, J.A. (eds), *Ecology of biological invasions of North America and Hawaii*, pp. 231–249. (Ecological Studies 58), Springer-Verlag, Berlin.

Smith, F.D.M., May, R.M., Pellew, R., Johnson, T.H. and Walter, K.R. (1993). How much do we know about the current extinction rate?. TREE, 8: 375–378.

Soulé, M.E. and Kohm, K.A. (eds). (1989). *Research Priorities for Conservation Biology*. Island Press, Washington, DC.

Stork, N.E. (1988). Insect diversity: facts, fiction and speculation. *Biol. J. Linn. Soc.*, 35: 321–337.

Stork, N.E. (1991). The composition of the arthropod fauna of Bornean tropical rain forest trees. *J. Trop. Ecol.*, 7: 161–180.

Stork, N.E. (1993). How many species are there?. *Biodiversity and Conservation*, 2: 215–232.

Stork, N.E. and Eggleton, P. (1992). Invertebrates as determinants and indicators of soil quality. *Amer. J. Alternative Agric.*, 7: 38–47.

Stork, N.E. and Lyal, C.J.C. (1993). Extinction or 'co-extinction' rates? *Nature*, 366: 307.

Swift, M.J. and Anderson, J.A. (1993). Biodiversity and ecosystem function in agricultural systems. In: Schultz, E.D. and Mooney, H.A. (eds), *Biodiversity and ecosystem function*, pp. 15–41.(Ecological Studies No. 99), Springer-Verlag, Berlin.

Tilman, D. and Downing, J.A. (1994). Biodiversity and stability in grasslands. *Nature*, 367: 363–365.

Tolba, M.K. and El-Kholy, O.A. (eds). (1992). *The World environment* 1972–1992. *Two decades of change*. Chapman & Hall, London.

Udvardy, M.D.F. (1975). A *classification of the biogeographical provinces of the world*, (IUCN Occ. Pap. 18). IUCN, Gland.

UNEP (1995). *Global Biodiversity Assessment*. Cambridge University Press, Cambridge.

UNEP. (1993). *Guidelines for country studies on biological diversity*, UNEP, Nairobi.

UNESCO (1994) *Biodiversity* No. 7. *Science, conservation and sustainable use*. UNESCO, Paris.

Viette, P. (1990). A provisional check-list of the Lepdioptera Heterocera of Madagascar. *Faune de Madagascar (supplément 1)*: 1–263.
Vitousek, P.M. (1988). Diversity and biological invasions of oceanic islands. In: Wilson, E.O. (ed), *Biodiversity*, pp. 181–189. National Academy Press, Washington, DC.
Vitousek, P.M. (1994). Beyond global warming: Ecology and global change. *Ecology*, 75: 1861–1876.
Vitousek, P.M., Ehrlich, P.R., Ehrlich, A.H. and Matheson, P.A. (1986). Human appropriation of the products of photosynthesis. *BioScience*, 36: 368–373.
Vitousek, P.M. and Hooper, D.U. (1993). Bioligical diversity and terrestrial ecosystem biogeochemistry. In: Schulze, E.D. and Mooney, H.A. (eds), *Biodiversity and ecosystem function*, pp. 3–14.(Ecological Studies No. 99), Springer-Verlag, Berlin.
WCMC. (1992). *Global biodiversity. Status of the Earth's living resources*. Chapman and Hall, London.
Walker, B.H. (1992). Biodiversity and ecological redundancy. *Conservation Biology*, 6: 18–23.
Walter, D.R., Du, R. and Horodyski, R.J. (1990). Coiled carbonaceous megafossils from the Middle Proterozoic of Jixian (Tianjin) and Montana. *Amer. J. Sci.*, 290: 133–148.
Watson, L. and Dallwitz, M.J. (1988). *Grass genera of the World*. The Australian National University, Canberra.
WCED. (1987). *Our Common Future The BrundHand Report for the World Commission on Environment and Development*, 1987. Oxford University Press, Oxford.
Weir, B.S. (1990). *Genetic data analysis*. Sinauer., Sunderland, Mass.
Wheeler, Q.D. (1990) Insect diversity and cladistic constraints. *Ann. Entom. Soc. Amer.*, 83: 1031–1047.
Whittaker, R.H. (1959). On the broad classification of organisms. *Q. Rev. Biol*, 34: 210–226.
Wiley, E.O. (1981). *Phylogenetics. The theory and practice of phylogenetic systematics*. John Wiley, New York.
Wilson, E.O. (1988). The current state of biological diversity. In: Wilson, E.O. and Peter, F.M. (eds.), *Biodiversity*, pp.3–18. National Acedemy Press, Washington D.C.
Wilson, E.O. (1989). Threats to biodversity. *Scient. Amer.* September: 108–116.
Wilson, E.O. (1992). *The diversity of life*. Penguin Press/Allen Lane, Harmondsworth.
Woodwell, G.M. (1983). The Blue Planet: of Wholes and Parts and Man. In: Mooney, H.A. and Godron, M, (eds), *Disturbance and ecosystems*, pp. 2–10. (Ecological Studies 44). Springer-Verlag, New York.
World Commission on Environment and Development. (1987). *Our Common Future*. Oxford University Press, New York.
WRI. (1994). *World Resources 1994–95. A guide to the global environment*. Oxford University Press, New York.
WRI/IUCN/UNEP. (1992). *Global biodiversity strategy*. WRI, Washington DC.
Woese, C.R., Kandler, O. and Wheelis, M.L. (1990). Towards a natural system of organisms: proposal for the domains Archawa, Bacteria and Eucaria. *Proc. Natal. Acad. Sci.*, USA, 87: 4576–4579.
Wuethrich, B. (1995). El Niño goes critical. *New Scientist* 1963: 32–35.
Zijlstra, J.J. and Baars, M.A. (1990). Productivity and fisheries potential of the Banda Sea ecosystem. In: Sherman, K., Alexander, L.M. and Gold, B.D. (eds.), *Large Marine Ecosystems; patterns, processes and yields*, pp. 54–65. Amer. Assoc. Advanc. Sci., Washington D.C.

LEGAL AND ETHICAL ASPECTS OF BIODIVERSITY ASSESSMENT:

The Convention on Biological Diversity

2.1 **Introduction**

2.2 **Intellectual Property Rights**

2.3 **The Convention on Biological Diversity**

2.4 **The role and responsibilities of the research institute and scientist**

2.5 **The importance of research agreements**

2.5.1 Collecting permits, research agreements and conservation

2.5.2 Who needs research agreements?

2.5.3 Who signs research agreements?

2.5.4 Biological damage versus economic benefits

2.5.5 Protecting biodiversity information

2.6 **Sharing the profits**

2.7 **Responsibilities of the commercial company or broker**

2.8 **References and bibliography**

Chapter authorship

This chapter has been drafted and edited by Clive Jermy with additional material and comments by D. Cutler, S. Linington, N. McGough, J. Ratter, M. Sands, N. Stork, M. Simmonds *and* P. Waterman.

2.1 Introduction

Chapter One discussed the framework for biodiversity assessment and emphasised the significance of the Convention on Biological Diversity in developing an international approach to the conservation and use of a country's biodiversity. Chapter Three develops the international aspects further. Increasing recognition of the value of biodiversity has also begun to crystallize international thought on sovereign rights over biological resources, both in the way they are exploited and how access to them should be controlled.

The collection of biological material for commercial purposes (termed bioprospecting) now covers a wide range of organisms and, although plants have been to the fore hitherto, microorganisms, marine invertebrates and insects are now becoming the focus of attention. It is not only those who collect material for bioassay and commercial ventures that are affected. Those who collect – and export – material for academic pursuits such as taxonomic studies can be involved, as can the guardians of the herbaria or museums in whose care those specimens may eventually reside. Similarly, in academic research areas of plant chemistry, studies elucidating species relationships, metabolic pathways, and effects of environmental stress on metabolism are providing a wealth of new information and claiming the attention of the chemical industry (see **Boxes 2.1** & **2.2**).

The ultimate trade results from such prospecting surveys (e.g. pharmaceutical; new strains of crop plants), although relevant to a country's economics, are not discussed here. The policies and programmes established by the leading pesticide and pharmaceutical companies are difficult to obtain other than in outline statements that rarely cover ethical questions in detail (but see **Box 2.16**). As Aylward (1993) points out, pharmaceutical prospecting in particular, could be a profitable endeavour, depending on a country's ability to use some combination of contractual arrangements, or property rights, to capture that pharmaceutical value. There is a tendency, however, perhaps understandably, to divorce prospecting and collection for chemical analysis, from the present (or future) ownership question because of the vast financial investment needed to realise the ultimate value. When this is removed from the equation the value of the source material seems unproportionately small.

The question of ethics in ethnobiological research is discussed clearly by Tony Cunningham (1993) in a report commissioned as part of the WWF/UNESC/RBGK '*Plants and the People*' Initiative, and in which he gives an account of the issues involved. In some cases traditional healers asked a research ethnobotanist not to report the aboriginal use of particular species. In other cases, traditional healers recognise that their knowledge has a wider value and want part of the benefits arising from its use. The three main issues which have been the focus of recent debate are: (i) local knowledge, farmers' rights and equity in the distribution and control of genetic resources from crop plants (Kloppenburg, 1988; Mooney, 1983); (ii) an increased interest in chemotaxonomy and the linked analysis of new plants for pharmaceuticals, waxes, oils and perfumes; (iii) the distribution of economic benefits from plants with a horticultural potential (e.g. *Saintpaulia*, *Hippeastrum*).

This whole issue of rights is not a new subject, and was being considered by some parties well before the Convention was established. It is complex and covers a range of aspects from ownership and access to exploitation and benefit sharing. Articles of the Convention that relate to these aspects are discussed in this Chapter together with other ethical considerations concerned with biodiversity assessment.

2.2 Intellectual Property Rights

Legislation governing Intellectual Property Rights (IPR) varies from country to country but the end effect is to create the legal right where potential users must seek the holder's permission before commercialising the intellectual property. The concept of 'common heritage' to wild plants and animals may be agreeable to some of those developing the potential of genetic resources but to remove legal ownership from the results of breeding (e.g. crop cultivars) would be counter-productive to the further development of those

Box 2.1 Biochemical studies at the Royal Botanic Gardens, Kew

Biochemical studies carried out in the Jodrell Laboratory at Kew are primarily fundamental studies in chemotaxonomy and into the role of secondary compounds in the taxonomy of plants. During this work the compounds are exposed to an array of bioassays to investigate their biological activity. Any extracts or compounds that show activity are studied further at Kew or in collaboration with pharmaceutical or academic partners. Nowadays a high proportion of studies are undertaken with students from other countries so that the techniques to isolate and identify compounds can be learnt and applied by these students in their home countries, thus developing the spirit of Article 18 of the Convention on Biological Diversity.

At Kew in the last 10 years phytochemical and biological studies have been carried out on over 4,050 species of plants out of the 25–30,000 species available in the living collection in the gardens. Random screening of 1,050 species identified 25 with potent anti-insect activity of which 10 species contained known anti-insect compounds. Out of the other 3,000 species selected because of taxonomic or biological interest, 860 species have been shown to have activity in insect, protozoa, HIV, cancer, diabetes or fungal screens. The active substances in these plants are currently being studied.

Biochemical research has shown that more needs to be learnt about the evolution of the metabolic pathways involved in producing compounds in different plant families. Nor is enough known about the variability of compounds within plant genera and the speed with which plants have been evolving compounds and discarding others. Plants under stress are known to produce secondary metabolites but we do not yet know how environmental stresses like ozone pollution will influence the metabolic pathways in plants. This could result in an increase in the diversity and concentration of secondary metabolites in plants. The level of specific compounds and the profile of compounds within a plant varies not only with the age of the plant or its parts but by the plant's locality (geographical factor) and its habitat (ecological factor). Leaf chemistry is known to be more variable than seed chemistry, thus chemotaxonomy studies undertaken on seeds can provide a clearer picture of the presence or absence of a type of compound in a plant species.

Chemotaxonomic studies have so far helped in the isolation of biologically active compounds that have, or are being, developed into drugs; e.g. indole alkaloids (Kutchan, 1989), polyhydroxy-alkaloids in legumes which have been shown to have potent activity against the HIV virus (Fellows *et al.*, 1989) and steroids (Dev, 1989). The time spent by the pharmaceutical industry investigating plants depends not only on the size of their research and development budgets but also on the leads given to them by those studying systematic plant chemistry.

(*Source*: *Monique S.J. Simmonds, Royal Botanic Gardens, Kew.*)

genetic resources. This is open to discussion. It is, for example, not always accepted by countries which have substantial genetic resources and considerable indigenous knowledge. Furthermore, the concept that was widely held that all wild organisms are freely available to all, and belong to no one, means there is little incentive to conserve either species or habitats. This is no longer so under the Convention on Biological Diversity which introduced the significant change that States have sovereignty over their own resources. Many indigenous societies depend on useful plants within a restricted resource catchment and therefore have strong incentives to protect both the plants and the knowledge concerning them.

There are many forms of IPR, including copyright on scientific publications, computer software and databases, which can affect technological transfer encouraged under the Convention on Biological Diversity. Several types of IPR are discussed by Gollin (1993) and Glowka *et al.* (1994) and from which the following has been summarised.

Box 2.2 Investigation of resins from Burseraceae

A characteristic feature of many species of the family Burseraceae is the production of resins, many of which are exuded from the bark, sometimes naturally and sometimes only in response to injury. A number of these resins have long been items of commerce, the most famous of these being frankincense (from *Boswellia* species) and myrrh (from *Commiphora* species).

The University of Strathclyde initiated a study of these resins and visited southern Ethiopia and northern Kenya where Burseraceae is a major component of the flora. In Ethiopia there is a well regulated commerce in olibanum, a fragrant resin originating from *Boswellia*, while in Kenya many resins are collected from the wild and simply sold in the market-place. The first project was to document, chemically, the resins from Kenyan species, in order to assess their similarities and differences to the 'benchmark' commercial products. It was hoped that this would allow a more structured and selective collecting programme leading to better quality (and hence more valuable) resins for the market. Studies revealed that within Kenya only a very low grade *Boswellia* resin was available (luban) while chemical variability in the myrrh-like resins from several *Commiphora* species meant that an extensive stock improvement programme would be necessary to produce a uniform product.

The isolation of hypolipaemic sterols from the resin of *Commiphora mukul*, an Indian species, prompted us to examine Kenyan resins for useful biological activity using the screens employed by the University's Institute for Drug Research. One species, eventually identified as *Commiphora kua*, yielded a resin which exhibited significant anti-inflammatory activity. From this initial observation we proceeded to separate and then identify the active substances. These turned out to be two novel chemicals which we have named mansumbinone (I) and mansumbinoic acid (II).

Unfortunately, while the activity present is significant, it was not sufficient to warrant development and the results of this investigation have been published in the journal *Planta Medica*.

H.D.V. Prendergast

***Boswellia sacra*, a Frankincense-producing species seen here growing in Dhofar, Oman.**

Another area where the resins have shown biological activity is in antimicrobials. This is not surprising as their probable function in the plant is to seal wounds and it would be beneficial in this context if they could also prevent infection. The very liquid resin produced by *Commiphora restrata* has been examined in some detail and shown to be very effective in preventing infection of that species after damage and many other species have been demonstrated to have antifungal and antibacterial properties. This activity is not restricted to species of the arid zones of Africa and the two attracting most interest in our laboratories currently are a *Canarium* species from West African rain forest and a *Protium* species from the Amazon.

(Source: Peter G.Waterman, Strathclyde Institute for Drug Research, University of Strathclyde.)

2.2.1 Utility patents

Patents are usually conceived as a protection order relating to any process, machine or composition of nature which is novel, useful and embodies an inventive step (Glowka *et al.*, 1994). Patent protection is, at the moment, a matter for national legislation. Species themselves cannot be patented unless biotechnology has been employed to develop a product, e.g. a genetic variation or a change in a metabolic byproduct. The USA was the first to extend patent protection to a living organism in 1980, thereby initiating a debate amongst other OECD countries as to whether they should do likewise. (See also 2.2.4 below).

2.2.2 Plant Breeders Rights

Plant Breeders Rights (PBRs) are recognised internationally through the 1961 *International Convention for the Protection of New Varieties of Plants* (UPOV), later amended in 1978, and again in 1991. They are granted to varieties which have shown to be Distinct, Uniform and Stable (DUS testing). PBRs are only granted for a limited period of time, at the end of which the variety passes into the public domain. PBRs are usually attached to varieties of plants with an agricultural or horticultural value that are commercially distributed. UPOV members are able to define the scope of the PBRs within their national legislation (Glowka *et al.*, 1994). Seed and other propagating material cannot be commercially marketed without prior authorisation of the holder of the PBR, who is then entitled to receive a royalty from that sale. However, the authorisation of the holder of the PBR is not required when the variety is used for research purposes, including use in the breeding of further new varieties. No plant variety can be monopolised through absolute ownership of subsequent generations although those propagated by vegetative means can be protected for many years (Cabinet Office, 1992).

Both patents and PBRs have an indirect and beneficial effect on plant genetic resource conservation, by encouraging investment in new technologies and the development of new varieties, and at the same time affording a mechanism through ownership/licensing agreements for owners of genetic resources to be recompensed when material is developed and exploited commercially.

2.2.3 Trade secrets

Trade secrets are used to protect subject matter which is either unpatentable or because the holder does not want to publish the subject matter publicly (Glowka *et al.*, 1994). Trade secrets can be applied to a wide range of subjects. For instance, information collected by an anthropologist/plant collector from, for instance, a traditional healer, and subsequently transcribed to a herbarium label attached to a voucher specimen, or otherwise published in a report, could be considered as a trade secret. Using such information to develop a new drug or crop without the permission of the originator may mean the informer has a claim on resulting profits. Enforcing a trade secret in law is difficult and disputes turn on the rigour of efforts made to maintain secrecy and the extent of public disclosure whenever duty of confidentiality is imposed – which may be considerable in the indigenous community (Gollin, 1993).

2.2.4 Trade related aspects of IPRs

The issue of intellectual property protection as it relates to trade was discussed as part of the Uruguay Round of GATT talks and included the topic of extending patent protection to living organisms. Members may, at their own discretion, offer patent protection to all eligible inventions using genetic resources, with mandatory protection for those involving microorganisms. Also under TRIPs (Trade-Related aspects of Intellectual Property), countries will have to consider plant varieties for protection using PBRs or straight forward patents where applicable. The TRIPs Agreement also requires members to protect trade secrets (Glowka *et al.*, 1994).

2.3 The Convention on Biological Diversity

The Objectives of the Convention on Biological Diversity are stated in **Article 1** (see **Box 2.3**). The *"conservation of biodiversity"* and its *"sustainable use"* are not the subject of this chapter but the third part, *"fair and equitable sharing of the benefits"* (such as those which come from applying the technology as to how to develop the potential of the plant, animal or microorganism, for example, for breeding or pharmaceutical use)

Box 2.3 The Convention on Biological Diversity
Article 1. Objective and Article 3. Principle

Article 1

The Objectives of this Convention, to be pursued in accordance with its relvant positions, are the conservation of biological diversity, the sustainable use of its components and the fair and equitable sharing of the benefits arising out of the utilization of genetic resources, including by appropriate access to genetic resources and by appropriate transfer of relevant technologies, taking into account all rights over those resources and to technologies, and by appropriate funding.

Article 3

States have, in accordance with the Charter of the United Nations and the principles of international law, the sovereign right to exploit their own resources pursuant to their own environment policies, and the responsibility to ensure that activities within their jurisdiction or control do not cause damage to the environment of other States or areas beyond the limits of national jurisdiction.

(*Source*: UNEP, *June* 1994.)

is an aspect considered here. The Article suggests three methods of achieving its aim (i) to give "*appropriate access to genetic resources*", (ii) to "*make relevant knowledge gained available to others*" (usually the owner country, taking into account all rights over these resources to technologies); and by "appropriate funding" (i.e. returning some of the gains). These are elaborated below.

Article 3 outlines the *Principles* of the Convention (see **Box 2.3**). As Glowka *et al.* (1994) point out the legal nature and significance of 'principles' in an international convention are a matter of controversy: there is no consensus on what distinguishes 'principles' from 'obligations and rights'. The principle regognises the 'sovereign right' of States to exploit their own resources according to their own environmental policies, providing that activities on their territory or under their control do not damage other States or areas beyond national jurisdiction, that is, the high seas, the deep sea-bed or the atmosphere. They must also observe the UN Charter and the principles of environmental protection which are part of international law.

The Convention is discussed in this chapter only in relation to ownership of biodiversity and related information which may impinge on scientific assessments and collection of data and specimens. All concerned with interpreting or acting within the Convention in any capacity at all should have at hand Glowka *et al.*, (1994) which discusses in detail the meaning and implication of each Article. It is, however, only those Articles that relate in depth to genetic resources and technical knowledge (**Articles 15–19**; see **Boxes 2.4–2.8**) that will be further discussed here. They are discussed in more detail by Glowka *et al.* and, in this same context but to a lesser extent, by Sands (1993).

Article 15(1) recognises the sovereignty of States over the access to genetic resources within its jurisdiction but it does not grant the State a property right over these resources. The latter is a matter for national legislation (which rarely exists) to determine the ownership of those genetic resources. Ownership, *per se*, is not addressed under the Convention.

Article 15(2) is an important clause in that it encourages Contracting Parties to cooperate with each other (i.e. 'facilitate access to genetic resources') thereby encouraging other States to join the Convention. This could include promoting the unrestricted exchange of genetic resources for research and non-commercial purposes (see also 15(6) below). Other aspects and its historical relationship to the 1983 FAO International Undertaking on Plant Genetic Resources (see below) are discussed by Glowka *et al.* (1994).

Article 15(4) allows for access to genetic resources on mutually agreed terms and subject to prior informed consent being given by the

Box 2.4 The Convention on Biological Diversity
Article 15. Access to genetic resources

M. Ambrose/John Innes Centre

'Purple Pod' a cultivar in the gene bank of the Pea Collection at JIC, Norwich.

1 Recognising the sovereign rights of States over their natural resources, the authority to determine access to genetic resources rests with the national governments and is subject to national legislation.

2 Each Contracting Party shall endeavour to create conditions to facilitate access to genetic resources for environmentally sound use by other Contracting Parties and not to impose restrictions that run counter to the objectives of this Convention.

3 For the purpose of this Convention, the genetic resources being provided by a Contracting Party, as referred to in this Article and Articles 16 and 19, are only those that are provided by Contracting Parties that are countries of origin of such resources or by the Parties that have acquired the genetic resources in accordance with this Convention.

4 Access, where granted, shall be on mutually agreed terms and subject to the provisions of this Article.

5 Access to genetic resources shall be subject to prior informed consent of the Contracting Party providing such resources, unless otherwise determined by that Party.

6 Each Contracting Party shall endeavour to develop and carry out scientific research based on genetic resources provided by other Contracting Parties with the full participation of, and where possible in, such Contracting Parties.

7 Each Contracting Party shall take legislative, administrative or policy measures, as appropriate, and in accordance with Articles 16 and 19 and, where necessary, through the financial mechanism established by Articles 20 and 21 with the aim of sharing in a fair and equitable way the results of research and development and the benefits arising from the commercial and other utilization of genetic resources with the Contracting Party providing such resources. Such sharing shall be upon mutually agreed terms.

(*Source*: UNEP, *June* 1992.)

contracting party. This assumes that efficient national systems, including appropriate institutions and administrative arrangements, are in place to cope with the demand for access to genetic resources. Previous to the Convention, the principle of free access to genetic resources prevailed and is recognised in the FAO Undertaking on Plant Genetic Resources 1983 (see Glowka *et al.*, 1994, p.78). Now, under the Convention, Article 15 recognises that the authority to determine access to genetic resources rests with the national governments and is subject to national legislation. However, no legal claim can be made by a Contracting Party for the past – *and future* – use of any genetic resource obtained before the Convention came into force.

Article 15(6) is particularly pertinent to assessing a State's biodiversity and genetic resources as it encourages joint programmes of scientific research into those resources with an aim to build scientific research capacity in the Parties providing the genetic resource. Such cooperation could also clarify the intellectual property rights of the research partners.

Article 15(7) requires that the results of research and development and the benefits arising from the commercial and other uses of

Box 2.5 The Convention on Biological Diversity
Article 16. Access to and Transfer of Technology

1 Each Contracting Party, recognising that technology includes biotechnology, and that both access to and transfer of technology among Contracting Parties are essential elements for the attainment of the objectives of this Convention, undertakes subject to the provisions of this Article to provide and/or facilitate access for, and transfer to, other Contracting Parties of technologies that are relevant to the conservation and sustainable use of biological diversity, or make use of genetic resources, and do not cause significant damage to the environment.

2 Access to and transfer of technology referred to in paragraph 1 above to developing countries shall be provided and/or facilitated under fair and most favourable terms, including on concessional and preferential terms where mutually agreed, and, where necessary, in accordance with the financial mechanism established by Articles 20 and 21. In the case of technology subject to patents and other intellectual property rights, such access and transfer shall be provided on terms which recognise and are consistent with the adequate and effective protection of intellectual property rights. The application of this paragraph shall be consistent with paragraphs 3, 4 and 5 below.

3 Each Contracting Party shall take legislative, administrative or policy measures, as appropriate, with the aim that Contracting Parties, in particular those that are developing countries, which provide genetic resources are provided access to and transfer of technology which makes use of those resources, on mutually agreed terms, including technology protected by patents and other intellectual property rights, where necessary, through the provisions of Articles 20 and 21 and in accordance with international law and consistent with paragraphs 4 and 5 below.

4 Each Contracting Party shall take legislative, administrative or policy measures, as appropriate, with the aim that the private sector facilitates access to, joint development and transfer of technology (referred to in paragraph 1 above) for the benefit of both governmental institutions and the private sector of developing countries and in this regard shall abide by the obligations included in paragraphs 1, 2 and 3 above.

5 The Contracting Parties, recognising that patents and other intellectual property rights may have an influence on the implementation of this Convention, shall cooperate in this regard, subject to national legislation and international law, in order to ensure that such rights are supportive of and do not run counter to its objectives.

(*Source*: UNEP, *June* 1992.)

genetic resources be shared in *"a fair and equitable way"*, and on mutually agreed terms. To be effective this would require an arrangement for sharing financial and other benefits derived from genetic resources that is acceptable to developed and developing countries. This is not a simple matter as it may be a decade or more before the benefit is realised. Furthermore, the genetic material used by a company or research consortium may have been collected from several sources. Article 15 does not address the contributions of indigenous people and farmers to the maintenance and development of genetic diversity through years of cultivation and husbandry (Glowka *et al.*, 1994). Futhermore, it clearly gives no rights to indigenous peoples with respect to royalties deriving from ethnobotanical materials (Barton, 1994). Such local knowledge is referred to, however, in **Article 8** (In-situ Conservation), where **para. 8j** includes the phrase, *"encourage the equitable sharing of the benefits arising from the utilization of such knowledge, innovations and practices"*.

Box 2.6 The Convention on Biological Diversity Article 17. Exchange of Information

1 The Contracting Parties shall facilitate the exchange of information, from all publicly available sources, relevant to the conservation and sustainable use of biological diversity, taking into account the special needs of developing countries.

2 Such exchange of information shall include exchange of results of technical, scientific and socio-economic research, as well as information on training and surveying programmes, specialised knowledge and indigenous and traditional knowledge as such and in combination with the technologies referred to in Article 16, paragraph 1. It shall also, where feasible, include repatriation of information.

Mark Hakansson/British Council Press Office

A user of the British Council's library in Johannesburg

(*Source*: UNEP, *June* 1992.)

Box 2.7 The Convention on Biological Diversity Article 18. Technical and Scientific Cooperation

1 The Contracting Parties shall promote international technical and scientific cooperation in the field of conservation and sustainable use of biological diversity, where necessary, through the appropriate international and national institutions.

2 Each Contracting Party shall promote technical and scientific cooperation with other Contracting Parties, in particular developing countries, in implementing this Convention, *inter alias*, through the development and implementation of national policies. In promoting such cooperation, special attention should be given to the development and strengthening of national capabilities, by means of human resources development and institution building.

3 The Conference of the Parties, at its first meeting, shall determine how to establish a clearing-house mechanism to promote and facilitate technical and scientific cooperation.

4 The Contracting Parties shall, in accordance with national legislation and policies, encourage and develop methods of cooperation for the development and use of technologies, including indigenous and traditional technologies, in pursuance of the objectives of this Convention. For this purpose, the Contracting Parties shall also promote cooperation in the training of personnel and exchange of experts.

5 The Contracting Parties shall, subject to mutual agreement, promote the establishment of joint research programmes and joint ventures for the development of technologies relevant to the objectives of this Convention.

(*Source*: UNEP, *June* 1992.)

Article 16 outlines some of the principles relating to technology transfer and its possible relationship to intellectual property laws. It recognises that the existing property rights (e.g. those set up by tradition or culture) may have to be limited. Similarly, some related international conventions, e.g. 1961 *International Convention for the Protection of New Varieties of Plants* (UPOV; see above), may have to be re-interpreted. Glowka *et al.* (1994) suggest that it is possibly the most complex and controversial article in the Convention. In the early stages of drafting

Box 2.8 The Convention on Biological Diversity
Article 19. Handling of Biotechnology and Distribution of its Benefits

1 Each Contracting Party shall take legislative, administrative or policy measures, as appropriate, to provide for the effective participation in biotechnological research activities by those Contracting Parties, especially developing countries, which provide the genetic resources for such research, and where feasible in such Contracting Parties.

2 Each Contracting Party shall take all practicable measures to promote and advance priority access on a fair and equitable basis by Contracting Parties, especially developing countries, to the results and benefits arising from biotechnologies based upon genetic resources provided by those Contracting Parties. Such access shall be on mutually agreed terms.

3 The Parties shall consider the need for, and modalities of, a protocol setting out appropriate procedures, including, in particular, advance informed agreement in the field of the safe transfer, handling and use of any living modified organism resulting from biotechnology that may have adverse effect on the conservation and sustainable use of biological diversity.

4 Each Contracting Party shall, directly or by requiring any natural or legal person under its jurisdiction providing the organisms referred to in paragraph 3 above, provide any available information about the use and safety regulations required by that Contracting Party in handling such organisms, as well as any available information on the potential adverse impact of the specific organisms concerned to the Contracting Party into which those organisms are to be introduced.

(*Source*: UNEP, *June* 1992.)

developed countries in particular were cautious of the provision on technology transfer.

Article 16(3) states that the source country should have access to the results of research and development of those resources, although **16(2)** reminds Parties that "*adequate and effective protection of intellectual property rights*" should be recognised. This has enormous implications for research institutes that collect material, to be named eventually and kept outside the source country.

Article 16(5) reaffirms that obligations under any other intellectual property rights that a country deems necessary should be supportive of the Convention.

Articles 17 (see **Box 2.6**) and **18** (see **Box 2.7**) refer to the exchange of information and technical and scientific cooperation respectively, and are discussed in the next section. **Article 19** (see **Box 2.8**) seeks to ensure that this participation of source countries in biotechnological research is encouraged by obvious financial input from the developed country. The well-quoted example of the Merck pharmaceutical company of the USA, which agreed to pay the Costa Rican government up to $1 million over two years and encourage, through donations of equipment and training, national participation in preliminary bioassays, is a step in the right direction. However, this must be counter-balanced against the very large, long-term research investment.

Article 20 (see **Box 2.9**) considers national and international responsibilities for financing action mandated by the Convention. It includes a commitment by all Contracting Parties to provide financial resources at the national level with a special obligation on the developed countries to provide new and additional finance to help developing countries implement their commitments.

2.4 The role and responsibilities of the research institute and scientist

Systematic research institutions in the UK have had, and continue to have, an important part to play in biodiversity assessments. The content and significance of the collections amassed in natural history institutions in the UK is

Box 2.9 The Convention on Biological Diversity
Article 20. Financial Resources

1. Each Contracting Party undertakes to provide, in accordance with its capabilities, financial support and incentives in respect of those national activities which are intended to achieve the objectives of this Convention, in accordance with its national plans, priorities and programmes.

2. The developed country Parties shall provide new and additional financial resources to enable developing country Parties to meet the agreed full incremental costs to them of implementing measures which fulfil the obligations of this Convention and to benefit from its provisions and which costs are agreed between a developing country Party and the institutional structure referred to in Article 21, in accordance with policy, strategy, programme priorities and eligibility criteria and an indicative list of incremental costs established by the Conference of the Parties. Other Parties, including countries undergoing the process of transition to a market economy, may voluntarily assume the obligations of the developed country Parties. For the purpose of this Article, the Conference of the Parties, shall at its first meeting establish a list of developed country Parties and other Parties which voluntarily assume the obligations of the developed country Parties. The Conference of the Parties shall periodically review and if necessary amend the list. Contributions from other countries and sources on a voluntary basis would also be encouraged. The implementation of these commitments shall take into account the need for adequacy, predictability and timely flow of funds and the importance of burden-sharing among the contributing Parties included in the list.

3. The developed country Parties may also provide, and developing country Parties avail themselves of, financial resources related to the implementation of this Convention through bilateral, regional and other multilateral channels.

4. The extent to which developing country Parties will effectively implement their commitments under this Convention will depend on the effective implementation by developed country Parties of their commitments under this Convention related to financial resources and transfer of technology and will take fully into account the fact that economic and social development and eradication of poverty are the first and overriding priorities of the developing country Parties.

5. The Parties shall take full account of the specific needs and special situation of least developed countries in their actions with regard to funding and transfer of technology.

(*Source*: UNEP, *June* 1992.)

discussed in Chapter Six. Programmes to build and maintain such collections, together with their associated locality data, initially for international academic research, raise a number of issues concerning the ownership and intellectual property rights of specimens and the information they contain.

Over the last 25 years many, especially tropical, countries have been increasingly aware that the scientific base for a taxonomic study and inventory of their biota resides abroad in institutions where the specimens are ultimately studied.

Over the past half century the larger international institutes have taken an increasingly responsible attitude to specimen acquisition. Before living or dead biological material is exported from a country the necessary permits will be cleared by the appropriate authorising department. In some countries this may mean clearances need to be obtained from several departments for different plant or animal groups collected. It has been the practice to leave a complementary set ('duplicates') of specimens in the country of origin in the appropriate National Collections Centre (usually a museum, botanic garden or

Brian Critchley/Natural Resources Institute

Biotechnology may help to control the cotton boll weevil in Nicaragua

national research institute). Unicates and specimens potentially new to science (i.e. hitherto un-named and undescribed) are usually allowed to be taken out with the understanding that they remain the property of the country of origin and must be sent back when the research is completed and potential new species named. In most instances type specimens (original specimens to which a scientific name is first attached) are now returned. Formerly this was not the case and types would remain in the institute holding the researcher's collections. It is natural for curators to want to hold such base-line material of organisms from their own territory. It is, of course, important that the host country institute has the resources to maintain those specimens in good condition, and if requested, to send them on loan to other research workers who wish to compare them with material from elsewhere.

Various *Codes of Ethics* or *Practice* have been drawn up to encourage collectors and curators to be more responsible on this matter. One for botanical collectors was launched in Perth, Western Australia, at the Botany 2000 Herbaria Curation Workshop (1990) and broadened to include all biota in 1992, in Manila, at the *7th Asian Symposium on Medicinal Plants and other Natural Products*. A similar *Code* was stated by Colvin (1992) and is the policy of the University of California Research Expedition Program.

These principles, already the policy for expeditions by the Natural History Museum (NHM), and the Royal Botanic Gardens at Kew (RBGK) and Edinburgh (RBGE), are recommended here for all British based expeditions and field projects involved in biodiversity work (see **Box 2.10**). It is recommended that other institutes should make clear in their publicity marketing brief their own policy on acquisitions and general ethical matters. Botanic gardens and others involved in ex situ conservation (e.g. setting up gene or seed banks) are (or should be) conscious of being the guardians of a source country's heritage.

Similar matters are being addressed by those involved in crop genetic resources (e.g Duvick, 1992). The International Plant Genetic Resources Institute (IPGRI) together with other international agencies have produced similar *Guidelines* for those collecting gene bank material (Guerino *et al.*, 1995, App.2.1).

The three major biodiversity research institutes in the UK (NHM, RBGE and RBGK) have become increasingly aware in recent years that they must be ethically responsible for their activities in such areas as intellectual property rights, ownership of biological resources, biological control and technology, and the transfer of skills to developing countries. They are each committed to the equitable sharing of benefits arising from their individual research activities, disseminating the data resulting from that research and applied conservation projects and to pursuing an active policy of training. RBGK and RBGE have adopted clear public positions on these issues (see **Box 2.14** for example) and have encouraged other botanic gardens to do likewise.

The RBGK, where taxonomic research involves the analysis of plant compounds (see **Box 2.1**), negotiates a Memorandum of Understanding with external collaborators, outlining the conditions of acceptance of plant material which on analysis may have the potential for commercial development (see **Box 2.11**).

2.5 The importance of research agreements

Janzen *et al.* (1993) discuss in detail the problems that can arise in collecting biological material and in granting biodiversity research agreements. The salient points from their paper will be briefly mentioned here. Research agreements help to establish a level of confidence between the visiting researcher and the national or local governments, host

Box 2.10 A Code of Practice for collectors of biological material

A visiting collector should:

1 Arrange to work with a local scientist(s) and institute(s).

2 Respect regulations of the host country; for example, by entering on the appropriate visa, and observe both national and international regulations for export of biological specimens, quarantine, the need for CITES licences, etc.

3 Enter into an official agreement (by an exchange of carefully worded letters, or by more elaborate *Research Agreements, Statements of Collaborative Intent* etc.) with the site managers and/or government department concerned which will include permission for all collecting, both in national parks or protected areas and elsewhere.

4 When budgeting (and applying for) a travel study or expedition grant, include equal expenses for local counterpart(s) and an amount to cover the cost of processing museum specimens or other costs of the visit to the host institute.

5 Where equipment to be used in the proposed scientific programme is difficult to obtain in the host country consider donating it to the appropriate department.

6 Leave a complete set of adequately labelled duplicate specimens with the institute before departing from the country. When material has to be left unidentified (e.g. a possible new species) send determinations to the host institute as soon as possible.

7 Ensure that 'type specimens' of new species described as a result of the research are deposited in the National Museum or Herbaria of the country of origin.

8 Inform the national collections repository in the country of origin where duplicate specimens are to be deposited.

9 Do not exploit the natural resources of the host country by removing high-value biological products; for example, collecting without prior permission plants with potential horticultural, medicinal, cultural, or other economic value.

10 When possible, and before your visit, obtain a list of rare and endangered species of the country visited and become familiar with their identity; collect these species only for a specific purpose, and then with permission.

11 Inform the host institute or appropriate organisation of the whereabouts of any rare or endangered species that are found. Leave photographs/slides that document your record.

12 Collect no more material than is strictly necessary; collect cuttings or seeds for live plant specimens, rather than uprooting whole plants; collect subsections rather than whole organisms, wherever possible, for marine specimens.

A. C. Jermy/NHM

A UK/Australian expedition collecting in Papua New Guinea

13 Collect identified reference voucher specimens for all biological products (e.g. textiles) to be exported.

14 Send copies of research reports and publications to collaborator(s) and host institute(s). Publish in journals of the host country and use the official language of that country or as joint publications whenever possible.

15 Acknowledge all collaborator(s) and host institutes(s) in research reports and publications.

16 Do not collect material for commercial use or development unless there is an existing Agreement with the host country.

scientists and members of the local community. The latter are particularly important when knowledge of indigenous peoples about the use of plants and animals is being collected.

2.5.1 Collecting permits, research agreements and conservation

Although large scale collecting has been the practice amongst the bigger 'international' institutes for some time, collecting may constitute a violation of national sovereignty. It is not sufficient to say that the reason for collecting is to build up the national collections in another country so that academic taxonomic research can proceed unabated. Pressure from the public domain concerning the effects of such collecting on ecosystems and, more recently, biodiversity (species) conservation – combined with the increasing costs of such exercises – have questioned the need for any large scale collecting. Physical samples (genes, tissues, specimens, populations) and the information about them both have a value (often not easy to quantify). Requests to visit a country and a site must make clear, both to the country and to the custodians of that site, the intended use of the physical or intellectual property removed. This is more than would be covered just by a 'collecting permit' and should be considered by all as a '*Research Agreement*'.

Increasingly complex legal or quasi-legal agreements are now being negotiated. Major long-term biodiversity research projects initiated by UK research institutes and universities tend to be governed by *Memoranda of Understanding* or similar official agreements between national research institutes and/or appropriate government departments e.g. between NHM and the Wallace Development Institute, Indonesia (see **Boxes 2.12** & **2.13**).

A similar Agreement has been signed between Cambridge University (Baritu Ulu Project) and the Indonesian government (LIPI).

2.5.2 Who needs research agreements?

A research agreement should be required by any researcher working on land conserved or used for its biodiversity. Differences in content will depend on the type of research and whether it will result in commercial exploitation or on what other agreements regarding intellectual and technical transfer may exist. Some different examples are given in **Boxes 2.11** and **2.12**. There may be different regulations for nationals of the host country than for foreigners.

Permissions granted to undergraduate and postgraduate expeditions may be difficult to incorporate into research agreements as the outcome is often hard to predict. However, students' curiosity about biodiversity should not be dampened and one would hope that host countries will encourage such groups to carry out joint projects with their own university student fraternity guided by appropriate senior staff. A developed country should, for its part, ensure supervision and input into project design before the student group leaves that country, an aspect considered seriously by the Royal Geographical Society when allocating funds for such projects. Such expeditions should follow the *Code of Practice* mentioned above (**Box 2.10**), and seek appropriate advice for the country concerned from the Expedition Advisory Centre at the Royal Geographical Society (see Chapter Five).

2.5.3 Who signs research agreements?

At a minimum, a research agreement should be evaluated and signed by the researcher (or a senior person representing his/her institute), the custodians of the areas to be studied, and a senior official of the government/institution/department responsible for it. The last two may be the same. In signing an agreement, the researcher (or institute) is making a commitment to carry out the project as planned and generate certain types of information. It is difficult to assign financial penalties in the case of non-compliance but a researcher's (or institute's) failure to comply with the terms of the agreement jeopardises future research contract applications, not only from that institute but by other national colleagues. This damage to personal or national reputation may be a more effective monitor of good practice.

In National Biodiversity Strategies prepared under the Convention, the role of taxonomists should be seen as a very beneficial form of technology transfer. In this context the UK Darwin Initiative encouraging involvement and training of host country personnel (see Chapter Five) can help. Any local or national laws that

Box 2.11 Conditions of acceptance of plant material from external collaborators for studies in the Biological Interactions Section of the Royal Botanic Gardens, Kew

Royal Botanic Gardens, Kew, Richmond, Surrey TW9 3AB, UK

Telephone *direct*: +44 - (0)181- 332 _ _ _ _ **Fax *direct*:** +44 - (0)181- 332 _ _ _ _

Memorandum of Understanding

1. Plant material should normally have been named by a botanist and be accompanied by a voucher specimen and also by any information in the Supplier's possession which will enable RBG to conduct mutually agreed studies in the most effective manner.

2. Unidentified material may be accepted for testing in exceptional circumstances but, should an identified sample of the same material be subsequently received from a second Supplier, then the second Supplier will be deemed to be sole Supplier for the purposes of publications or patents.

3. All information and/or material will be held by RBG in strict confidence and not passed to a third party without the Supplier's written permission.

3a. If the information at time of receipt has already been supplied to us by another collaborator under a confidentially agreement, then the Supplier will immediately be informed of this in writing and his material will be returned to him/her.

3b. If any information held in confidence by RBG becomes public knowledge through no fault of staff of RBG, then the commitment to confidentially by RBG with respect to that particular information will cease.

4. The Supplier should state at the time of submission of the material what tests are requested and RBG will indicate what in its opinion is feasible. All studies are carried out at discretion of RBG.

5. If RBG finds itself unable to work on material received for any reason that material will be returned promptly to the Supplier. Confidentiality will be maintained.

6. Results of work carried out at RBG will be sent to the Supplier as soon as available. The Supplier agrees to keep the results confidential until the method of publication and/or patent protection is agreed with RBG. Should RBG wish to carry out confirmatory or follow-up studies, then the Supplier will use his/he best endeavours to supply further samples of the material of interest.

7. Should the results of studies at RBG be judged by RBG to be of possible commercial interest, then it will attempt to develop the work in collaboration with a commercial partner, with the Supplier's agreement. According to the publicly declared policy of RBG, any net profits derived by RBG from such collaboration will be shared equally between RBG and the Supplier, or as otherwise determined by pre-existing agreements and/or national legislation.

8. Should RBG not wish to approach a commercial company, the Supplier is free to do so at his/her own expense. In this case RBG will expect to share in any net profits derived from the commercial exploitation of data generated at RBG, proportionately in relation to the contribution made by RBG.

Email: _ _ _ _ _ _ _ _ _ _ _ _ _ _ _ _ @rbgkew.org.uk

Telephone *central*: +44 - (0)181- 332 5000 24hr recorded information: +44 - (0)181- 940 1171 Fax *central*: +44 - (0)181- 332 5197 Telex: 296694 KEWGAR

Box 2.12 An Endorsement of the Memorandum of Understanding between the Wallace Development Institute, Indonesia and The Natural History Museum

**AN ENDORSEMENT
OF THE
MEMORANDUM OF UNDERSTANDING
BETWEEN
WALLACE DEVELOPMENT INSTITUTE, INDONESIA
AND
THE NATURAL HISTORY MUSEUM**

We the undersigned wish to endorse the Memorandum of Understanding signed between Dr H Ibnu Sutowo, Chairman of the Wallace Development Institute (WDI) of the Republic of Indonesia, and Dr S Blackmore, Associate Director of The Natural History Museum (NHM) of the United Kingdom in which the two institutions extended to one another their mutual respect and recognised their shared commitment to the investigation and maintenance of biological diversity.

We support their aims of fostering scientific development and understanding and entering into a true spirit of co-operation by collaborating in the development of WDI's activities in research, training and the building of educational resources, and finding of resources together to underpin such collaboration.

We recognise that, as formal and official collaborators in Indonesian research programmes, the specialists of the NHM will use their best endeavours to work in close collaboration with their Indonesian counterparts, and that the WDI will facilitate the participation of the NHM by making available the collaborative use of their field station and other facilities.

We exhort international funding agencies and potential commercial sponsors to view favourably joint applications from the WDI and the NHM for financial support to underpin their collaborative initiatives which are designed to help to facilitate, or to develop resources necessary to understand mechanisms for the sustainable utilisation of Indonesian biological diversity as outlined in the Convention on Biological Diversity signed by representatives of the Governments of both Indonesia and the United Kingdom in Rio de Janeiro in 1992.

Wallace Development Institute

Dr Ing BJ Habibie

Founder and Adviser

The Natural History Museum

Dr Neil R Chalmers

Director

The Natural History Museum Cromwell Road London SW7 5BD Telephone 071-938 9123

Box 2.13 A Memorandum of Understanding between the UK Government and that of Malaysia

The Government of Malaysia and the Government of the United Kingdom of Great Britain and Northern Ireland, Considering both countries' deep concern for the global environment and for the conservation of natural resources, within a balanced perspective as expressed in the Langkawi Declaration, according due emphasis to the need for economic growth and sustainable development, including the eradication of poverty, provision for basic needs, and enhancement of the quality of life,

Recognising that the responsibility for ensuring a better and stable environment should be equitably shared in accordance with the abilities of individual countries to respond, and with their related needs or capabilities to receive or to provide assistance,

Reaffirming the sovereign rights of countries to exploit their own resources pursuant to their own environmental policies, in accordance with Principle 21 of the Stockholm Declaration,

Acknowledging the deepening global concern over the future of all forests, following the earlier and extensive loss of temperate and boreal forests, and the unique global significance of tropical rainforests as the major haven of terrestrial biodiversity, of incalculable potential benefit to present and future generations,

Having in mind the importance of sustainable management of forests for the welfare of local people, including the forests' contributions to household resources and rural livelihoods, and to the stability of local and regional environments, soils and water resources,

Taking account of the vital and multiple roles of tropical forests as a major resource for socio-economic development, and their potential for greatly enhanced and sustainable contributions of both timber and non-timber forest products, under appropriate management, with the generation of additional employment, income and value through local processing and equitable conditions of international trade,

Considering the International Tropical Timber Organisation's target of the Year 2000 for all trade in tropical timber to come from sustainably managed sources, and both countries' commitment to that target,

Considering the potential value of enhanced cooperation, particularly among Commonwealth countries, towards the twin goals of sustainable development and environmental conservation of tropical forests, in pursuance of the objectives of the Langkawi Declaration, and

Recognising the significance of the Malaysian forests and forestry experience in relation to the sustainable management and conservation of the tropical moist forests,

Have reached the following understandings:

Section I – General Objectives

1. To promote wider and deeper understanding of approaches for the sustainable management of forests; conservation of biological diversity; and of the essential links between the rational harvesting of timber and other forest products and the conservation of both the productive and the environmental values and benefits of forests, taking into account the needs of the local population as appropriate;
2. to facilitate the acquisition, adoption, development and utilisation of additional technologies and systems for sustainable forest management, within a strengthened system for international cooperation among Commonwealth countries, and more widely.

Section II – Programme of Cooperation

The Governments to the present Memorandum will decide on specific projects and programmes for collaborative action in accordance with the relevant laws, regulations, powers and responsibilities of their respective governments, to develop demonstration models of sustainable management in relevant forest areas. The programme of cooperation will encompass sites for associated research and training activities, and will include promoting the integration of environmental and social objectives in economic decision-making.

The Governments will also consider together, and in consultation with the Commonwealth Secretariat, possible joint action to encourage and, where appropriate, to assist the development of similar demonstration areas in other Commonwealth countries, as part of an integrated network, and as a contribution to wider global action in this field, with particular attention to the strengthening of the local capacities and human resources in the countries concerned, and to the role of TCDC (Technical Cooperation among Developing Countries) in this connection.

Section III – Activities of Cooperation
The programme of cooperation will include the following activities:

i Deployment and exchange of scientists, technicians, experts and specialists in appropriate fields;

ii Exchange of scientific, technical and technological information;

iii Development or replication of appropriate systems of sustainable forest management and harvesting practices;

iv Transfer of appropriate technology, with particular attention to the development of forest-based processing industries and small-scale enterprises, and the manufacture, marketing and export of higher value-added forest products;

v Provision for academic training, and for the joint organisation of seminars, symposiums, courses and academic interchange, and for the promotion of public understanding;

vi Joint exploration and development of systems for monitoring and evaluation of environmental and social impacts in sustainable forest management, and for the adoption of national systems of forest resource accounting;

vii Interchange and supply of equipment, instruments, specimens, products, data and other material necessary for the efficient conduct of projects and programmes of cooperation jointly decided on; and

viii Joint research programmes, incorporating activities already agreed under the link arrangements between the Forest Research Institute of Malaysia (FRIM) and the Oxford Forestry Institute (OFI), and with particular attention to the better understanding of forest ecology, autecology and the functioning of tropical forest ecosystems, the conservation of genetic resources and the development of non-timber forest production.

Section IV – Participants
These programmes will be open to both Federal and State institutions in Malaysia and to appropriate collaborative institutions throughout the United Kingdom of Great Britain and Northern Ireland. The Governments will jointly identify as those participating in the cooperative programme, scientists, technicians, specialists and experts linked to official government agencies, academic institutions and other non-governmental bodies, in accordance with the principles already established under the bilateral arrangements between the two countries, and under Commonwealth auspices, for example within the programme 'Institutional Development for Environmental Action' (IDEA). The Governments may jointly consider the value of the proposed demonstration areas and management models may be further extended through the promotion of wider participation of individuals and organisations from other countries, both within the Commonwealth and more widely.

Section V – Finance
The costs of the development and implementation of the cooperative programme will be shared equitably between the two Governments to the Memorandum of Understanding in accordance with the existing bilateral agreements and understandings between them.

Section VI – Implementation
The implementation of the programme of cooperation decided upon within the ambit of the present Memorandum will be subject to the norms applied to technical scientific and technological cooperation current in each country. The two Governments will meet annually to review progress of the programme of cooperation as set out in Section III of this Memorandum.

Section VII – Validity, duration, amendments and deletions

1. The present Memorandum will come into operation on the date of its signature and will continue until it is terminated either by the joint decision of both Governments or as provided for in sub-section VII. 3 below.
2. Amendment to the present Memorandum may be made at any time by written decision of the two Governments.
3. Either Government may terminate the Memorandum at any time by giving written notice to the other Government. Such termination will have immediate effect without prejudice to the projects and programmes already started and not completed during the period in which the Memorandum was in operation.

hamper collection and the free flow of information about a species should be discouraged and replaced by a thorough system of research agreements. In this context the memoranda of understanding between large institutions in the developed world and governments of developing countries should prove useful. So should the training facilities in the developed countries for the developing country staff (see Chapter Five).

2.5.4 Biological damage versus economic benefits

Wildlife custodians may be concerned about the damage researchers may cause. Often the custodians may not be aware of the techniques to enable them to evaluate biological damage. Scientific study or routine monitoring of an ecosystem rarely does lasting damage (except perhaps in a few fragile montane systems). National park authorities should be aware of the relative dangers of use in an ecosystem. The effect of restrictions on free collecting should be balanced against the mass use of a park for tourism and a larger income.

Hard cash payments to visit and work in national parks (as licence fees) are frequently paid to central government but, unless these get back to the custodians, the incentive for them to charge such fees is low. Cash grants by the researcher paid directly into the conservation system should be encouraged and other benefits (e.g. from the information generated and passed on through collaborative efforts) should be taken into consideration. All of this should be weighed against the amount of biodiversity damage which is acceptable (e.g., whether the habitat can quickly recover).

2.5.5 Protecting biodiversity information

Research agreements will often contain a clause about confidentiality of the information collected. Where plants with a pharmacological potential are involved, not only the site manager but the government may want to put an embargo, for instance, on drawing attention to their presence in a protected area (e.g. publishing in scientific journals).

Conflicts over confidentiality may have far-reaching consequences. For example, publishing information on where a particular plant or animal has been seen may be unwise from a conservation point of view. Whatever the legal viewpoint on the ownership of such 'records' (and there is much debate on this subject), there rests a moral obligation on the part of a visiting scientist to comply with the wishes of a host site manager. Even though an embargo on publishing could jeopardise a researcher's career development, the interests of the host country should always be borne in mind. Publicising the presence of highly valued plants, for instance, before a country has developed its own technology, could benefit neighbouring countries at the expense of in-country commercialisation of that product. Commercial companies signing research agreements will similarly want embargoes on announcing information by land managers. Guarantees which help a country protect its biodiversity as a commercial resource, already commonplace in mining and biotechnology, will be required more and more in the field of biodiversity.

2.6 Sharing the profits

The world's herbaria and museums are full of 'free samples' for the use of pharmacological companies and, even if material is not assayed, the location of plants with potential medicinal properties will be clearly displayed on the herbarium label. Research agreements may want to address these topics and state that the specimens are strictly for taxonomic study. Where public funds are used to amass and curate such collections it may not be easy to restrict access to such material and information especially if there is a policy to loan specimens. Requests to collect herbarium material of a particular family or the plants of a given area for a specific chemical assessment (or simply to copy the ethnobotanical notes from the herbarium labels), could present herbarium managers with a dilemma, especially when the investigator is prepared to pay for the information received.

Those who are custodians of living plant collections will have to take a more guarded approach especially if they have a policy of distributing (free or on exchange) seed (see Linington, 1994) or other propagated material from those collections. RBG Kew has a clear policy on this (see **Box 2.14**) and all recipients of its free seed distribution have to sign that they agree to the following:

a) They are used for the common good in the areas of research, trialling, breeding, education and the development of public botanic gardens.

b) If the recipient seeks to commercialise either the genetic material, its products or research derived from it, then written permission must be sought from the Royal Botanic Gardens, Kew. Such commercialisation will be subject to a separate agreement embodying Kew's policy that a *proportion of net profits be distributed to the country from which the seed was collected.* [Editor's emphasis].

c) The genetic material, its products or research derived from it are not passed onto a third party for commercialisation, without written permission from the Royal Botanic Gardens, Kew.

Similarly, a policy statement of intent to share profits derived from chemical screening (see Section 2.1 and **Box 2.1**) is made clear to all collaborative organisations (see **Box 2.14**).

2.7 Responsibilities of the commercial company or broker

Pharmaceutical companies and other commercial organisations screen extracts of plants, animals or microorganisms for bioactive principles. Suffness & Douros (1979) summarises the three objectives of pharmaceutical exploration as follows:

- the discovery of active agents which can be developed as drugs;
- the discovery of agents which can be modified through analogue studies to yield useful drugs;
- the discovery of agents with novel structures and mechanisms of action which can be used in the study of disease states.

Rarely are such substances drugs in themselves but if the potency, selectivity and specificity is acceptable, isolation, purification and structural elucidation follows. This can be a lengthy and expensive programme. In most cases companies have three main concerns regarding the supply of natural products:

(i) the supply of samples must continue and be available at a fair price;

(ii) the market must be exclusive for the end products (patent protection) (West,V. in Henderson, 1993, p.15); to which one can add a third, wider concern:

(iii) that royalties paid should be directed at the right area of the community and, when through governments, to be directed to the declared aims of the Convention on Biological Diversity (biodiversity conservation, sustainable use and benefit sharing).

Such companies have the ability to obtain intellectual property rights through legal patents but a patent may run out before the candidate drug has been tested and clinically developed. Nevertheless, patent protection is an essential component of biodiversity prospecting and the lack of this in a developing nation can severely limit the value of that prospecting. Understandably, ethical drug companies are dependent on patents because of their need to finance expensive future research (it can take up to 12 years and $700 million to develop a new drug from laboratory to marketplace; West, *loc.cit.*). The success rate for the discovery of a new drug can range from 3:100 to 1:40,000! (Aylward, 1993). Some countries (e.g. Brazil, India, Turkey) do not allow patents for pharmaceutical compounds (Baxter & Sinnott 1992), and may have strict rules about the exportation of any living germplasm. India cites the example of *Rauwolfia serpentina*, the annual sales of which are US $260 million, but India says it gets none of that profit (Jayaraman, 1994).

Materials should come from a sustainable and accessible source and endangered species must not be collected. Each company or broker should agree with its suppliers or the source-country institute on a royalty of which a 'fair proportion' would benefit the people or peoples at the collection source. Some such input might be in the form of

Box 2.14 Policy statement of the Royal Botanic Gardens, Kew, with regard to intent to share chemical derivatives

Royal Botanic Gardens, Kew, Richmond, Surrey TW9 3AB, UK

Telephone *direct*: +44 - (0)181- 332 ____ **Fax *direct*:** +44 - (0)181- 332 ____

POLICY STATEMENT OF INTENT TO SHARE PROFITS DERIVED BY CHEMICAL SCREENING

"The mission of the Royal Botanic Gardens Kew is to ensure better management of the Earth's environment by increasing knowledge and understanding of the plant kingdom - the basis of life on Earth."

In urgent pursuit of this mission, Kew, a non-profit making organisation with an international reputation for scientific integrity, needs to expand its programme of collaborative research on plants gathered from across the world. As always the findings will be disseminated worldwide by publication and training.

The field programmes of RBG Kew are always covered by collaborative bilateral or multilateral agreements with counterpart organistations which conform to appropriate national and international legislations. New agreed joint programmes with overseas research institutes are being sought in order to benefit fundamental science and training, conservation and resource management for the benefit of all parties.

Among the many areas of plant research at Kew is the screening of species for novel chemicals. Joint programmes with chemical companies will provide opportunities both to increase the scope of the screening and to obtain further funds for the purposes indicated above.

Kew clearly states its intention to share equally any net revenue it derives from the commercialisation through chemical screening of any plant accessions of known provenance. The form of this benefit will either be funds, materials or services as determined by agreements with collaborators or governments.

The above policy is operative from 1 April 1992 and will remain in force until superseded by relevant international agreements.

Professor Sir Ghillean Prance FRS
Director

Email: _______________ @rbgkew.org.uk

Telephone *central*: +44 - (0)181- 332 5000 24hr recorded information: +44 - (0)181- 940 1171 Fax *central*: +44 - (0)181- 332 5197 Telex: 296694 KEWGAR

The Royal Botanic Gardens, Kew has charitable status

100% recycled paper

local training or education. The idea of commercial companies paying a research licence fee to local people for conservation of useful ecosystems sounds attractive but the market value of wild material at source is not large and companies are likely to pay only token fees. The idea of returning money from ultimate profits of product development back to the community or government for future conservation, as in the Merck-INBIO project in Costa Rica, is one positive way of carrying out this recommendation but it remains to be seen whether the sums involved are fair and equitable.

The concept of 'Rain Forest Harvest' originated in the USA to develop the idea that indigenous people could "actively use and manage their forest resources in an economic way for consumers in the developed world, and thereby promote rain forest conservation". It has been criticised by Corry (1993) who points out the over-exploitation through encouragement of cash income and the difficulty of balancing that price against the 'subsistence value' of the rain forest to those that live in and on it. Nevertheless, the Convention on Biological Diversity encourages sustainable use and the fair and equitable sharing of benefits. An interesting and laudable example is the project by the UK-based cosmetics company, The Body Shop, whereby trade for Brazil nut oil with the Kayapo Indians of Brazil has been set up to bring them economic returns (see Box **2.15**).

Box 2.15 The Body Shop trade links with the Brazilian Kayapo

The Body Shop first made contact with the Kayapo of A-Ukre, in the Xingu River basin, Brazil, in 1989, when the community was looking for an economic and sustainable alternative to logging and mining developments that were threatening their rain forest home. After some preliminary investigations by The Body Shop laboratories it was discovered that Brazil nut oil was a very effective ingredient for hair conditioner.

Brazil nuts are proliferate in the area but transport is difficult so methods were designed whereby the Kayapo could use hand-operated machinery to press out the oil. Body Shop International agreed to fund the outlay needed. Once collected and the hard husk broken, the nuts are washed in the river, put into sacks and taken back to the village where they are laid out on tarpaulins and dried in the sun for three weeks.They are then heated over boiling water for twenty minutes and pressed to extract oil which is put into 50 kg plastic drums for transport to Redençao, a nearby town. No samples have had to be rejected because of rancidity or spoiling. Nearly everyone in the community is involved and in their first years A-Ukre was producing about 1,500 kg annually.

A second village in the Kayapo community, Pukanuv, asked The Body Shop to set up a similar trade link with them, which they did in 1992. They learnt from the experiences of their sister community and refined their techniques. They produced in their first year some 2,000 kilos which is probably the limit that each community can produce without disturbing traditional life practices.

Thomas Kelly/The Body Shop

Kayapo Indians breaking husks of Brazil nuts.

One of the main challenges of these trade links has been to help the Kayapo Indians learn basic numeracy skills, managing money and investing the profits for the benefit of the communities. With help from The Body Shop and its franchisees the Kayapo raised enough money to buy a second-hand Cessna airplane to provide emergency flights to hospital. In a more ambitious project involving the Federal indian Agency in Brazil (FUNAI) and the Brazilian Health Ministry, the Body Shop Foundation (a registered charity) provides health care research in the middle Xingu.

The Body Shop is now researching new raw materials to try to assist other Kayapo communities.

(*Source: The Body Shop plc, Littlehampton* BN17 6LS.)

Box 2.16 Glaxo Group Research policy on plant acquisition

Policy for the acquisition of natural product source samples

Glaxo Research and Development (GRD) is aware of, and sensitive to, issues relating to biodiversity and conservation. In particular, GRD recognises the importance of matters considered by the International Biodiversity Convention at the United Nations Conference on Environment and Development in Rio in 1992. GRD understands the impact that unauthorised habitats can have on the environment and economy of a country.

GRD recognises that various natural materials, such as plants, microorganisms, algae and marine invertebrates, are a valuable source of novel biologically-active molecules that may serve as templates from which new therapeutic drugs can be derived. GRD works with small quantities of natural materials to discover bio-active principles that, in turn, allow lead compounds to be identified. In the vast majority of cases, further supplies of such lead compounds and derivatives are synthesised by Glaxo's own medicinal chemists.

In seeking access to natural materials, GRD's policy is to collaborate with organisations that possess the expertise and the authority to obtain such materials from whatever source. Agreements will be concluded with prospective sample suppliers only when they can provide documentary evidence that they have permission from appropriate government authorities to collect such samples.

Samples of plants and other organisms must be classified taxonomically and their supply must be reproducible and sustainable. GRD will neither seek, nor knowingly support, the collection of endangered species.

In collaborating with *bona fide* suppliers, GRD's practice is to reimburse them for the costs incurred in collecting natural product source samples and to reward their expertise (e.g. in taxonomic classification). All costs of freight of natural materials are borne by GRD.

***Catharanthus roseus*, the source of two important anti-cancer medicines.**

GRD's Material Transfer Agreements may make reference to intermediate forms of compensation, and may involve a financial benefit payable to the supplier in the event that GRD is able to develop a commercial product as a consequence of screening the natural products supplied. The magnitude of this payment will recognise the relative contribution of the discovery of the bio-active principle to the subsequent development of the commercial products. This does not normally entail a transfer or sharing of intellectual property rights by Glaxo. In addition, the Agreements require that a significant portion of this payment will be returned to the source country to support scientific training and education at the community level.

A distinction is drawn between supply of natural materials for drug discovery and the broader philanthropic support of efforts to conserve resources of which these materials are a part. GRD as a company with interests in basic research applications of natural materials, negotiates purchase terms on the basis described above for sample acquisition. Conservation support *per se* is a matter for consideration by the Appeals Committee of Glaxo Holdings p.l.c.

(Source: Glaxo Research and Development document. Amended 6th January 1994.)

Throughout that programme frequent contact was essential to build up mutual respect between the two parties. The Body Shop hired an experienced Brazilian indigenist to liaise with the Kayapo. The Kayapo believe that their gathering of the Brazil nuts does not disrupt propagation patterns of the trees as they gather the nut carapaces which fall to the forest floor.

In some other communities there may also be different levels of local authority and the laws that emanate from each. An example given by Boom (1990) is the Kuna Indians of Panama who control access to research sites and ask a payment for scientists to visit. They further insist on researchers taking a paid Kuna assistant/guide and request a report on the investigation.

A number of academic institutions in the UK have set up companies, part of whose remit is to broker the relevant research of associated departments. More and more commercial companies are collaborating with biodiversity institutes and university programmes (e.g. Glaxo; see **Box 2.16**), to develop projects for all parties to use plants on a fair and sustainable basis.

2.8 References and bibliography

Aylward, B. (1993). The economic potential of genetic and biochemical resources. In Henderson, C. (ed.) p. 6. *qv*.

Aylward, B., Echeverría, J., Fendt, L. and Barbier, E.B. (1993). *The economic value of species information and its role in biodiversity conservation: Costa Rica's National Biodiversity Institutes*. LEEC Paper DP 93.06. IIED, London.

Barton, J.H. (1994). Ethnobotany and intellectual property rights. In Chadwick, D.J and Marsh, J. (eds). *Ethnobotany and the search for new drugs*, pp. 214–221. (CIBA Frdn Symp. 185). John Wiley, Chichester.

Baxter J.W. and Synnott, J.P. (1992). *World patent law and practice*. Matthew Bender, New York.

Boom, B.M. (1990). Ethics in ethno-pharmacology. *Proc. 1st Internat. Congr. Ethnobiology*: 147–153.

Cabinet Office. (1992). *Intellectual property in the public sector research base*. HMSO, London.

Colvin, J.G. (1992). Editorial: a Code of Ethics for research in the Third World. *Conservation Biology*, 6: 309–311.

Colchester, M. (1993). The perspective of indigenous peoples and the risks of commercialising traditional knowledge. In Henderson, C. (ed.), p. 9, *qv*.

Corry, S. (1993). *Harvest moonshine taking you for a ride: a critique of the 'Rain Forest Harvest' – its theory and practice*. Survival International, London.

Cunningham, A.B. (1993). *Ethics, ethnobiological research and biodiversity*. WWF, Gland.

Dev, S. (1989). In Swaminathan, M.S and Kochhar, S.L.(eds). *Plants and Society*, pp. 267–292. Macmillan, London.

Duvick, D. (1992). Possible effects of Intellectual Property Rights on erosion and conservation of plant genetic resources in centers of crop diversity. *Proc. Internat. Crop Sci. Congr.*

Fellows, L.E., Kite, G.C.. Nash. R.J., Simmonds, M.S.J. and Scofield, A.M. (1989). In Poulton, J.E., Romeo J.T and Conn, E.E. (eds), *Plant nitrogen metabolism*, pp. 395–427. Plenum Press, New York.

Findeisen, C. and Laird, S. (1991). *Natural products research and the potential role of the pharmaceutical industry in tropical forest conservation*. Rainforest Alliance, New York.

Glowka, L., Burhenne–Guelmin, F. and Synge, H. (1994). *Guide to the Convention on Biological Diversity*. IUCN, Gland and Cambridge.

Gollin, M.A. (1993). An intellectual property rights framework for biodiversity prospecting. In Reid W.V. *et al.*, (eds), pp 159–197, *qv*.

Guerino, L., Rao, R. and Reid, R. (1994). *Collecting plant genetic diversity: technical guidelines*. IPGRI/FAO/IUCN, Nairobi.

Hamilton, A. (1993). The role and responsibilities of scientists and international brokers (3). In Henderson, C. (ed.), pp. 10–11, *qv*.

Henderson, C. (ed.) (1993). *Intellectual property rights, indigenous culture and biodiversity*. Green College Centre, Oxford.

Janzen, D.H., Hallwachs, W., Gámez, R. Sittenfeld, A. and Jiminez, J. (1993). Research management policies: permits for collecting and research in the tropics. In Reid, W.V. *et al.*, (eds), pp. 131–157, qv.

Jayaraman, K.S. (1994). India set to end 'gene robbery'. *Nature*, 370: 587.

Kloppenburg, J.R. (1988). *First the seed*. Cambridge University Press, Cambridge.

Kutchan, S.L. (1989). Plants as stimuli for exploration and exploitation. In Swaminathan, M.S. and Cochar, S.L., pp. 44-85. Macmillan, London.
Linington, S. (ed.). (1994). *List of seeds* 1994. Royal Botanical Gardens, Kew.
Mooney, P.R. (1983). The law of the seed. *Development Dialogue* 1983(1–2): 1–172.
Posey, D.A. (1995). *Indigenous peoples and traditional resource rights: a basis for equitable relationships*. Green College Centre, Oxford.
Rainforest Alliance (1992). *Tropical forest medical resources and the conservation on biodiversity*. Proceeding R.A. Symposium, New York.
Reid, W., Laird, S.A., Meyer, C.A., Gámez, R., Sittenfeld, A., Janzen, D.H., Gollin, M.A. and Juma C. (eds) (1993). *Biodiversity prospecting: using genetic resources for sustainable development*. World Resources Institute, Baltimore.
Sands, P. (1993). The Biodiversity Convention: legal issues and mechanisms for equitable trade in biological resources. In Henderson, C. (ed.), pp. 7–8, *qv*.
Siebeck, W.(ed.) with Evenson, R.E., Lesser, W. and Braga, C.A.P. (1990). *Strengthening protection of intellectual property in developing countries*. World Bank Discussion Papers 112, Washington, DC.
Suffness, M. and Douros, J. (1979). Drugs of plant origin. *Methods in cancer research*, 16: 73–125.
UNEP (1992). *Convention on Biological Diversity*. United Nations Environment Programme, Nairobi.

INVENTORIES, MONITORING AND MANAGEMENT:

International approaches

3.1 **Introduction**

3.2 **National approaches to describing and monitoring biodiversity**

3.3 **International coordination of biodiversity inventory and monitoring**

3.3.1 Diversitas

3.3.2 Systematics Agenda 2000

3.3.3 Species 2000 and the International Organisation of Plant Information (IOPI)

3.3.4 The European Science Foundation Systematic Biology Network

3.3.5 BioNET-International

3.4 **Identifying and monitoring biodiversity for conservation and sustainable management**

3.4.1 The International Union for the Conservation of Nature – the World Conservation Union

3.4.2 Monitoring vegetation cover and land use

3.4.2.1 Monitoring forests

3.4.2.2 Tropical forest organisations

3.4.3 CoORdination of INformation on the Environment (CORINE)

3.4.4 Monitoring marine systems

3.5 **Global monitoring of biotic and abiotic environmental variables**

3.5.1 The Global Terrestrial Observing System and the Geographical Observatories Programme

3.5.2 The Global Ocean Observing System

3.5.3 The Global Climate Observing System

3.6 **Sites for biodiversity conservation**

3.6.1 Protected Areas

3.6.2 Determining future priority areas

3.7 **Translating survey into policy**

3.8 **References and bibliography**

Chapter authorship

This chapter has been researched and drafted by Harriet Eeley *and* Nigel Stork *with additional contributions and comments by* I. Gauld, M. Gillman, P. Harding, C. Humphries, C. Jermy, S. Knapp, D. Long, J. Robertson-Vernhes, M. Sands, K. Sherman, K. Singh, R. Smith, P. Williams *and* S. Winser.

3.1 Introduction

Much of what has been so far recorded about the nature and distribution of life on Earth is the result of historical expedition and exploration rather than any planned attempt to describe systematically the Earth's biological diversity. Collections of animals and plants from North America and all over the world were assembled between the 18th and 19th century, concomitantly with European exploration and global colonisation. The early herbalists and plant breeders had an enormous effect, sponsoring collectors to send back living and preserved material. These collections are now mostly housed in the museums and herbaria of Europe. The formal description of the Earth's flora and fauna has been almost entirely carried out by taxonomists from Europe and North America. However, in spite of their effort, it seems that taxonomists have named probably less than 15% of all organisms.

Natural History Museum Photo Library

Sloane's viper fish *(Chaliodus sloani)*, of a cosmopolitan benthic species and one of the earliest specimens in the UK National Collections.

Until now, there have been no attempts to coordinate efforts to inventory the world's biota. Some groups of organisms have received a disproportionate amount of attention (May, 1988), while others have been almost entirely neglected. Groups such as terrestrial and deep-sea invertebrates and microorganisms will require greater effort on the part of systematists in order to redress the balance.

Although international initiatives, such as the International Biological Programme (1962–1972) and the World Conservation Strategy (1976–1977) have helped to create a baseline understanding of natural systems and their conservation, the problems of cataloguing and monitoring biodiversity have been largely ignored.

Describing and monitoring biological diversity at national, regional and global levels requires substantial systematic and ecological infrastructure, economic input and human resources. Such requirements exceed the capabilities of even the most scientifically-advanced nation. This, along with the global shortage of taxonomists for many groups of organisms (Gaston & May, 1992; Systematics Agenda 2000, 1994), means that it is important to share and coordinate expertise and capacity, particularly for inventories and monitoring of diverse and poorly known groups, and collectively to agree priorities for future programmes of inventorying. Furthermore, the processes and patterns of the elements of biodiversity are not restricted by political, social or economic boundaries and international cooperation and coordination is essential if resources (time, money and personnel) are not to be wasted. Without such coordination some areas will be left unresearched and others will be analysed repeatedly using different and incompatible methodologies.

This chapter reviews current attempts to provide national and international coordination for the inventorying and monitoring of biodiversity. It aims to provide an international context for this biodiversity manual. A range of initiatives are outlined, including Diversitas, a programme which aims to provide an international framework for biodiversity studies, as well as Systematics Agenda 2000 and Bio-NET International, both of which address the lack of adequate local and global resources for systematics. Also discussed are international programmes concerned with describing and monitoring terrestrial and marine biodiversity (such as FAO's Forest Resources Assessment 1990, CORINE and the Large Marine Ecosystem networks), international organisations involved in identifying and monitoring diversity for conservation (the World Conservation Union and the World Conservation Monitoring Centre) and global programmes concerned with monitoring physical variables (the Global Terrestrial Observing System, the Global Ocean Observing System and the Global Climate

Observing System). The choice of appropriate sites for biodiversity studies world-wide is also considered. For further details on many of the programmes and initiatives outlined here see Stork and Samways (1995).

Until recently there was little call for biodiversity information from policy makers. In the past scientists have often failed to transmit new information resulting from their discoveries and research both to non-scientists and, more importantly, to policy makers. The final section of this chapter addresses this problem and suggests ways in which survey information can be transferred to policy and implemented.

3.2 National approaches to describing and monitoring biodiversity

No country has yet completed a full inventory of all their species. Indeed, some of the most well-resourced and well-developed countries still have a very poor knowledge of their biota. This is perhaps not surprising as few countries support a sufficient number of taxonomists and often the range of organisms that they are required to cover is enormous. Which groups have been inventoried will have been determined by the present and past availability of taxonomic expertise. For example, although groups such as harvestmen, mutillid wasps, aphids, butterflies and dragonflies have been well inventoried in South Africa, little is known about millipedes, mosses or Protozoa.

As Parties to the Convention on Biological Diversity, countries are now committed to identify and monitor biodiversity for both conservation and sustainable use. Action plans have been produced by a number of countries, but, although many countries are now attempting to make inventories of their biota, few appear to have cohesive programmes. Often the responsibility for national inventories rests on poorly resourced national museums, herbaria and a handful of individuals. Few countries have coordinated and planned programmes of action. For many countries the lack of even basic inventory data prevents studies of populations of species and the monitoring of biodiversity at the species level or below.

Box 3.1 The Australian government's Environmental Resources Information Network (ERIN)

The Environmental Resources Information Network (ERIN) was set up by the Australian federal government in 1989 to provide an environmental database, bringing together information on the distribution of endangered species, vegetation types and heritage sites, to aid planning and conservation at a governmental level. ERIN operates as a programme within the Department of Arts, Sport, the Environment, Tourism and Territories and is administered by the Australian National Parks and Wildlife Service. The network includes only agencies within the department, both as suppliers and users of the data, although it has links with various other agencies such as the National Resources Information Centre and specimen-based data from State and Federal herbaria, museums and official expeditions. ERIN incorporates both networked computer technology and Geographical Information Systems. The data sets being assembled include the standard names and descriptions of Australian fauna and flora, specimen observations and managed areas, as well as a directory of environmental information, experts and references (FINDAR) which gives users access to 130 environmental databases, over 1000 taxonomic experts and an extensive bibliography of publications and maps.

Source: ERIN *and* Burnett *et al.* 1994

The long history of taxonomic work in Europe means that many western European countries have a good understanding of their biotas. In contrast, elsewhere in the world few countries have such a history and their biotas are poorly recorded. Some countries, such as Australia (see **Box 3.1**) and Canada, have made major advances in their biodiversity inventories over the past few decades. More recently, countries such as the USA, Mexico and Costa Rica have all commenced national inventory programmes. The National Biological Survey (now the National Biological Service) of the

USA (NRC, 1993), established in 1993, will coordinate the making of inventories and monitoring of biodiversity for the United States. In the United States, The Nature Conservancy's Biological and Conservation Data System (BCD), operating through the National Heritage and Data Center Network and involving all fifty US states as well as Canada, Latin America and the Caribbean, also provides a range of computerised environmental information, including species biogeography, population trends and ecology.

Arguably the most impressive advances in inventorying of developing nations have taken place in Mexico and Costa Rica through the formation of the National Commission for the Knowledge and Use of Biodiversity (CONABIO) and the Instituto Nacional de Biodiversidad (INBio), respectively (see **Box 3.2** and **Box 3.3**). The approaches taken by these two countries are different, reflecting their different scientific and cultural heritages. Both approaches are built on the skills and enterprise of local peoples. However, as in almost all national efforts they are still dependent, to some extent, on expertise from other nations.

3.3 International coordination of biodiversity inventory and monitoring

In this section we outline a range of different international programmes providing a framework for biodiversity studies, inventories and monitoring. This list of programmes and initiatives is by no means comprehensive. However, those highlighted here are felt to be of most relevance in this context. In future what may be needed is a mega-database of organisations with consistent information on each one that could be edited and updated continually. This is discussed more fully in Chapter Seven of this volume.

Box 3.2 The Mexican National Commission for the Knowledge and Use of Biodiversity (CONABIO)

CONABIO was created by the Mexican Government in March 1992, to coordinate and promote the biodiversity efforts of Mexican universities, NGOs, research centres and other government agencies. CONABIO's mandate is to advance the inventory of the Mexican biota by:

- supporting projects for the computerisation and networking of national museum collections;
- agreeing with foreign museums on the repatriation of information from Mexican specimens;
- field work in selected areas.

CONABIO now maintains 80 databases with more than half a Gygabyte of information from the labels attached to specimens collected in Mexico, a large proportion of which has come from more than 40 foreign museums and herbaria. As agreements with these source institutions proceed, this information will be added to CONABIO's Gopher system. The specimen data, together with maps and socio-economic data, are being organised in a single information system for government, non-government and research users and will provide the factual basis for the National Biodiversity Strategy that Mexico is obliged to complete as a signatory of the Convention on Biological Diversity.

In addition, CONABIO supports the sustained use of tropical species, the eradication of invading species, and projects for furthering public awareness. Since it's inception, nearly 200 projects have been selected for support. CONABIO also acts as an advisor in assessments of environmental impact studies, contracts between pharmaceutical companies, indigenous peoples and Mexican universities and in multinational negotiations.

CONABIO funding comes mainly from donations by the Federal Government to a private fund which channels these resources to specific projects. For this, the Mexican government has committed an annual budget of about U.S.$ 6 million.

(*Source*: CONABIO, *Mexico*)

Box 3.3 Costa Rica's Instituto Nacional de Biodiversidad (INBio)

INBio was created in 1989 and is a non-governmental, non-profit institution dedicated to the conservation of wildland biodiversity via the facilitation of its non-destructive intellectual and economic use (Reid *et al.*, 1993b). INBio's activities focus on biodiversity inventory, prospecting, and information dissemination and management, and are concentrated largely on Costa Rica's National System of Conservation Areas, a network of government-owned parks and protected areas covering 25% of the country and containing an estimated 500,000 species.

I. Gauld/Natural History Museum

An INBio parataxonomist training workshop.

INBio's national inventory and biodiversity prospecting are well under way for plants and insects and spreading rapidly to other taxa. This inventory is mainly conducted by parataxonomists (fully trained field collectors) who collect and bring specimens and other information to INBio from regional biodiversity offices in or near the Conservation Areas. Their field work is guided by INBio's inventory managers working in coordination with national and international taxonomic specialists. Detailed information on species identities, ranges and natural histories are accumulating rapidly and are being made publicly accessible via the Internet and various publications. INBio is also establishing field guides, reference collections and electronic identification services. The knowledge gained through inventory and information management is being shared through on-site workshops. Collaborative agreements for biodiversity inventorying also are being developed between INBio and various tropical countries and biodiversity institutions.

The aim of INBio's biodiversity prospecting programme is to identify potential users and to facilitate the contracted flow of wildland biodiversity research samples and information to commercial research and development institutions. Profits from this collaboration will contribute to the management of Costa Rica's conserved wildlands and, in time, may make a serious contribution to the national economy (Reid *et al.*, 1993b). The current focus of biodiversity prospecting is to discover chemicals from plants, insects and micro-organisms that may be used in the pharmaceutical and medical industries. In the future this may be expanded to include using organisms as sources of genes for biotechnology, pesticides and biological control. Two successive two-year contracts have now been signed between INBio and the pharmaceutical company, Merck & Co., to prospect for biologically active compounds. In each contract Merck & Co. will pay INBio U.S.$ 1.3 million, in return for sample collection, identification, data management and personnel training. Samples, collected by INBio within the government owned Conservation Areas are managed under agreement with the Ministry of Natural Resources, Energy and Mines and are processed in INBio's laboratory. Ten percent of the payments received (and 50% of any royalties) will go towards conservation efforts, while the remainder will be spent on biodiversity development at INBio and in the Conservation Areas. A similar 5-year contract is now in progress between INBio, Cornell University and Bristol Meyers Squibb.

(*Source*: INBio, *Costa Rica*.)

3.3.1 Diversitas

The Diversitas programme was launched in 1991 by the International Union of Biological Sciences in collaboration with UNESCO and the Scientific Committee on problems of the Environment to provide a conceptual framework and direction for the rapidly growing number of studies on biodiversity. It arose in part out of the 'Decade of the Tropics' programme of IUBS in the 1980s. Diversitas is one of the first attempts to coordinate activities with respect to all levels of biological diversity and involves a range of scientific disciplines, including genetics, physiology, population biology, ecology and systematics. One main theme of Diversitas is to promote international cooperation in inventorying and monitoring biodiversity (di Castri & Younès, 1990; di Castri *et al.*, 1992a, 1992b). Diversitas has the following themes:

- Ecosystem function and biodiversity;
- Origins, maintenance and loss of biodiversity;
- Inventorying and monitoring of biodiversity;
- The biodiversity of wild relatives of cultivated species;
- Marine biodiversity;
- Microorganism diversity; and
- Human dimensions of biodiversity.

The proposed framework for biodiversity inventory and monitoring comprises a few sites that will be intensively studied for all organisms and, in parallel, a range of selected taxa to be more extensively inventoried at a larger number of sites (Coddington *et al.*, 1991;

Box 3.4 Diagramatic representation of an intensive and extensive sampling programme
(Source: after di Castri et al. 1992 IUBS, Paris)

Ecosystem	Sites	Taxa										
		a	b	c	d	e	f	g	h	i	j	k
Moist Tropical Forest	1											
	2											
	3											
	4											
	5											
Dry Tropical Forest	6											
	7											
	8											
	9											
	10											
Dessert	11											
	12											
	13											
	14											
	15											

Inventories would be made of all groups of organisms (a-k) at a few selected sites (4, 8, 13) and at a wider range of sites (1-15) inventories would be made of only selected focal taxa (c, f, j).

di Castri *et al.*, 1992a, b; and see **Box 3.4**). The UNESCO Man and the Biosphere Reserve network (see **Box 3.12**) has been promoted by Diversitas to provide the focal sites for the inventory programme. Diversitas activities have been initiated in several countries, including Brazil, China, Japan, France and the USA, but as yet there is little coordination between them. The 'intensive sampling' theme of Diversitas has been expanded by Janzen and Hallwachs (1994) in their All Taxa Biodiversity Inventory (ATBI) (see Volume Two, Chapter One).

3.3.2 Systematics Agenda 2000 (SA2000)

SA2000 aims to 'discover, document and classify the world's species within the next two to three decades (Systematics Agenda 2000, 1994). The programme has been proposed jointly by three organisations; the American Society of Plant Taxonomists, the Society of Systematic Biologists and the Willi Hennig Society, in cooperation with the Association for Systematic Collections, and has three main missions:

- To discover, describe and inventory global species diversity;
- To analyse and synthesise the information derived from this global discovery effort into a predictive classification system that reflects the history of life; and
- To organise the information derived from this global programme in an efficiently retrievable form that best meets the needs of science and society.

The main goals of each of these missions are outlined in **Box 3.5.**

SA2000 recognises the importance of national, scientific, systematic research centres, such as CONABIO in Mexico and INBio in Costa Rica (see **Box 3.2** and **Box 3.3**), staffed by professional systematists with taxon-based expertise. A coordinated action plan has been proposed to increase training in systematics and help individual countries to establish the capability to survey and inventory their biological diversity and develop collection-based infrastructures (including museums, herbaria and repositories for microorganisms and genetic resources). To prevent duplication of collections or taxonomic expertise, SA2000 also proposes that systematists with a global knowledge of particular taxa be funded through international centres and conduct research at sites worldwide. Cooperative research, database linkages and partnerships between institutions are regarded as essential. SA2000 currently plans to hold three workshops within the next two years, in South America, South Africa and South east Asia.

Although originally a North American initiative, SA2000 has been widely discussed in Europe. In May, 1995, the European Science Foundation Network in Systematic Biology (see below), the Rjiksherbarium in Leiden, the Linnean Society of London, and the UK Systematics Association organised a conference to examine how Europe should contribute to the goals of SA2000. The support of this conference was instrumental in SA2000 being adopted as a global initiative under the auspices of the International Union of Biological Sciences.

3.3.3 Species 2000 and the International Organisation for Plant Information (IOPI)

Comparable to SA200 are two other organisations which are now making considerable progress. The International Organisation for Plant Information (IOPI) in 1991 began work on a global plant checklist and is now also engaged in a Species Plantarum Project, initiated by the Royal Botanic Gardens, Kew, to prepare a Flora of the World. In a wider biodiversity context there is also the development of a programme, *Species* 2000, for a global master species database (see **Box 3.6**).

3.3.4 The European Science Foundation Systematic Biology Network

The European Science Foundation (ESF) Systematic Biology Network recognises the global importance of Europe's biological and palaeontological collections which, supported by an unrivalled concentration of systematic expertise, place Europe in a unique position to contribute in the worldwide response to the current 'biodiversity crisis' (Blackmore, 1994, 1995; Donlon, 1995). The European systematic community has a fundamental role to play in improving our knowledge of the species on Earth and the phylogenetic relationships between them. However, systematics by tradition has been split along taxonomic lines,

Box 3.5 The Missions and Goals of Systematics Agenda 2000

Mission 1 To discover, describe and inventory global species diversity

Mission goals:

- to survey marine, terrestrial and fresh-water ecosystems;
- to achieve a comprehensive knowledge of global species diversity;
- to determine the geographic and temporal distributions of these species;
- to discover, describe and inventory species living in threatened and endangered ecosystems;
- to target groups critical for maintaining the integrity and function of the world's ecosystems, for improving human health and for increasing the world's food supply;
- to target the least-known groups of organisms.

Mission 2 To analyse and synthesise the information derived from this global discovery effort into a predictive classification system that reflects the history of life

Mission goals:

- to determine the phylogenetic relationships among the major groups of organisms, thus providing a conceptual framework for basic and applied biology;

- to discover the phylogenetic relationships of groups of species that are critical for applied biology, targeting species that are important for human health and food production, as well as for conservation of the world's ecosystems;
- to discover the phylogenetic relationships of groups of species that are of critical importance for the basic biological sciences, such as those having broad relevance for experimental science and those critical for maintaining the integrity and function of ecosystems;
- to develop more powerful techniques and methods for systematic data analysis.

Mission 3 To organise the information derived from this global programme into an efficiently retrievable form that best meets the needs of science and society

Mission goals:

- to develop systematic, biogeographic and ecological databases of species information based on species housed in the world's natural history collections;
- to integrate data from specimens housed in systematic collections with information contained in GIS databases, thus providing a means to monitor past and present effects of global change on species distributions and extinction;
- to develop linkages among databases for the efficient retrieval of all available information about species and the places they occur;
- to develop and implement an information system that can be accessed efficiently by a broad international user community;
- to develop data dictionaries of taxonomic names, geographic localities and other information basic to all systematics databases;
- to develop data products, including guides, keys, electronic floras and faunas and monographic works;
- to develop mechanisms for maintaining and updating databases and information networks including continuing hardware and software support.

(*Source*: SA2000, 1994.)

a fact which is now limiting its ability to meet these new scientific challenges.

The ESF Systematic Biology Network, running for three years from September 1994, aims to establish a framework within which Europe's unique resources in systematics can be coordinated (Blackmore, 1994; Donlon, 1995). It will mobilise activity, possibly through initiatives such as Systematics Agenda 2000 and the All Taxa Biodiversity Inventory concept, but also through new projects arising from European collaboration. The primary purpose of the ESF Network is to stimulate cross-disciplinary communication and collaboration between the scientists. The specific objectives of the Network include:

- To unite European systematists through good communication;
- To define and mobilise Europe's contribution to tackling the scientific challenges of biodiversity;
- To consider new developments in systematic theory, practice and application;
- To explore the relationship between morphological and molecular systematics;
- To produce briefing papers on systematics for policy makers; and
- To establish priorities for and consider ways of promoting the training of systematists.

The Network's activities include a series of five workshops being held between September 1994 and 1997, a Network newsletter and an Internet information service about the Network and European systematics to help improve communication between European systematists and the institutions in which they work. One workshop, held in May 1995 (see above), included an assessment of European resources in systematic research and proposed ways of developing European collaboration. There is no fixed membership of the network, it is open to all members of the European systematic biology community, including those in Central and Eastern Europe, and it aims to distribute information as widely as possible.

3.3.5 BioNET-International

BioNET-International was established in 1993 by CAB International to develop biosystematic self-reliance within developing countries (Jones & Cook, 1993). The initiative aims to maximise the use of existing resources and improve these through the transfer of skills, knowledge and

Box 3.6. Global Master Species Databases: Species 2000

Species 2000, Indexing the World's known Species is a new International Union of Biological Sciences project, adopted and operating in cooperation with ICSU Committee on Data for Science and Technology (CODATA) and the International Union of Microbial Societies (IUMS). Its objective is to census all known species of plants, animals and microbes on Earth, providing a baseline dataset for global biodiversity studies and a common medium for global communication about biotic resources and facilitating the implementation of the Convention on Biological Diversity. It will develop existing global master species databases, as well as those for new groups until all species of all groups of organisms are recorded. The resulting index will contain accepted names and synonyms, sources and classificatory position. The implementation of this programme will involve forming a federation of the existing global master species databases, establishing a common access framework for member databases, accelerating the development of global master species databases, both existing and new ones, developing access amongst the federated databases to provide a virtual index of all known species.

Global master databases are currently available or being established for several major groups of organisms; namely fossil plants, bacteria, protists, arthropods, molluscs, fish, birds and fungi (Bisby, 1994). While many remain very incomplete (especially for speciose groups such as the arthropods), these databases already demonstrate the feasibility of such forms of data collection. Some are however reaching completeness, listing all known species, and having full internal classificatory coordination and many give accepted scientific names, synonyms and common names. Master species databases could thus prove very useful for a number of inventory initiatives worldwide as a quick method for verifying species lists, for checking species names for uniformity, for checking species taxonomy and as a source of prepared electronic species checklists (Bisby, 1994).

scientific expertise from institutions in developed countries. BioNET-International originally had particular concern for arthropods, nematodes and microorganisms (Jones & Cook, 1993) but has now expanded to include all invertebrates (Jones, 1994) and other taxa may be included in the future.

The proposed framework for BioNET-International is a series of interlinked, regional networks of countries (known as Locally Organised and Operated Partnerships or LOOPs) (see **Box 3.7**), supported by a consortium of expert institutions from developed countries and a central technical secretariat. Two LOOPS have been developed since 1993, one in Europe (EuroLOOP) and one in the Caribbean (CARINET). Four other LOOPS, in East Africa (EAFRINET), South-east Asia (ASEANET), the South Pacific (PACINET) and Southern Africa (SAFRINET), are currently being established (Jones & Cook, 1993; Jones, 1995). Initial work will be focused on four areas; information and communication services, training of personnel, rehabilitation of collections, and the development and application of new resources, such as electronic, interactive, taxonomic keys.

Box 3.7 Diagram of the interrelated LOOPs of BioNET INTERNATIONAL

BioNET-INTERNATIONAL – Global Network for Biosystematics

- Technical Secretariat
- Network Coordinating Institutes
- Consortium of pertinent Biosystematic Institutions, International Organisations and Donors (BIOCON)

Planned Loops: Caribbean, East Africa, Europe, Latin America, North America, Pacific, South Asia, South-east Asia, West Africa, and others.

3.4 Identifying and monitoring biodiversity for conservation and sustainable management

3.4.1 The International Union for the Conservation of Nature – the World Conservation Union

The International Union for the Conservation of Nature and Natural Resources (IUCN) was established in 1948 and under its present title, the World Conservation Union, aims to encourage and assist nations worldwide to conserve the integrity and diversity of their environment and to ensure the ecologically sustainable use of natural resources. The organisation comprises over 800 members, including governments, non-governmental organisations, research institutions and conservation agencies in around 125 countries. A worldwide network of scientists and experts support IUCN's six Commissions which focus on; threatened species, protected areas, ecology, sustainable development, environmental law and environmental education and planning. From time to time IUCN maintains thematic programmes which have included tropical forests, wetlands, marine ecosystems, Antarctica and population and natural resources.

IUCN, along with several other national and international organisations, has played a fundamental role in the establishment of treaties and conservation initiatives worldwide. In addition to the Convention on Biological Diversity, two international treaties of particular importance for the conservation of biodiversity are the Convention on Wetlands of International Importance especially as Waterfowl Habitat (also known as the Ramsar Convention), and the Convention on International Trade in Endangered Species of Wild Fauna and Flora (CITES).

Originally signed in 1971, the Ramsar Convention provides a mechanism for international coordination of the identification and conservation of wetland sites, including their monitoring. Each signatory to the treaty designates at least one wetland area of

Ouse Washes. A wetland and designated Ramsar Site in eastern England.

The Green Turtle, Barracuda Point, Grand Turk Is., Caribbean.

international importance for preservation and reports any change in the ecological character of the site, as well as promoting the conservation and wise use of wetlands. As of 1994, 81 states were contracting parties to the Convention and 654 Ramsar wetlands had been designated, covering an area of over 43 million hectares (Navid, 1994).

CITES provides a legal framework that aims to "control, reduce or eliminate the international trade in threatened species whose numbers and conditions suggest further removal of individuals from its natural habitat would be detrimental to species survival" (Favre, 1989). Species in danger of extinction and banned completely from international trade are listed in Appendix I of the Convention, while those which might become threatened and for which effective regulation of trade is important are listed in Appendix II. The Convention was first signed in 1973 and there are currently 128 signatory nations. These countries all submit annual reports of their trade in listed species to the Convention Secretariat. Global levels and trends in the trade of listed species are assessed from these reports and monitored relative to the state of populations in the wild.

IUCN's Species Survival Commission (SSC) was established in 1949 to promote action to "arrest the loss of the world's biological diversity and to restore threatened species to safe and productive population levels". It now has approximately 3,500 members in 135 countries (WCMC, 1992) and provides a forum for both amateurs and professionals to make a direct contribution. The SSC is divided into some 104 Specialist Groups, covering different taxonomic groups or geographical areas. The taxon-based Specialist Groups are responsible for monitoring the status and requirements of their groups and producing Action Plans for effective long-term conservation.

The World Conservation Monitoring Centre (WCMC) was founded jointly by IUCN, UNEP and WWF, to provide information and technical services for the conservation and the sustainable use of species and ecosystems, as well as supporting others in the development of their own information management services. WCMC has global databases on national parks and protected areas, threatened plant and animal species and habitats of conservation concern. The Centre also undertakes and publishes research into the status, management and utilisation of these resources and provides information to its founder organisations and many other groups, including scientific institutions, conservation and development aid organisations, governments, commercial businesses and the media. IUCN's World Trade Monitoring Unit, based at WCMC, plays a fundamental role in monitoring the trade in CITES listed endangered species and species products. The resources offered by WCMC are more fully described in Chapter Seven of this volume.

The SSC has initiated the setting up of the Biodiversity Conservation Information System (BCIS), a consortium of IUCN Commissions, BirdLife International, International Waterfowl and Wetlands Research Bureau, and WCMC. The goal of BCIS is to support environmentally-sensitive decision-making and management practices through the provision of data and information (Smith, 1995). WCMC is the data-management partner of this consortium.

The IUCN Red Data Books are produced by the Species Survival Commission in collaboration with WCMC. The volumes are divided on a taxonomic or geographical basis and provide a range of authoritative information on the current status, distribution, population levels, habitat, ecology and conservation requirements of many globally threatened taxa. WCMC, together with the SSC and BirdLife International, also publish Red Lists for threatened animals, generally at two year intervals. Using information supplied by the SSC Specialist Groups, national Red Data Books and scientists and naturalists working in the field, these volumes present a global inventory of species or subspecies that are known to be, or are suspected of being, extinct or threatened with extinction in the wild. The 1994 Red Data List contains information on a total of nearly 6,000 species (WCMC, 1993). Provisional assessments cover all birds, just over 50% of mammals (but many insectivores, rodents and microbats are not included), 20% of reptiles, 12% of amphibians and under 10% of fish. Only a small proportion of invertebrate species are included, such as some dragonflies, butterflies and molluscs. A Red List for plants should appear in 1996. Both Red Data Books and Red Lists provide a useful method for monitoring the threatened status of species. IUCN has recently completed a revision of the categories of threat used to classify species in both the Red Data Books and Red Lists. (See Mace and Stuart, 1994).

Malcolm Penn/NHM

GIS satellite image showing areas of forest and logging, Belize.

3.4.2 Monitoring vegetation cover and land use

3.4.2.1 Monitoring forests

The Food and Agriculture Organisation (FAO) has played a fundamental role in assessing and monitoring the distribution of natural vegetation types. The FAO 1990 Forest Resources Assessment (FAO, 1993) investigates the distribution of forest ecosystems worldwide and changes in forest distribution from 1980 to 1990. The study covers three areas; temperate forests in developed countries (UNECE/FAO, 1993), tropical forests in developing countries and non-tropical forests of developing countries (FAO, 1993). The analysis was undertaken at a scale which also allows the investigation of extent and change of forest cover at the sub-national and national levels.

The FRA 1990 tropical forest assessment covered in total 10% of the tropics and incorporated both a statistical analysis of existing forest inventory data from different countries, and multi-date observations of forest cover from high resolution satellite images for 1980 and 1990 at statistically chosen sample locations. The remaining two studies were based solely on the compilation of national forestry statistics. The tropical forest assessment also included demographic and ecological information in order to model forest cover change, to standardise national data to the two base years (1980 and 1990) and to reduce problems introduced by variation in the data availability, timeliness and quality from country to country (FAO, 1993; WRI, 1994). To model cause and effect of deforestation FRA 1990 compiled three geo-referenced data-sets for the tropical countries depicting (i) the forest extent around 1980 using available national maps; (ii) eco-floristic zones; and (iii) sub-national boundaries with time series of population data.

As a result of FRA 1990, FAO found that there was considerable variation in the current national capacity for monitoring forest change and that data on change, produced from national statistics, are not adequate to meet the needs of policy makers and scientists. Forest resource assessments are among the most neglected aspects of forest resource management, conservation and development in the tropics. In cooperation with other interested organisations,

Box 3.8 Vegetation and land use monitoring programmes and their coverage

Programme	Coverage
The International Geosphere Biosphere Programme (IGBP)	Global landcover
The U.S. Geological Survey	Global land cover
The Commission of European Communities TREES project	Tropical moist forests
NASA's PATHFINDER programme	Global tropical moist forests
FAO's AFRICOVER project	African land cover
The Woods Hole Research Centre	South American land cover

(*From*: ENRIC, 1994; D. *Lantieri, pers. comm.*; *Stone et al.*, 1994).

FAO now intends to assess forest cover on an annual basis, using both a 10% sampling scheme and coarse resolution remote sensing, to build up a spatially referenced picture of tropical forest cover for the world (K.D. Singh, pers. comm.). The FRA sampling approach will be continued in order to assess trends in deforestation and also extended to cover all the forests of the world, both tropical and temperate, developing and developed world.

Several other organisations are also applying remote sensing techniques to digitally map forest or land cover at global or regional levels (see **Box 3.8**).

With the exception of the Woods Hole South America map, these projects are either ongoing or at the planning stages. NASA's PATHFINDER project uses high resolution satellite data, while the TREES, IGBP, US Geological Survey and Woods Hole programmes rely on low-resolution (AVHRR) imagery. In all cases, supplementary ground information will be necessary to provide detailed data and to validate the satellite imagery. Only NASA's PATHFINDER and the FAO's FRA provide more than a snap-shot view of land cover. For more details on remote sensing and satellite imagery see Chapter Seven, this volume.

Mapping land cover at the national level is, however, more advanced and several countries, including India, Thailand, Indonesia and Brazil, are using high resolution imagery to record and monitor changes in their national vegetation (WRI, 1994). An example of a project using Geographical Information System integrating satellite data is described in Chapter Seven. This project is examining the effects of management practises on some elements of the fauna and flora in lowland tropical forests in Trinidad and Guyana.

3.4.2.2 Tropical forest organisations

There are a number of organisations involved in coordinating the international research and monitoring of forest resources and two are considered here: the Centre for International

David Sutton/NHM

Mahogany logging, Belize.

Forestry Research (CIFOR), and the International Union of Forestry Research Organisations (IUFRO). In addition, two international initiatives concerned with tropical deforestation and providing international assistance in the national regulation of forest resources are the Tropical Forestry Action Plan (TFAP) and the International Tropical Timber Agreement (ITTA), the latter administered by the International Tropical Timber Organisation (ITTO).

The Centre for International Forestry Research (CIFOR)

CIFOR is one of 18 international agricultural research centres supported by the Consultative Group on International Agricultural Research (CGIAR). CIFOR is based in Bogor, Indonesia, and is the focal point for strategic forestry research and training in the CGIAR. It has a global mandate but is currently focusing on tropical regions. CIFOR has four broad areas of research:

- Social sciences, economics, policy analysis and development;
- Ecology, conservation and management of natural resources;
- Rehabilitation of degraded and depleted forest lands; and
- Utilisation and marketing of forest goods and services.

The centre is also involved in the coordination of research objectives and methodologies at long term sites of ecological research in tropical forests worldwide.

The International Union of Forestry Research Organisations (IUFRO)

IUFRO is an association of nearly 710 private, government, and academic research organisations representing 15,000 scientists from 111 countries. The purpose of IUFRO is to give forestry researchers more opportunity to communicate with their counterparts in other countries. Two of the subject groups within IUFRO deal with forest resource inventory and monitoring and remote sensing technology.

The Tropical Forestry Action Plan (TFAP)

The Tropical Forestry Action Plan, adopted in 1985, was developed by FAO in collaboration with numerous international organisations and NGOs. Its aim is to provide a mechanism for the coordination of international aid to tropical timber countries, with a view to halting deforestation and promoting the sustainable use of forest resources. The role of the TFAP is to help countries to develop national forest management strategies (national TFAPs) to be used as a basis for increasing international investment in tropical forestry and, thus, to provide information to enable donor countries to make informed choices as to where best to direct funding. By end of March 1993, 90 countries, accounting for 60% of worlds remaining tropical forests, were involved in the TFAP but none had yet completed a full national TFAP (WCMC, 1992). As result of continuing rise in rates of deforestation, the TFAP has been reviewed and a restructuring of the system recommended.

The International Tropical Timber Organisation and the International Tropical Timber Agreement

ITTO is an intergovernmental organization comprising a total of 53 Members, 26 tropical timber consuming countries plus the European Union, a Member in its own right, collectively responsible for 95% of the import trade and 26 producing countries whose territories enfold more than 85% of the world's closed moist tropical forest. It was founded under the International Tropical Timber Agreement (ITTA), 1983, which came into force in 1985.

Statutory objectives now include the expansion and diversification of international trade to tropical timber from sustainable sources with equitable prices, improved structural conditions and better transparency in the market, consultation on promotion of non-discriminatory trade practices, support for research and development to improve forest management. Currently this is highlighted by ITTO's Year 2000 Objective, which is a commitment by Members that all internationally traded tropical timber products shall originate from sustainably managed resources by 2000.

Policy work towards this objective has included the development and publication of guidelines on sustainable management of both natural and planted tropical forests, the conservation of biological diversity in tropical production forests, and criteria and indicators of sustainable management.

ITTO has, since 1987, approved and funded over 200 projects, pre-projects and other activities with budgets totalling US$110 million in the fields of Reforestation and Forest

A. C. Jermy/NHM

Tundra areas in the northern hemisphere, like here in central Iceland, can hold a rich biodiversity.

L. Gill/Scottish National Heritage.

This hay meadow at Brerachan, Tayside, Scotland, is protected under UK law.

L. Gill/Scottish National Heritage

Man's forest cutting and grazing practises have reduced the diversity of this former boreal forest here at Loch Torridon, Scotland.

R. Ormond/TMRU, York

Ras Mohammed N.P., Egypt, is increasingly pressurised by tourism – snorkellers, dive boats and reef walkers.

Management, Economic Information and Market Intelligence, and Forest Industry. These enterprises take a wide variety of forms, from research projects through pilot demonstration plots of sustainable management to major field projects, including establishment of national parks as well as forest management for timber and other products. Most are nationally oriented, but there are also many regionally and globally active projects.

The UK Tropical Forest Forum

In the UK, matters relating to tropical forests are debated in the UK Tropical Forest Forum which was launched in February, 1991, and acts as a UK focal point for the EC European Tropical Forest Research Network. It is open to all British-based governmental and non-governmental public and private organisations, companies and individuals concerned with the sustainable utilisation and conversation of tropical forests. The Forum aims to improve the understanding between such groups through the exchange of views and information and thus to facilitate the coherent support of forest conservation and management. It meets regularly to discuss issues, both through main meetings twice a year and working groups and workshops or specific issues including the timber trade and biodiversity. In the course of debate, the Forum has also helped to identify areas for potential research initiatives, several of which have since been carried forward by the appropriate organisations. For example, projects concerned with trade-related incentives to tropical forest conservation have subsequently been funded by the International Tropical Timber Organisation (see above).

3.4.3 CoORdination of INformation on the Environment (CORINE)

CORINE was set up in 1985 to gather, coordinate and improve the consistency of information on the state of the environment in the European Communities. This aim is being achieved through the development of "procedures for the collection, standardisation and exchange of data in the Community, as well as the establishment of a Geographical Information System, to provide policy related information on the community environment" (CEC, 1989). CORINE uses the ARC/INFO GIS programme and is based jointly at Birkbeck College, London, and the Directorate General for the Environment in Brussels. The system contains a geographic database including information on coastline, national boundaries, administrative regions, water pattern, slopes and settlements. Projects for which data are being gathered include water quality, air, biotopes, coastal erosion, land cover, marine environments, soil erosion, water resources and land quality.

As part of the biotopes project, an inventory of sites considered worthy of conservation at the community level has been established. The criteria for site selection include the presence of vulnerable species of plants or animals, the presence of vulnerable habitats, the richness of the site for a particular taxonomic group and the richness of the site for a syntaxon of phytosociologial units. The information is currently only in a provisional state and the data vary in their consistency and completeness. Recently, attention has been drawn to the fact that the vegetation classification used by CORINE ignores some important habitats and as a consequence some of these are not covered by the EU Habitats Directive. Around 7,000 registered sites are currently contained within the CORINE database and these are expected to increase to between 8,000 to 9,000 in the future. Some of these sites already have protected status but others do not. CORINE also holds information on designated areas, sites which are recognised by national legislation as being under formal protection for the purpose of nature conservation. To date, this includes about 12,000 sites but again the number is expected to change as existing data are validated and new information becomes available.

Rich reef communities, Grand Turk Islands, Caribbean.

The relatively diversity-poor community of an intertidal Atlantic shore, Isle of Mull, Scotland.

In addition, CORINE holds mapped information on the distribution of vegetation types on scale of 1:3 million. This was published in 1988 as a map with over 100 vegetal classes. However, it depicts only climax and plagioclimax vegetation, not the actual existing vegetation types (CEC, 1989). The system also holds data on soils, climate (including rainfall, temperature humidity and snow cover), air (atmospheric emissions and air quality), water (freshwater quality and bathing water quality) and socio-economic data (including air traffic and nuclear power stations). For the Mediterranean region only it also has information on land resources (soil erosion risk and land quality), coastal erosion, seismic risk and stream discharge.

The future of the CORINE project as a whole and habitat classification in particular are bound up with the European Environment Agency and the development of the Nature Conservation Topic Centre, an international consortium led by the Muséum National d'Histoire Naturelle, Paris, and of which the Institute of Terrestrial Ecology and the Joint Nature Conservation Committee are partners.

3.4.4 Monitoring marine systems

Marine stations are ideally placed in terms of their facilities, expertise, long data sets and geographical location to make a major contribution to identifying and monitoring biodiversity (Grassle *et al.*, 1991) and many have already published inventories which describe the distribution patterns of marine species. Because of the nature of marine systems and the high cost of some marine research, international cooperation is essential and there has been a common tradition of such collaboration for investigating marine biodiversity (Lasserre *et al.*, 1994). Recent regional initiatives include the Caribbean Coastal Marine Productivity Programme set up in 1985, which includes 21 marine laboratories in 16 countries of the Caribbean, and the European Marine Stations network, set up in 1991 to collaborate on biodiversity research and now including 80 marine stations. A useful manual for coral reef monitoring in the Caribbean is that of Rogers *et al.*, (1994).

It is important to expand the geographical scope of this coordinated work because of the multiple habitat requirements, the possible wide distribution of many marine organisms (planktonic larval life and many invertebrates and fishes) and the ocean-wide migrations of species such as tuna, turtles, seabirds, seals and whales. A second area for future international research is the study of the dynamics of marine biodiversity, as ecological processes operate over wide areas and on long time scales (Lasserre, 1993)

The Large Marine Ecosystems (LMEs) network is a regional approach to this problem, LMEs are regions, in the order of 200,000 km^2 of ocean space, characterised by distinct bathymetry, hydrography, productivity and trophically dependent populations. These areas include coastal areas from river basins and estuaries to the seaward margins of coastal current systems (see **Box 3.9**).

LMEs allow useful comparisons of the various processes influencing large-scale changes in ecosystem status, health and biomass yields (Bax & Laevastu, 1990). For example, a recent marked decline in biodiversity among principal fish stocks of the Black Sea is believed to be partly due to the introduction of a 'foreign' species of ctenophore ('comb jelly') which preys on both the early life-stages of fish and their prey.

Efforts are now underway to place a greater emphasis on the linkage between the needs of society and the scientific community and long-term, broad-scale coastal ocean assessment and monitoring aimed at improving the sustainability of marine resource species. Biodiversity studies can contribute more towards the achievement of marine sustainability and the restoration of degraded biodiversity when they are conducted within a multidisciplinary, scientific framework and focused on populations, habitats and ecosystems of large spatial scales. The driving forces of variability in biodiversity and biomass yields are currently being examined in several LME's, along with observations on the changing states of 'health' of these systems. The GEF, in collaboration with the National Oceanic and Atmospheric Organisation, the Intergovernmental Oceanographic Commission, UNEP, FAO, NERC (Plymouth Marine Laboratory; see Chapter Four), the Sir Alister Hardy Foundation for Ocean Science and scientists from national marine resource agencies of several of the more developed countries, is helping developing nations to implement coastal ecosystem assessment, monitoring and mitigation programmes, aimed at providing a scientific basis for improving the prospects for the long-term sustainable development of marine resources (Sherman *et al.*, 1992). These programmes include one for the Gulf of Guinea ecosystem, involving five countries of the region, and one for the Yellow Sea large marine ecosystem,which is being developed jointly by specialists from China and Korea (Wu & Qui, 1993).

3.5 Global monitoring of biotic and abiotic environmental variables

3.5.1 The Global Terrestrial Observing System (GTOS) and the Geographical Observatories Programme (GOP)

GTOS is a joint initiative by FAO, the World Meteorological Organisation, UNEP, UNESCO and the International Council of Scientific Unions which aims to improve our understanding of the ways in which terrestrial systems respond to global change. By integrating and expanding existing monitoring sites and systems, GTOS aims to provide a coordinated, permanent, observational framework with adequate spatial coverage and temporal continuity to produce data for:

- Detecting and understanding local, regional and global change in terrestrial and freshwater ecosystems;
- Assessing ecosystem responses to change and the role of ecosystems in causing change;
- Quantifying and mapping change at all scales;
- Assessing the consequences of changes for land use, biodiversity, biogeochemical cycles and climate; and
- Developing, calibrating and validating models of change (Norse, 1994).

GTOS will also support the integrated assessment of changing socio-economic forces and ecosystem responses.

A four level system of sites is envisaged for GTOS, the levels varying in the number and size of the sites they include, the number of variables they monitor and the intensity (or temporal scale) of measurement (Norse, 1994). thus GTOS is designed to meet both the practical needs of planners and policy makers and the more comprehensive requirements of the research community. It will provide a global framework for national ecosystem monitoring schemes, a method for relating local

Box 3.9 Map of Large Marine Ecosystems (from Sherman, 1994, GBA Workshop)

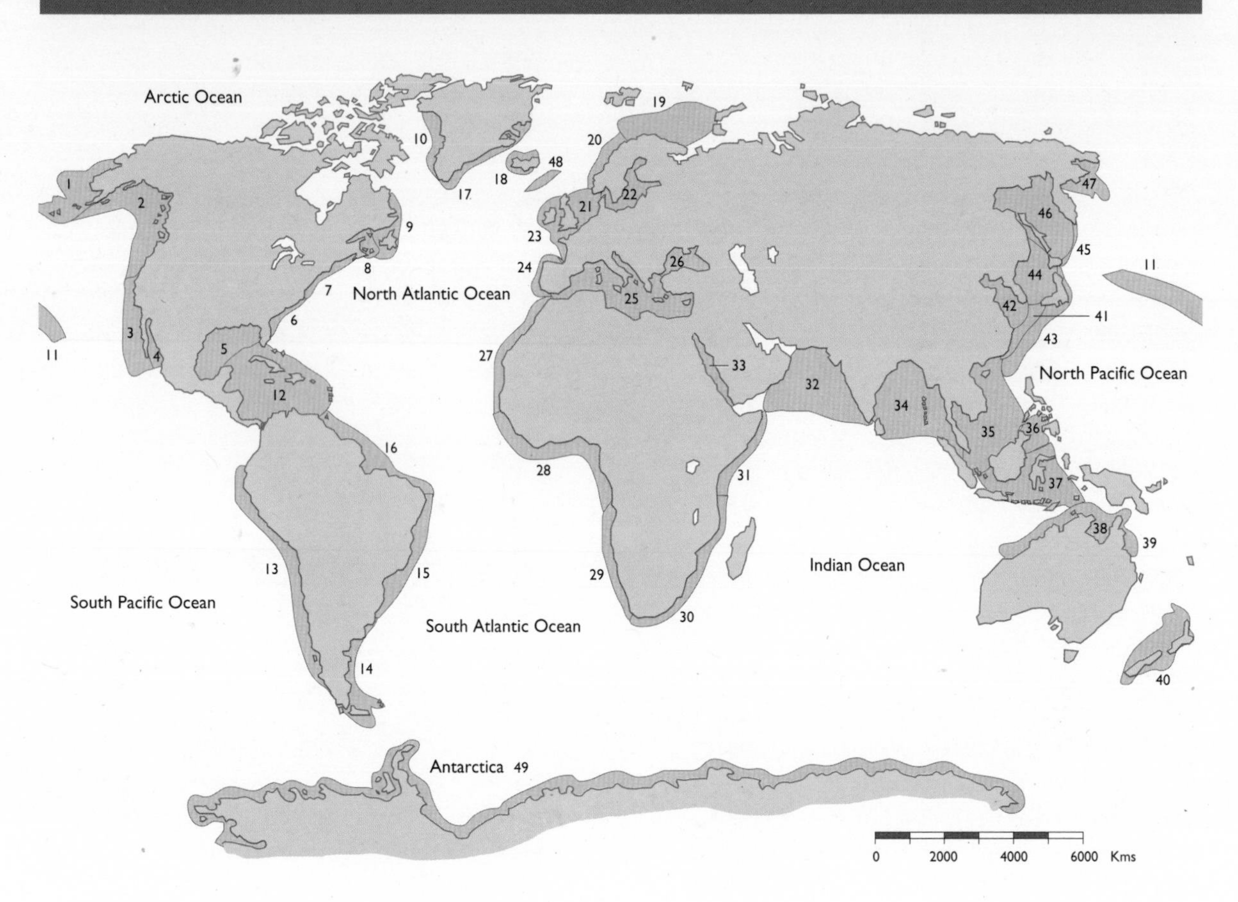

1. Eastern Bering Gulf
2. Gulf of Alaska
3. California Current
4. Gulf of California
5. Gulf of Mexico
6. Southeast U.S. Continental Shelf
7. Northeast U.S. Continental Shelf
8. Scottish Shelf
9. Newfoundland Shelf
10. WestGreenland Shelf
11. Insular Pacific-Hawaiian
12. Caribbean Sea
13. Humbolt Current
14. Patagonian Shelf
15. Brazil Current
16. Northeast Brazil Shelf
17. East Greenland Shelf
18. Iceland Shelf
19. Barents Sea
20. Norwegian Shelf
21. North Sea
22. Baltic Sea
23. Celtic-Biscay Shelf
24. Iberian Coastal
25. Mediterranean Sea
26. Black Sea
27. Canary Current
28. Guinea Current
29. Benguela Current
30. Agulhas Current
31. Somali Coastal Current
32. Arabian Sea
33. Red Sea
34. Bay of Bengal
35. South China Sea
36. Sulu-Celebes Seas
37. Indonesian Seas
38. Northern Australian Shelf
39. Great Barrier Reef
40. New Zealand Shelf
41. East China Sea
42. Yellow Sea
43. Kuroshio Current
44. Sea of Japan
45. Oyashio Current
46. Sea of Okhotsk
47. West Bearing Sea
48. Faroe Plateau
49. Antarctic

monitoring to global processes and permanent ground truthing for remote sensing activities. Many of the sites and some regional and global ecosystem monitoring networks are already established, which, combined with recent advances in data integration and communication, should enable GTOS to be 80% operational within five years.

The Royal Geographical Society in London is currently developing a Geographical Observatories Programme (GOP) which highlights the importance of long-term research sites. GOP will establish six permanent field stations to monitor global environmental change and provide a platform for local and regional environmental monitoring, research, training and education (see **Box 3.10**). GOP will adopt global protocols, where these exist and will be integrated with the GTOS monitoring sites.

3.5.2 The Global Ocean Observing System (GOOS)

GOOS was set up by the Intergovernmental Oceanographic Commission and is an international system for gathering, processing and analysing oceanographic observations from the open ocean and coastal and shelf seas. It will regularly provide descriptions and predictions of the state of the world oceans, as well as data on ocean-atmosphere and other global products for global distribution. GOOS will facilitate the study, monitoring, understanding and prediction of global climate change and global climate monitoring and will be integrated with regional and national activities. The main benefits at these levels will be improved operational and design data for marine industries services and environmental management.

GOOS recognises the importance of long-term monitoring and will integrate physical, chemical and ecological data. The system gathers information by remote sensing, sea-surface and sub-surface instrumentation and provides data contributing to the assessment and monitoring of changes in the biodiversity of fish, shellfish and plankton species groups and ocean conditions.

Global scientific programmes currently underway or planned include; the World Ocean Circulation Experiment, the Tropical Ocean Global Atmospheric Project and the Joint Global Ocean Fluxes Study. These provide the basis for the design of GOOS. However, work is needed in equipment specification, testing and trials development of operational procedures, procurement of equipment and services, and progressive merging and testing parts of the whole system. Equipment, vessels satellites and personnel will be dedicated to the GOOS by participating nations and will remain under national control.

3.5.3 The Global Climate Observing System (GCOS)

GCOS was established jointly by the International Council of Scientific Unions, the World Meteorological Organisation, UNESCO's Intergovernmental Oceanographic Commission and UNEP. It aims to develop an effective system to monitor the world's climate, detect climate change and predict climatic variation and change (Spence, 1994). The importance of an integrated and comprehensive approach is recognised, incorporating all climatic components (the global atmosphere, world oceans, the land surface, the cryosphere and the biosphere). The proposed initial operational system (IOS) for GCOS includes:

- Existing, essential observational components for climate;
- Necessary improvements which can be immediately identified and implemented;
- A comprehensive data system, all of which should be operational over the next decade.

Five advisory panels (covering atmosphere, oceans, land/ecosystem, data and space-based observations) have been established to consider how best to implement the IOS (Spence, 1994). The importance is also recognised of integrating GCOS with existing research programmes (such as the World Climate Research Programme and the International Geosphere Biosphere Programme) and operational activities (such as World Weather Watch, the Global Atmosphere Watch and the Integrated Global Ocean Services System), as well as with developing programmes such as GTOS and GOOS.

Contributing to the establishment of GCOS, a baseline upper-air network of weather observing stations and a reference surface network are

Box 3.10 The Royal Geographical Society's Geographical Observatories Programme

In partnership with leading national and international research organisations the Royal Geographical Society is planning to establish six permanent field research centres, Geographical Observatories, one in each of the world's major biomes. The primary objective of these centres is to monitor global environmental change and to provide a platform for local and regional environmental monitoring, research, training and education priorities. Such a facility will be of importance in collecting reliable, long-term, ground-based, geographical data which are essential for understanding the patterns and processes involved in current environmental changes and hence in providing an effective response. Key features of the Global Observatories Programme (GOP) include:

1. Ground-based direct observation

Advances in satellite remote sensing techniques have revolutionised environmental science. However, as the complexity and intricacy involved in environmental systems becomes apparent field-based activities also are becoming increasingly relevant. Being on the ground is the only way to provide the necessary resolution to study processes and to gather data for validating remotely sensed information.

2. Global scale

Global changes with natural causes have been happening throughout the Earth's history, but only comparatively recently have human activities been implicated at this scale. The consequences are barely understood and predictions are invariably controversial. In order to ensure a unity of research aims and results, environmental scientists require coordination across national and continental boundaries. The GOP aims to achieve this by defining an all-site core set of monitoring, research and training activities in consultation with the UN, ICSU and other international research programmes.

3. Long-term change

Data collection has traditionally been limited to specific research projects and funding availability. The consequent lack of consistent long-term results is now a serious constraint in understanding environmental change. In creating Geographical Observatories, the RGS will offer its own commitment to establishing long-term partnerships (aimed initially at 50 years) with national governments and research organisations.

4. Intensity of research and monitoring

By focusing at single locations over long time periods, each Observatory will be able to give its environment an unprecedented level of scientific scrutiny. The long-term programmes will set targets for each phase of operation, starting with a GIS of site descriptions and species inventories and developing to regular statements of change and the production of descriptive and process-based models of the site.

5. Local, national and regional priorities

As permanent field facilities with international backing and cooperation, Geographical Observatories will also be ideal platforms for monitoring, research, training and education activities determined on a site-by-site basis according to local, national and regional circumstances. The priority will be to use environmental science to help answer specific questions to the benefit of the community (for example environmental degradation, resource utilisation and biodiversity assessment).

6. Developments in remote sensing

Remote sensing provides a way of gathering data, frequently and globally, about any phenomenon with a distinct spectral response and thus has a key role in environmental science. However, such data is limited in resolution and there is no intrinsic indication of what a certain signal, or change of signal, represents on the

Box 3.10 *continued*

ground. Geographical Observatories offer a platform for directly observing surface phenomena in order to correlate them with remotely sensed signals, both at single points in time and for observing long-term changes.

7. Training and capacity building

Fieldwork offers an unrivalled opportunity for first hand experience and involvement for training. It provides practical skills in planning and data management and stimulates further interest, enthusiasm and commitment. Geographical Observatories will have the necessary access, accommodation and scientific facilities for students visiting from local and foreign institutions. On-going monitoring and research activities, using standardised field techniques, will provide many training opportunities.

8. Linking traditional and scientific approaches

Any approach to solving socio-environmental problems must have the fullest participation of local people. Not only are they the most important participants, but traditional systems of managing fragile or extreme environments and adapting to change are often highly effective and sustainable. Being field-based, the GOP is in a strong position to cooperate closely with local people and organisations. This approach has been extremely productive in previous RGS field projects and the GOP will build on these experiences.

9. Beneficial results

Ultimately, the GOP is committed to helping to improve the environment and the quality of people's lives. Research and monitoring at Geographical Observatories will provide the basis for environmental understanding, modelling and prediction. These in turn may be used, for example, in the development of 'green technologies' to ameliorate environmental change, to provide a foundation for environmental policy and legislation and to improve renewable resource management and conservation. Training and education activities will have direct benefits also for national capacity building and in developing public awareness. In these ways GOP activities will provide tangible social and environmental benefits, locally, nationally and globally.

being developed, both of which will optimise the geographical coverage, reliability and quality of records. GCOS is also cooperating with ocean programmes to increase the numbers and coverage of drifting buoys which provide surface observations, and is helping to ensure the continuation of the Tropical Ocean Global Atmosphere Programme as a source of climate data for research and operational prediction. Ultimately, GCOS will provide an international mechanism for the coordination of climate observations and will contribute to capacity building and training projects for climatic research in participating countries.

3.6 Sites for biodiversity conservation

3.6.1 Protected Areas

The world's protected areas play a fundamental role in the conservation of species and ecosystem diversity worldwide. In the past, areas with varying levels of conservation status (such as National Parks, Forest Reserves and Nature and Game Reserves) were established for a variety of different reasons, many of which

B. Clayton-Jolly/Raleigh International

Montane forest at 2,000m in the Ruwenzori Mountains, western Uganda.

A.C. Jermy/NHM

Lower montane forest, Tjibodas National Park, Java, Indonesia.

A.C. Jermy/NHM

Tourists are confined to board walkways here in G. Mulu NP, Sarawak, to protect the fragile soils in the alluvial forests.

were non-biological and some of which were purely financial or socio-economic. Furthermore, Governments also use a variety of different legal and administrative mechanisms to manage national areas for the conservation of biodiversity. This variation between countries in the mechanisms used to create and maintain systems of protected areas has prompted the IUCN Commission on National Parks and Protected Areas to establish a system for the classification of Protected Areas, based on management objectives, in order to facilitate international comparisons. Ten categories are recognised, including eight national and two (ix and x below) international categories (IUCN, 1984):

- I Scientific Reserve/Strict Nature Reserve;
- II National Park;
- III Natural Monument/Natural Landmark;
- IV Managed Nature Reserve/Wildlife Sanctuary;
- V Protected Landscape or Seascape;
- VI Resource Reserve;
- VII Natural Biotic Area/Anthropological Reserve;
- VIII Multiple-Use Management Area/Management Resource Area;
- IX Biosphere Reserves; and
- X World Heritage Sites.

One hundred and sixty nine countries worldwide now support protected areas of 1,000 hectares or more in IUCN categories I to IV, a total of 7,734,900km^2 or 5.19% of the Earth's land area (WCMC 1992).

Two international conventions and one international programme include provision for the designation of internationally important sites: the Convention Concerning the Protection of the World Cultural and Natural Heritage, the Convention on Wetlands of International Importance especially as Waterfowl Habitat (the Ramsar Convention) and UNESCO's Man and the Biosphere (MAB) programme.

The Convention Concerning the Protection of the World Cultural and Natural Heritage was adopted in 1972 and came into force in 1975. It provides for the designation of areas of outstanding universal value as World Heritage Sites and its aim is to promote international cooperation to ensure the preservation of these areas. World Heritage Sites are nominated by the states party to the Convention and pass

Box 3.11 Map of World Heritage sites (from WCMC, 1992, with updates)

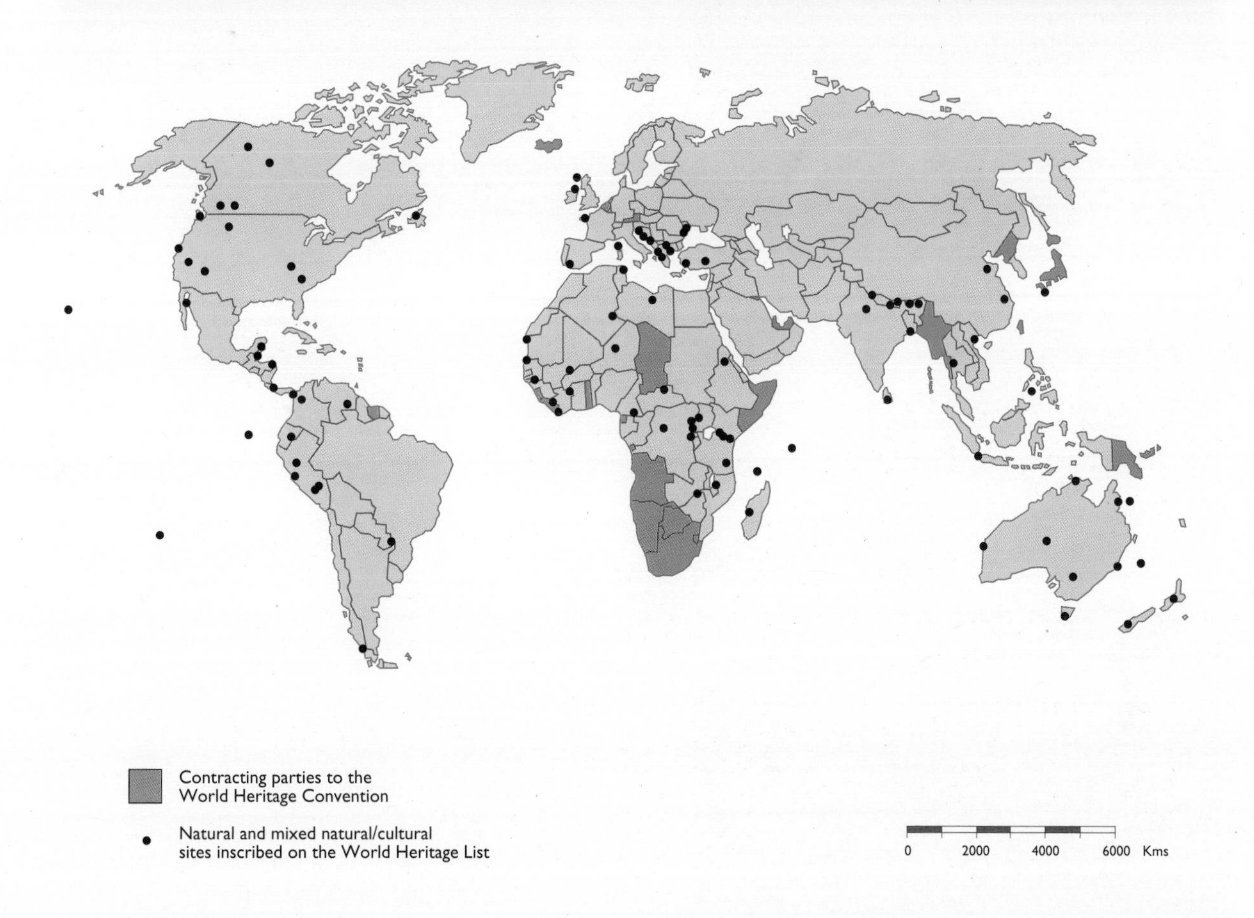

through a review procedure prior to being listed by the World Heritage Committee. As of 1992, 97 natural and 17 mixed natural and cultural areas were listed as World Heritage Sites worldwide (see **Box 3.11**). The Ramsar Convention was outlined in Section 3.4.1 and, as stated there, 654 Ramsar wetlands, covering an area of over 43 million hectares, in 81 states had been designated as of 1994 (Navid, 1994).

UNESCO's Man and the Biosphere (MAB) Programme was launched in 1971 to improve the links between people and the biosphere. It now involves some 110 countries, coordinating their national contributions under MAB National Committees. The backbone of the MAB programme is the International Biosphere Reserve Network (see **Box 3.12**)which was established with the aim of coordinating studies of natural systems at national, regional and international levels. Biosphere Reserves are alternative types of protected areas with a combination of functions including *in situ* conservation of natural and semi-natural areas, sustainable management of natural resources for local people, scientific research and monitoring and environmental education and training. As of mid-1995, there were 328 biosphere reserves located in 82 countries, covering an area of over 217 million hectares.

The 1984 Action Plan for Biosphere Reserves refers specifically to actions for inventorying and monitoring of flora and fauna. One effort to promote such work has been the preparation of a directory of contacts and environmental databases for Biosphere Reserves in Europe and North America (EuroMAB, 1993). Of the 175 Biosphere Reserves in this directory, over 88% had undertaken a biological inventory and possessed collections of local fauna and flora. 86% also monitored climatic variables, 76% monitored various data on vegetation and 76% monitored wildlife population dynamics. These high figures fall considerably for Biosphere Reserves in some other parts of the world, where the Reserves range from long-established sites with good inventories of flora and fauna and monitoring programmes, to new sites which have not had the time to develop

Box 3.12 Map of UNESCO Man and Biosphere Reserve Network (from WCMC, 1992)

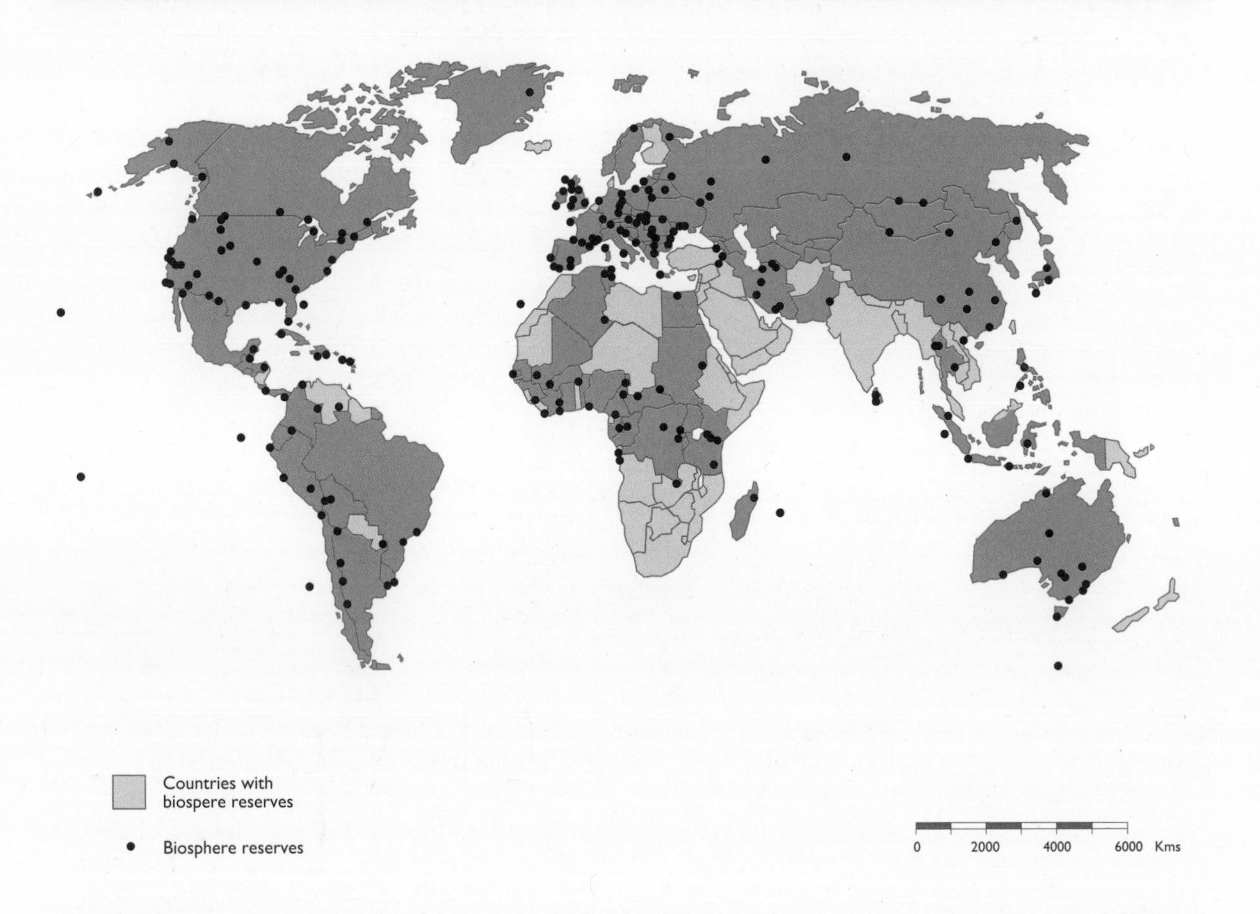

Box 3.13 Map of Endemic Bird Areas (from Bibby *et al.*, 1994.)

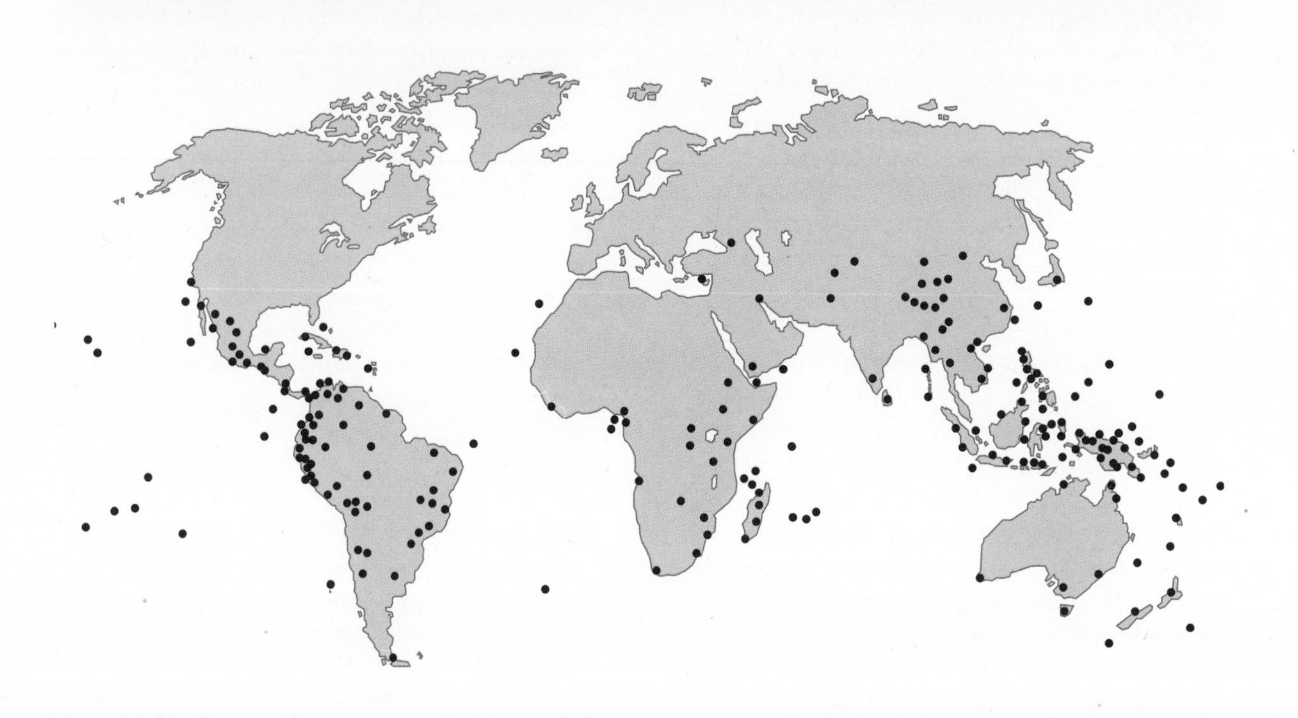

such activities, to those that do not possess the capacity to collect such data.

At the regional level, a number of sub-networks of Biosphere Reserves exist which undertake biodiversity inventories and monitoring as part of their activities. These include:

- The *Observatoire du Sahara et du Sahel* which monitors desertification in 20 countries of the Sahara and Sahel region of Africa, emphasising the human uses of biological resources (OSS, 1994).
- The Ibero-American Network of Biosphere Reserves organised under CYTED (*Programma Ibero-americano de Ciencia y Technologia para el Desarrollo*) which involves 19 countries of Latin America and the Caribbean, plus Portugal and Spain, and specifically addresses the inventorying and monitoring of biodiversity in the Biosphere Reserves of these countries (Halffter, 1992).
- The Biosphere Reserves Integrated Monitoring system, set up for 175 Biosphere Reserves in 32 countries in Europe and North America (EuroMAB), is a compilation of existing inventorying and monitoring activities in these sites that facilitates collaboration among participating scientists (EuroMAB, 1993).

As a complement to the terrestrial element of the Biosphere Reserve Network and the growing number of Biosphere Reserves that encompass both terrestrial and marine parts, UNESCO also operates a COstal and MARine programme (COMAR) for scientific research and training in coastal and marine systems.

The figures quoted here for MaB Reserves and WHSs are regularly updated on UNESCO's information page on the Internet.

3.6.2 Determining future priority areas

In order to make the best use of limited resources for *in situ* biodiversity conservation two basic questions must be addressed, namely 'how to do it' and 'where to do it'. Leaving aside socio-political considerations, methods for conserving biodiversity depend largely on applied ecology and systematics, while where to concentrate efforts depend on both systematics and biogeography. In 1988, Norman Myers focused attention on the plight of biodiversity in so-called 'hotspot' areas (Myers, 1988, 1990). He identified a total of 18 areas worldwide (totalling 746,400km^2 or 0.5% of the Earth's land surface) of high species richness which support 49,955 endemic plant species or 20% of all plant species on Earth. Similarly, Mittermeier (1988) identified between six and twelve 'megadiversity' countries (most importantly Mexico, Colombia, Brazil, Zaire, Madagascar and Indonesia) which together account for 50–80% of the world's total biological diversity and where conservation efforts should be concentrated.

Bibby *et al.* (1992) used data on the distribution of bird species worldwide and, defining restricted-range species as those with breeding ranges less than 50,000 km^2 (in total 2,609 species or 27% of all bird species), identified 221 Endemic Bird Areas (EBAs) (see **Box 3.13**). Each EBA incorporated the ranges of at least two restricted-range birds and in total the EBAs identified include the ranges of 2,484 species (95% of all restricted-range birds). It is suggested by Bibby *et al.*, that these areas of high avian endemism are of importance also for plants and other animals and the countries concerned should consider designating them as conservation areas.

A similar review of the distribution of endemic species of plants has been provided by the Centres for Plant Diversity project (Davis *et al.*, 1994), an international initiative coordinated by IUCN and WWF. The objectives of the project are:

- to identify areas around the world which, if conserved, would preserve the greatest number of plant species;
- to record the economic and scientific benefits the conservation of these areas would bring; and
- to outline a conservation strategy for each area.

Two hundred and thirty four priority areas have been selected, considered to be the most important sites for plant biodiversity and conservation by authorities. The results will be published in three volumes, the first, covering Europe, Africa and South West Asia (Davis *et al.*, 1994) the second, covering Asia, Australasia and the Pacific (Davis *et al.*, 1995), and the third, the Americas, to be published in the near future. These provide a unique global

N. M. Collins/A. C. Jermy/N. deN. Winser/RGS

Biodiversity in a tropical rain forest, Gunung Mulu National Park, Sarawak.

and regional analysis of the nature and distribution of plant diversity worldwide and a guide to the practical and cost-effective conservation of these areas.

Although there can be no doubting the significance of such findings (and the threat to these areas and their organisms), these approaches on birds and plants take little account of the rarity and threatened status of other taxa. In part, this is due to lack of information for other groups. The problem is exemplified by an analysis of detailed information on the distributions of a range of taxa including breeding birds, butterflies, dragonflies, liverworts and aquatic plants in Great Britain (Prendergast *et al.*, 1993; Lawton *et al.*, 1994). Species richness and rarity are the most frequently cited criteria for area selection by conservationists (Usher, 1986; Scott *et al.*, 1993). However, the data for Great Britain indicate that there may be little overlap between areas of high species richness for different taxa. Furthermore, many areas of high species richness do not contain any rare species and about 17% do not contain any uncommon species. There is, therefore, a danger in selecting areas for management on the basis of information for only one or a few taxa and 'a limited number of species-rich areas do not guarantee effective conservation for rare and restricted organisms' (Prendergast *et al.*, 1993).

A number of organisations and individuals are concerned about how to use distributional data for prioritising areas for conservation. It should be emphasised that Myers' definition of 'hotspots' is different from that of Prendergast *et al.* (1993). Williams *et al.* (in press) have redefined 'hotspots' as 'range size rarity hotspots', distinct from 'species richness hotspots', to clarify this difference. They emphasise the need for appropriate methods to suit the job and point out that biodiversity may be measured as: i) taxic richness (e.g. species richness), ii) character contributions and character richness (taxic measure), and iii) endemism (range size rarity). In particular, area selection involves developing methods for integrating and analysing data on the distribution and relationships of taxa, land classes, or environmental domains (Pressey *et al.* 1993). The computer programme, WORLDMAP (Williams, 1994), offers an accessible and easily understood example which allows such biodiversity information to be incorporated into algorithms for conservation evaluation (see **Box 3.14**). Witting and Loeschcke (1995) suggest that conservation efforts should be directed explicitly towards a minimisation of the future loss of biodiversity and that there is a need to consider both phylogenetic relationships and the degree of vulnerability of species in order to optimise reserve selection.

On a practical level, Conservation International (CI) has now organised two workshops with the aim of bringing together the world's leading experts on the species and ecosystems of specific regions. Together, their knowledge and experience can be used to develop an understanding of the region as a whole (CI, 1992). The first of these workshops was held in 1990 in Manaus, Brazil, to discuss biological priorities for the Amazon Basin. The second was held in Madang, Papua New Guinea and was able to build on the experiences of the first. Prior to the Papua New Guinea workshop, six teams were organised; covering birds and mammals, fishes, reptiles and amphibians, insects and other invertebrates, freshwater ecosystems and marine ecosystems. These teams held discussions and prepared maps of the distribution of their groups prior to the meeting. The preliminary maps, along with base maps generated by CI's geographic information system showing the geographical data needed to set conservation priorities (rivers, roads, political boundaries etc.), were used as a focus for discussion at the meeting (CI, 1992). One of the main problems with these maps is that they often just show the differences in the distribution and intensity of collecting. It is difficult to determine whether gaps in data for some areas are due to low biodiversity or whether little or no collecting has been carried out there.

In spite of these problems, these maps are already being used to help government leaders, non-governmental organisations and funding agencies to decide where to focus conservation resources. They can also help in a variety of land-use planning activities, including selecting new sites for protected areas, targeting environmental education programmes and helping to assess the selection of sites for the exploitation of resources (CI, 1992).

3.7 Translating survey into policy

The provision of comprehensive and extensive information from biodiversity inventories and monitoring and the expertise to analyse this information is essential for identifying key issues with respect to biodiversity policy and management goals, as well as determining whether those goals are being met. Such data help in assessing priorities and conservation needs in land use and planning, and in environmental impact assessments for inclusion in national and other reports, and also for informing policy makers and the general public. Actions affecting biodiversity and natural resources are often based on inadequate information. There are three main reasons for this:

- The data necessary for informed decision-making are unavailable, incomplete or unreliable;
- The data are not presented in a format which policy-makers and managers can understand or use; and
- The data are incorrectly interpreted.

Defining the audience and determining their needs is critical for knowing what should be assessed or monitored in the future. There is a need to ensure that the results and implications of exploration and scientific study become translated into policy and action. Reid *et al.* (1993a), for example, explain in simple terms how a set of bioindicators can be measured and used to help set priorities. Their bioindicators provide measures that can be monitored over time in order to detect changes in biodiversity. The criteria, or set of indicators used in biodiversity planning, may differ depending on the perspective and the needs of the individual or organisation. For example, NGOs will have different perspectives to national bodies and sometimes compromises are necessary.

How best then to maximise the effectiveness of scientific endeavour? Too often, scientists write

Box 3.14 WORLDMAP and prioritisation for conservation: integration of systematic data for conservation evaluation

WORLDMAP is a PC-based graphical tool for the fast, interactive assessment of priority areas for the conservation of taxonomic diversity. The programme provides a specialised platform for developing biodiversity evaluation methods and is designed to evaluate data for one or more taxonomic groups, with the distribution of each member taxon (family, genus, species etc.) recorded separately on a gridded map (an equal-area projection map of the world or a customised regional, national or local grid). Each cell represents one specified, separate region of land or sea. The interrelationships between the taxa are also specified, according to a given or chosen taxonomic hierarchy. The individual maps for each taxon can then be integrated to show a 'biodiversity surface', according to richness of species, richness of endemic species or predicted richness of individual characters among species (based on the taxonomic relationships) (IUCN, 1980; Faith, 1994; Williams, in press).

Basic principles have been established for the analysis of systematic data for the effective evaluation of priority areas,

Paul Williams/NHM

Fig. 1 Map showing 140 1° x 1° grid sqs, selected as high priority in the Flora Neotropica are on the basis of species richness (in this case, of 729 spp of five signficant flowering plant families, and the genus *Panopsis*). Key: Red are unique and irreplaceable; yellow, selected in this complement as essential for the minimum set; those which could be replaced, in a reassessment, by light blue sqs; the remaining species distributions are dark blue.

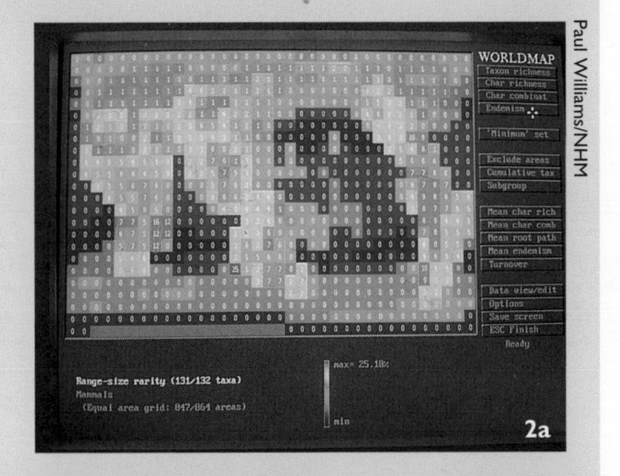

Paul Williams/NHM

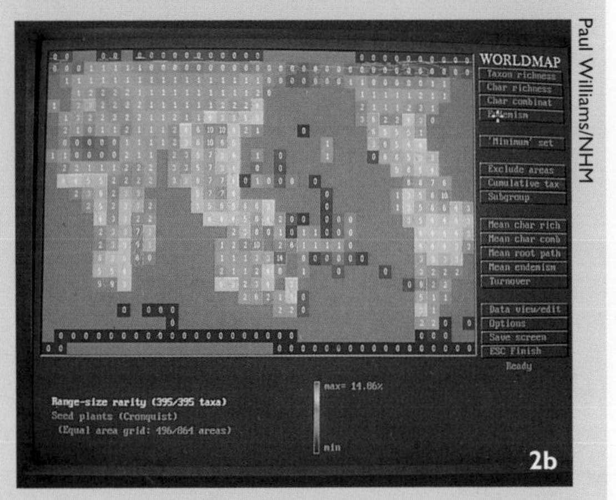

Paul Williams/NHM

Fig. 2 map of plant family range-size rarity scores. Key: Red (high scores) to blue (low scores). 2a Flowering plants; 2b Mammals.

including efficiency, flexibility, viability and vulnerability (Pressey *et al.*, 1993). WORLDMAP currently implements the first two of these and is being developed to take account of other factors. Efficiency depends on complementarity (Vane-Wright *et al.*, 1991), i.e., the degree to which one or more areas contribute to reaching an explicit representation goal (such as all the birds of Britain or all vegetation types in Africa). WORLDMAP implements complementarity through both step-wise analyses and near-minimum sets. A third procedure, for maximising the number of attributes represented by sets of areas smaller in number than the minimum set, is also being explored.

In the step-wise procedure, WORLDMAP can evaluate the additional complement to be gained by the selection of any area in the study system, in relation to the attributes

Box 3.14 *continued*

(taxa) represented by areas already chosen (such as existing reserves or perhaps non-threatened areas). Iteration of this procedure, to the point where all attributes are accounted for, results in a fully representative set. However, this is not necessarily the most efficient set for full representation, even if maximally additive areas are chosen at each step. Nevertheless, this approach can be of great practical value when faced with the need to evaluate an existing reserve system and then to consider recommendations for one or a small number of additional areas.

To represent all focal taxa in the minimum number of sites is an almost prohibitively time-consuming calculation when working with real data sets involving thousands of area comparisons and hundreds of taxa, especially if the practical need for interactivity and flexibility is acknowledged (Pressey *et al.*, 1995a). WORLDMAP employs a rapid, heuristic, three-stage procedure to find near-minimum sets for 1–10 representations of all taxa in the data set. This also implements flexibility by identifying irreplaceable areas (areas which cannot be omitted if the goal is to be achieved) and flexible areas (those areas which can be replaced by others without compromising the overall goal or efficiency).

Other systems such as the algorithms described by Margules *et al.* (1994), CODA (Pressey *et al.*, 1994) and ERMS (Pressey *et al.*, 1995b) take account of such factors as modelled probability of occurrence, area contiguity (trading efficiency to increase adjacency or connectivity of chosen areas) and degrees of area irreplaceability. The DIVERSITY package (Faith & Walker, 1994) includes procedures for sampling diversity within environmental space and can also evaluate choices in relation to competing land-use suitabilities or variable costs. A very different 'workshop' approach involving a Geographical Information System (GIS) platform and participatory consensus, is described by Mittermeier *et al.*, (1994). GAP analysis (Scott *et al.*, 1993), based on a full scale GIS, is being developed to implement a wide range of analytical options.

No system currently handles all the measures, principles and systematic procedures necessary for a fully integrated approach to biodiversity evaluation, but tools such as WORLDMAP can already be very helpful for conservation managers and planners. All such procedures, however, are limited or constrained by a general lack of comprehensive and consistent data. This inequality includes partiality by taxonomic group and unevenness of sampling effort across regions, the available data being often old and uncertain. However, this does not mean that existing data are valueless (one priority is to mobilise existing data so that they can be used), but the choice of a surrogate to represent biodiversity then becomes a critical factor (Ryti, 1992; Kremen, 1992; Pressey, 1994; Vane-Wright, 1994; Williams *et al.*, 1994;). A major effort is needed to gather data on a far wider and more consistent basis than previously. However, before this is done careful attention must be given to the design of economical methods and statistically appropriate protocols, to ensure that such data will serve evaluation needs more effectively in future.

(*Source*: *P.H. Williams, C.J. Humphries, Natural History Museum, London.*)

the results of their studies for other scientists alone, ignoring the general public and policy makers. Those writing reports of field studies should ensure that the reports are timely and appropriately presented. They should include a section on policy implications resulting from their studies. Reports that are too long often do not get read and a summary of the main findings, written in plain language is therefore essential.

One of the greatest concerns for global biodiversity is the apparent lack of action at most levels (local, national and international) to deal with many biodiversity issues. This is perhaps not surprising as little information had been assembled on the plight of biodiversity prior to the UNCED meeting in Rio de Janeiro in 1992, nor had the scientific issues been fully debated. This is in contrast to the Convention on

Climate Change, where an international scientific review was available beforehand. In this respect, the *Global Biodiversity Assessment* (GBA) recently published by UNEP should help fill this gap by providing a comprehensive, peer-reviewed document on many of the scientific issues concerned with biodiversity. Perhaps the most important section of the GBA is the Summary for Policy Makers, which is provided to help those involved in decision-making to assess where and how to act in order to best utilise and maintain biodiversity (UNEP, 1995).

In the UK there is much evidence of scientists, NGOs and government bodies cooperating. For example, several scientists acted as chapter editors for Biodiversity: The UK Action Plan (DoE, 1994) which was generally well received by non-governmental organisations. For their part, the UK non-governmental organisations also have produced a plan for action for the voluntary conservation sector (Wynne *et al*, 1995). Representatives from UK NGOs now work with government and regional officials on the UK Biodiversity Steering Group for joint action (see DoE, 1995). The publication, '*Conservation Issues in Strategic Plans*' produced by English Heritage, English Nature and the Countryside Commission (Anon., 1993), shows how coordinated actions by different organisations can result in policy and management decisions which satisfy the cultural, historical, economic and biological backgrounds of the areas concerned.

3.8 References and bibliography

Anon. (1993). *Conservation issues in strategic plans*. Countryside Commission, English Heritage and English Nature. Northampton.

Bax, N.J. and Laevastu, T. (1990). Biomass potential of Large Marine Ecosystems. In: Sherman, K., Alexander, L.M. and Gold, B.D. (eds.), *Large Marine Ecosystems: patterns processes and yields*, pp.188–205. American Association for the Advancement of Sciences Press, Washington, DC.

Bibby, C.J., Collar, N.J., Crosby, M.J., Heath, M.F., Imboden, Ch., Johnson, T.H., Long, A.J., Stattersfield, A.J. and Thirgood, S.J. (1992). *Putting biodiversity on the map: priority areas for global conservation*. ICBP, Cambridge.

Bisby, F.A. (1994). Global master species databases and biodiversity. *Biology International*, 29: 33–38.

Blackmore, S. (1994). ESF scientific networks: Systematic Biology. *Communications*, 31: 14–15.

Blackmore, S. (1995). Welcome to the network. *Syst. Biol. Network Newsl.* 1: 1.

Burnett, J., Copp, C. and Harding, P. (1994). *Biological recording in the United Kingdom: Present practice and future development*. Volume 1: *Report*. The Department of the Environment, London.

Castri, F. di, and Younès, T. (1990). Ecosystem function of biodiversity. *Biology International Special Issue No.* 22. IUBS, Paris.

Castri, F. di, Robertson Vernhes, J. and Younès, T. (1992a). The network approach for understanding biodiversity. *Biology International*, 25: 3–9.

Castri, F. di, Robertson Vernhes, J. and Younès, T. (eds). (1992b). Inventorying and monitoring biodiversity. *Biology International Special Issue No.* 27. IUBS, Paris.

CEC (Commission of the European Communities). Directorate General for the Environment, Nuclear Safety and Civil Protection. (1989). CORINE *Database Manual Version* 2.2. Compiled by Whimbrel Consultants CORINE Central Team, CDXI, CEC, Brussels. Laboratory of Land Management, K.U.L. University, Belgium.

CI (Conservation International). (1992). *Mapping biodiversity, computers and conservation priorities. Lessons from the Field* 1. Conservation International, Washington, DC.

Coddington, J., Hammond, P. M., Olivieri, S., Robertson, J., Sokolov, V., Stork, N.E. and Taylor, E. (1991). Monitoring and inventorying biodiversity from genes to ecosystems. In: Solbrig, O. (ed.), *From genes to ecosystems: A research agenda for biodiversity*. pp. 83–117. IUBS, Cambridge, Mass.

Davis, S.D., Heywood, V.H. and Hamilton, A.C. (eds) (1994). *Centres of plant diversity: A guide and strategy for their conservation. Volume* 1: *Europe, Southwest Asia and the Middle East*. WWF and IUCN, Cambridge, UK.

Davis, S.D., Heywood, V.H. and Hamilton, A.C. (eds) (1995). *Centres of plant diversity: A guide and strategy for their conservation. Vol. 2: Asia, Australasia and the Pacific*. WWF and IUCN, Cambridge, UK.

DoE (Department of the Environment) (1994). *Biodiversity, the UK Action Plan*. HMSO, London.

DoE (Department of the Environment). (1995). *Biodiversity: The UK Steering Group Report*, Vols. 1 & 2. HMSO, London.

Donlon, N. (1995). A European network for systematic biology. *Soc. General Microbiol. Quarterly*, 22: 62.

ENRIC (Environment and Natural Resources Center) (1994). *A source book on tropical forest mapping and monitoring through satellite imagery: The status of current international efforts*. Datex Inc., Arlington, Virginia.

European Community. (1991). *CORINE Biotypes: The design, compilation and use of an inventory of sites of major importance for nature conservation in the European Communities*. Office for Official Publications of the European Community, Luxembourg.

EuroMAB (1993). ACCESS. *A Directory of contacts, environmental data bases and scientific infrastructure on 175 Biosphere Reserves in 32 countries*. Department of State Publication 10059, Bureau of Oceans and Environmental and Scientific Affairs, Springfield.

Faith, D.P. (1994). Phylogenetic diversity: a general framework for the prediction of feature diversity. In Forey, P.L., Humphries, C.J. and Vane-Wright, R.I. (eds), *Systematics and conservation evaluation*. 251–268. Oxford University Press, Oxford.

Faith, D.P. and Walker, P.A. (1994). DIVERSITY. *A software package for sampling phylogenetic and environmental diversity: reference and user's guide (version 2.0)*. CSIRO Division of Wildlife and Ecology, Canberra.

Favre, D.S. (1989). *International trade in endangered species; A Guide to CITES*. Martinus Nijhoff, Dordrecht.

FAO (1993). *Forest Resources Assessment 1990: Tropical Countries*. FAO, Rome.

Gaston, K.J. and May, R.M. (1992). Taxonomy of taxonomists. *Nature*, 536: 281–2

Grassle, J.F., Lasserre, P., McIntyre, A.D. and Ray, G.C. (1991). Ecosystem function of marine biodiversity. *Biology International Special Issue No.* 23. IUBS, Paris.

Halffter, G. (ed.) (1992). *La diversidad biologíca de Iberoamerica* I. Acta Zoológica Mexicana, (Volumen especial 1992), Mexico City.

IUCN (1980). *World Conservation Strategy: Living resource conservation for sustainable development*. IUCN/UNEP/WWF, Gland, Switzerland.

IUCN (1984). Categories and criteria for protected areas. In: McNeely, J.A. and Miller, K.R. (eds), *National parks, conservation and development. The role of protected areas in sustaining Society*, pp. 47–53. Smithsonian Institution Press, Washington, DC.

IUCN (1994). IUCN *Red List Categories*. IUCN, Gland.

Janzen, D.H. and Hallwachs, W. (1994). *All Taxa Biodiversity Inventory (ATBI) of terrestrial systems. A generic protocol for preparing wildland biodiversity for non-damaging use*. Report of a NSF workshop, 16–18 April 1993, Philadelphia, Pennsylvania.

Jones, T. (1994). *Announcement of the establishment of EuroLOOP*. TECSEC, CAB International, Wallingford.

Jones, T. (1995). *BioNET-International Southern Africa LOOP (SAFRINET), A feasibility study*. TECSEC, CAB International, Wallingford.

Jones, T. and Cook, M.A. (eds). (1993). *Proceedings of the First BioNET-INTERNATIONAL Consultation, London, June 1993*. CAB International, Wallingford.

Kremen, C. (1992). Assessing the indicator properties of species assemblages for natural areas monitoring. *Ecological Applications*, 2: 203–217.

Lasserre, P. (1993). The role of biodiversity in marine ecosystems. In: Solbrig, O.T., van Oordt, P. and van Emden, H. (eds), *Biological diversity and global change*, pp 105–130. IUBS Monograph Series, Paris.

Lasserre, P., McIntyre, A.D., Ogdon, J., Ray, G.C. and Grassle, J.F. (1994). *International Marine Biodiversity Programme: Marine laboratory networks for the study of biodiversity function and management of marine ecosystems*. Biology International Special Issue No. 31. IUBS, Paris.

Lawton, J.H., Prendergast, J.R. and Eversham, B.C. (1994). The numbers and spatial distributions of species: analysies of British data. In: Forey, P.L., Humphries, C.J. and Vane-Wright, R.I. (eds), *Systematics and conservation evaluation*, pp.177–195. Oxford University Press, Oxford.

Mace, G.M. and Stuart, S. (1994). Draft IUCN Red List Categories, Version 2.2. *Species*, 21/22: 13–24.

Margules, C.R., Cresswell, I.D. and Nicholls, A.O. (1994). A scientific basis for establishing networks of protected areas. In: Forey, P.L., Humphries, C.J. and Vane-Wright, R.I. (eds), *Systematics and conservation evaluation*, pp.327–350. Oxford University Press, Oxford.

May, R.M. (1988). How many species are there on Earth? *Science*, 241: 1441–1449.

Mittermeier, R.A. (1988). Primate diversity and the tropical forest: case studies from Brazil and Madagascar and the importance of the megadiversity countries. In: E.O. Wilson (ed.), *Biodiversity*, pp. 145–154. National Academy Press, Washington DC.

Mittermeier, R.A., Bowles, I.A., Cavalcanti, R.B., Olivieri, S. and da Fonseca, G.A.B. (1994). *A participatory approach to biodiversity conservation: the regional priority setting workshop*. Conservation International, Washington.

Myers, N. (1988). Threatened biotas: 'hot spots' in tropical forests. *The Environmentalist*, 8: 187–208.

Myers, N. (1990). The Biodiversity Challenge: expanded hotspots analysis. *The Environmentalist*, 10: 4, 1–14.

Navid, D. (1994). The legal development of the Convention on Wetlands: Getting it right, or the importance of proper legal drafting. *Ramsar Newsl*., Special Issue April 1994: 1–4.

Norse, D. (1994). Global Terrestrial Observing System (GTOS). *The Globe*, 22: 6–7.

NRC (National Research Council) (1993). *A biological survey for the Nation*. National Academy Press, Washington DC.

Prendergast, J.R., Quinn, R.M., Lawton, J.H., Eversham, B.C. and Gibbons, D.W. (1993). Rare species, the coincidence of diversity hotspots and conservation strategies. *Nature*, 365: 335–337.

Pressey, R.L. (1994). Land classifications are necessary for conservation planning but what do they tell us about fauna? In: Lunney, D., Hand, S., Reed, P. and Butcher, D. (eds), *Future of the fauna of Western New South Wales*. Royal Zoological Society of New South Wales, Sydney.

Pressey, R.L., Humphries, C.J., Margules, C.R., Vane-Wright, R.I. and Williams, P.H. (1993). Beyond opportunism: key principles for systematic reserve selection. *Trends in Ecology and Evolution*, 8: 124–128.

Pressey, R.L., Bedward, M. and Keith, D.A (1994). New procedures for reserve selection in New South Wales: maximising the chances of achieving a representative network. In: Forey, P.L., Humphries, C.J. and Vane-Wright, R.I. (eds), *Systematics and conservation evaluation*, pp.351–373. Oxford University Press, Oxford.

Pressey, R.L., Possingham, H.P. and Margules, C.R. (in press). Optimality in reserve selection algorithms: when does it matter and how much? *Biological Conservation*, 10:

Pressey, R.L., Ferrier, S., Hutchinson, C.D., Siversten, D.P. and Manion, G. (1995). Planning for negotiation: using an interactive geographic information system to explore alternative protected area networks. In: Saunders, D., Craig, J. and Mattiske, L. (eds). *Nature conservation: the role of networks*. Surrey Beatty, Sydney.

Reid, W.V., McNeely, J.A., Tunstall, D. B., Bryant, D.A., Winograd, M. (1993a). *Biodiversity indicators for policy-makers*. WRI, Washington DC.

Reid, W.V., Laird, S.A. Laird, Elmez, R.G., Sittenfeld, A., Janzen, D.H., Gollin, M.A. and Juma, G. (eds). (1993b). *Biodiversity Prospecting*. WRI, Washington DC.

Rogers, C.S., Garrison, G., Grober, R., Hillis, Z-M. and Franke, M.A. (1994). *Coral reef monitoring manual for the Caribbean and Western Atlantic*. U.S. Virgin Isles, National Parks Service, St. Johns, USVI.

Ryti, R. (1992). Effect of the focal taxon on the selection of nature reserves. *Ecological Applications*, 2: 404–410.

Scott, J.M., Davis, F., Csuti, B., Noss, R.F., Butterfield, B., Grovers, C., Anderson, H., Caicco, S., D'Erchia, F., Edwards, Jr., C., Ulliman, J. and Wright, R.G. (1993). Gap analysis: a geographic approach to protection of biological diversity. *Ecological Monographs*, 123: 1–41.

Sherman, K., Jaworski, N. and Smayda, T. (1992). *The northeast shelf ecosystem: stress, mitigation and sustainability*, 12–15 August 1991, *Symposium Summary*. U.S. Dep. Commer., NOAA Tech. Mem. NMFS-F/NEC-94.

Smith, A.T., (1995).Progress in the development of a species conservation data management system for the SSC. *Species* 24: 9–10.

Spence, T. (1994). Global Climate Observing System. *The Globe*, 22: 2–4.

Stork, N.E. and Samways, M. (1995). Inventorying and monitoring of biodiversity. Section 5, UNEP *Global Biodiversity Assessment*. Cambridge University Press, Cambridge.

Stone, T., Schlesinger, P., Houghton, R. and G. Woodwell. (1994). A map of the vegetation of South America based on satellite imagery. *Photogrammetric Engineering & Remote Sensing* 60: 541–551.

Systematics Agenda 2000 (1994). *Systematics Agenda 2000: Charting the biosphere. Technical Report*. Systematics Agenda 2000, Washington, DC.

UNECE/FAO (1993). *The forest resources of the temperate zones. Volume 1: General forest resource information*. UNECE/FAO, Geneva.

UNEP (1995). *Global Biodiversity Assessment*. UNEP, Nairobi.

Usher, M. B. (1986). *Wildlife conservation evaluation*. Chapman and Hall, London.

Vane-Wright, R.I. (1994). Systematics and the conservation of biodiversity: global, national and local perspectives. In Gaston, K.J., New, T.R. and Samways, M.J. (eds), *Perspectives on insect conservation*. Intercept, Andover.

Vane-Wright, R.I., Humphries, C.J. and Williams, P.H. (1991). What to protect – Systematics and the agony of choice. *Biol. Cons.*, 55: 235–254.

WCMC (1992). *Global biodiversity: Status of the Earth's living resources*. Chapman and Hall, London.

WCMC (1993). 1994 IUCN *Red List of Threatened Animals*. WCMC, Cambridge.

Williams, P.H. (1994). *Using* WORLDMAP: *priority areas for biodiversity, Version* 3.08/3.18 [software user document; unpublished]. Natural History Museum, London.

Williams, P.H. (in press). Biodiversity value and taxonomic relatedness. In: Hochberg, M., Clobert, J. and Barbault, R. (eds), *Aspects of the genesis and maintenance of biological diversity*. Oxford University Press, Oxford.

Williams, P.H., Humphries, C.J. and Gaston, K.J. (1994). Centres of seed-plant diversity: the family way. *Proc. Roy. Soc. Lond.* (B), 256: 67–70.

Williams, P.H., Gibbons, D., Margules, C., Rebelo, A., Humphries, C. and Pressey, R.L. (1996) Comparison of richness hotspots, rarity hotspots an complementarity areas for conserving diversity using British birds. *Conserv. Biol.*, 10:.

Williams, P.H., Prance, G.T., Humphries, C.J. and Edwards, K.S. (in press). Promise and problems in applying quatitative complementary areas for representing the diversity of some Neotropical plants (families Dichapetalaceae, Lecythidaceae, Caryocaraceae, Chrysobalanaceae and Proteaceae). *Biol. J. Linn. Soc.*

Witting, L., and Loeschcke, V. (1995). The optimization of biodiversity conservation. *Biol. Cons.*, 71: 205–207.

WRI (1994). *World Resources* 1994–95. Oxford University Press, New York.

Wu, B. and Qui, J. (1993). Yellow Sea fisheries: from single and multi-species management towards ecosystems management. *J. Oceanogr. Huanghai and Bohai Seas*, 11: 13-17.

Wynne, G., Avery, M., Campbell, L., Gubbay, S., Hawkswell, S., Juniper, T., King, M., Newbery, P., Smart, J., Steel, C., Stones, T., Stubbs, A., Taylor, J., Tydeman, C., and Wynde, R. (1995). *Biodiversity Challenge* (2 edn.). RSPB, Sandy.

THE CONTRIBUTION OF UK RESEARCH INSTITUTES TO BIODIVERSITY ASSESSMENT

4.1 **Introduction**

4.2 **Natural History Museum**

4.3 **Royal Botanic Gardens, Kew**

4.4 **Royal Botanic Garden, Edinburgh**

4.5 **Office of Science and Technology**

4.5.1 Natural Environment Research Council

4.5.1.1 Centre for Ecology and Hydrology
Institute of Freshwater Ecology
Institute of Hydrology
Institute of Terrestrial Ecology
Institute of Virology & Environmental Microbiology

4.5.1.2 Centre for Coastal and Marine Sciences

4.5.1.3 Southampton Oceanography Centre

4.5.1.4 British Antarctic Survey

4.5.1.5 Sea Mammal Research Unit

4.5.2 Biotechnology and Biological Sciences Research Council

4.5.2.1 Babraham Institute

4.5.2.2 Institute for Animal Health

4.5.2.3 Institute of Arable Crops Research

4.5.2.4 Institute of Grassland and Environmental Research

4.5.2.5 John Innes Centre

4.5.2.6 Roslin Institute

4.6 **Ministry of Agriculture Fisheries and Food**

4.6.1 Directorate of Fisheries Research

4.7 **Scottish Office**

4.7.1 Scottish Crop Research Institute

4.7.2 Macaulay Land Use Research Institute

4.7.3 Other agricultural institutes

4.7.4 Fisheries Research Services

4.8 **Forestry Commission Research Division**

4.9 **Foreign & Commonwealth Office: Overseas Development Administration**

4.9.1 Natural Resources Institute

4.10 **CAB International**

4.10.1 International Institute of Biological Control

4.10.2 International Mycological Institute

4.10.3 International Institute of Parasitology

4.10.4 International Institute of Entomology

4.11 **Bibliography and references**

Chapter authorship

This chapter has been researched and edited by Shane Winser with additional material contributed by M. Ambrose, J. Anderson, J. Bodrell, G. Boxshall, S. Cable, M. Chase, N. Donlon, M. Cheek, N. Garwood, M. Gee, R. Grimble, P. Harding, C. Humphries, C. Jermy, D. Long, M. McBride, C. Milner, K. Pipe-Wolferstan, D. Rae, P. Ryan, P. Sanders, M. Sands, M. Schultz, S. Seal, K. Shawe, M. Simmonds, M. Tebbs, R. Warwick, A. Watt, M. Wilkinson and G. Young.

4.1 Introduction

The United Kingdom has contributed to the study of biological diversity through its government-financed research centres for over two centuries. In its early form, this study consisted of comparisons of natural objects as scientists became increasingly aware of the variety of living organisms. Early inventories of plants and animals and the listing of their known distribution stimulated the beginning of major private museum collections. In England, the extensive collection of natural objects amassed by Sir Hans Sloane and left to the nation on his death became, in 1753, the nucleus of the internationally famed British Museum (now The Natural History Museum). This was the government's first acceptance of responsibility for the study of natural science. There rapidly followed in 1759, the establishment of a Royal Botanic Garden at Kew Palace, exhibiting exotic plants in the living form brought back by early travellers. Indeed the cultivation of plants useful in medical practice was an essential part of the teaching of apothecaries and the first university botanic garden was founded at Oxford in 1621. The Royal Horticultural Society was one of the first to send plant collectors abroad and national collections, as well as more local ones increased in number and size as more natural scientists travelled abroad and research facilities using these collections were developed at Kew, Edinburgh and the British Museum. It was another 150 years before government-sponsored research into biodiversity was to develop further. It was obvious that the increasing knowledge of biological science should be directed towards producing food and, as an island nation, it is not surprising that the study of the variation and ecology of marine fish should be one of the first to be tackled, with the formation of the Fisheries Board in 1883, and the Board of Agriculture in 1911.

The utilisation of wild plants and animals as crops and domestic beasts has a longer history, but it was only this century that biological diversity was scanned for new varieties to use in breeding programmes and organisations like the John Innes Centre and the Plant Breeding Institute in Cambridge were established. The Agriculture Research Council (later the Agriculture and Food Research Council and

Box 4.1 Diagram showing relationship between U K Research Insitutes and Government

Foreign Office	Scottish Office	Ministry of Agriculture Fisheries and Food	Office of National Heritage	Office of Science and Technology	
Overseas Development Administration				Natural Environmental Research Council	Biotechnology & Biological Sciences Research Council
	Royal Botanic Garden, Edinburgh	Royal Botanic Gardens, Kew	Natural History Museum	Centre for Ecology & Hydrology	Institute of Arable Crops Research
Natural Resources Institute	Scottish Crop Research Institute	National Institute of Agricultural Botany		Centre for Coastal & Marine Sciences	Institute of Grasslands & Environmental Research
	Macaulay Land Use Research Institute	Forestry Commission		Southampton Oceanographic Centre	Roslin Institute
	Other agricultural research institutes	Directorate of Fisheries Research		British Antarctic Survey	Institute for Animal Health
	Scottish Agricultural Science Agency and Statistic Service			Sea Mammal Research Unit	John Innes Centre (formerly Institute of Plant Science Research)
	Fisheries Research Services			British Geological Survey	Horticulture Research Institute

now incorporated into the Biotechnology and Biological Sciences Research Council) was established in 1931 to direct research on this wider issue and these genetic institutes came under its aegis.

As recently as 1949, with the move to develop national parks and nature conservation, greater research into the environment, and the plants and animals that populate it, became a higher priority in government research planning.

The majority of research institutes are a networked group within the Research Council system through which the majority of government research funds are channelled by the Office of Science and Technology. The Research Councils also fund specialist research groups and individual postgraduate studentships in UK universities (see Chapter 5). Two Research Councils are involved in biodiversity initiatives, the Natural Environment Research Council (NERC) and the Biotechnology and Biological Sciences Research Council (BBSRC).

The Ministry of Agriculture, Fisheries and Food and its counterpart in the Scottish Office also partly fund biodiversity programmes that relate plant biodiversity and fish biodiversity with plant and animal breeding and gene-banking. Particular projects relating to overseas research, including biodiversity, may be funded by the Foreign Office through the Overseas Development Administration, the appropriate Directorates of the European Union, Commonwealth organisations, or in partnership with other nations' aid agencies. This chapter also includes a description of CAB International and its component institutes, for although an inter-governmental agency, it has its main base in the UK, and arose from a Commonwealth initiative.

The Department of the Environment encourages and funds science and technology in the UK and overseas dependencies in support of its policy, statutory, operational, regulatory and procurement responsibilities. Research projects relating to biodiversity are carried out under the Department's air quality, global atmosphere, environmental protection, water and countryside research programmes, as well as by the statutory conservation agencies and the National Rivers Authority. The Department's countryside research and long term monitoring programme seeks to ensure that countryside and wildlife policies are based on the best scientific evidence available. This includes monitoring change, assessing and predicting the impacts of new policies and practices, and producing practical prescriptions for conservation. Much of the programme involves collaboration with a wide range of other Government departments, research councils, statutory conservation agencies and partners in the voluntary sector. Its role as administrator of Government funds for an international biodiversity programme under the Darwin Initiative for the Survival of Species is discussed in Chapter Five.

The country conservation agencies (English Nature (EN), Countryside Council for Wales (CCW), Scottish Natural Heritage (SNH) and the Environment and Heritage Service of the Department of the Environment (Northern Ireland)) are the statutory advisers responsible for nature conservation in the UK. The Joint Nature Conservation Committee (JNCC) is the statutory committee through which EN, CCW and SNH exercise their special national and international responsibilities. Whilst outside the remit of this manual, research reports and publications resulting from those agencies will be of direct interest to those working in similar bodies abroad. An overseas visitor wishing to make contact should do so through the JNCC (see Appendix 2 for address). The Department of the Environment's Biodiversity Secretariat co-ordinated the UK Biodiversity Action Plan and Biodiversity Steering Group Report.

Competition for research funding generally has increased in recent years. Government departments often put out research contracts to tender in their endeavours to get value-for-money. This has led to many research establishments and universities setting up commercial departments to market their services. This is also enabling them to bid for commercial contracts for environmental impact assessments, monitoring programmes, policy and law, audit and conservation management. These projects may involve biodiversity studies. Cooperative projects and joint funding agreements are now commonplace and industrial sponsorship is not unusual, especially in the field of biotechnology.

The diverse range of Government departments currently involved in biodiversity-related

Box 4.2 Seedling Flora Project of Panama – a guide for forest practice

Accurate identification of seedlings and saplings is a critical stumbling block to the success of many ecological studies aimed at improving the management and conservation of tropical forests or restoring damaged forests to productive use.

The *Seedling Flora Project* is a long-term collaborative project between The Natural History Museum in London, the Field Museum of Natural History in Chicago, and the Smithsonian Tropical Research Institute in Panama. The US National Science Foundation and the NERC in the UK funded its early phases.

Current work focuses on Barro Colorado Island, Panama, because it is the ecologically and floristically best known neotropical forest. Our broad survey of about 700 species in over 100 families includes canopy and subcanopy trees, shrubs, lianas, strangling figs, and herbs. Because a large and varied sample of species has been studied, many of which are widespread throughout the Neotropics, the results of work in Panama will be immediately useful throughout the neotropics. The programme will contribute to a long-range goal of identifying the seedlings of all neotropical genera. The illustrated identification manuals will be published in Spanish and English to reach a wide audience.

The best way to identify tropical seedlings and saplings, but also the most time-consuming, is to collect seeds from fruiting individuals, of which voucher specimens are made to confirm the identification, germinate the seeds, then grow the seedlings to a range of sizes in sun and shade environments. For nearly five years, seedlings were grown in this manner in Panama. Over 40,000 juvenile specimens were grown. Illustrations, descriptions, and keys to the seedling and sapling stages are now being produced from this material at the NHM. About 60 seedlings per species were studied, the specimens eventually contributing significantly to representation of tropical seedlings in major herbaria.

The project has concentrated on forest species that germinate and establish on the ground, both in the shaded understory and in canopy gaps, as these are the key groups in forest management and restoration, but also include the hemi-epiphytic strangling figs because these are often significant components of the canopy and keystone species in the community. Examples of weedy species common in permanent clearings and epiphytes (but not the orchids) were included to broaden the taxonomic base of *The Seedling Flora*, as these are frequently from families rare in the forest.

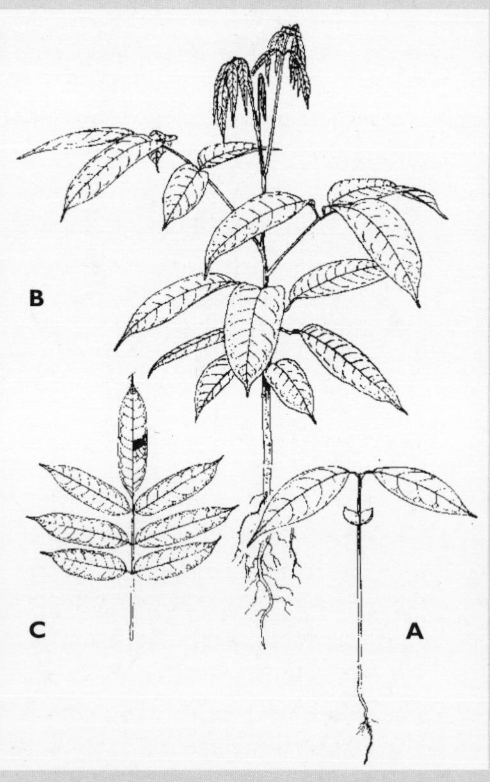

Tetragastris panamensis: Seedling (A), sapling (B) and mature leaf (C).

The initial impact of *The Seedling Flora* will provide the means

- to document patterns of forest regeneration after natural disturbance or planned manipulations;
- to developed new techniques of transplanting identified seedlings to preserve or increase genetic diversity;
- increase our knowledge of the neotropics and their biodiversity.

(*Source*: N. *Garwood* & M. *Tebbs*, *The Natural History Museum*.)

research are shown in **Box 4.1**. Current science and technology programmes and 5–10 year perspectives are published annually in *Research Councils and Departments Forward Look Statements* (Cabinet Office *et al.*, 1995). In the following account the research programmes of the three major biodiversity institutes, the Natural History Museum, the Royal Botanic Gardens, Kew, and the Royal Botanic Garden, Edinburgh are discussed first, followed by the Research Councils (Office of Science and Technology) and other government departments.

4.2 The Natural History Museum

This is one of the leading institutes in the UK for biodiversity research, funded primarily by a Parliamentary grant through the Department of National Heritage. The Museum's extensive specimen collections, library and databases are described in Chapter Six. The Museum employs about 300 scientific staff in its five scientific departments: Botany, Entomology, Mineralogy, Palaeontology and Zoology. As well as the collections and libraries, Museum researchers have access to an excellent range of modern facilities and equipment, including molecular biology and analytical chemistry laboratories, electron microscope and microprobe facilities, on-line databases and electronic mail facilities, a GIS system and advanced biometric services.

The scientific work of The Natural History Museum is structured around seven major programmes that emphasise the Museum's contribution to issues of contemporary human concern. The programmes also serve to foster interdisciplinary cooperation and collaboration between scientists with experience on diverse groups of animals, plants and minerals.

Natural History Museum

The South Kensington buildings of the Natural History Museum, opened in 1881.

The current programmes are: Biodiversity, Environmental Quality, Living Resources, Mineral Resources, Biomedical Sciences, Human Origins and Curation.

The Biodiversity Programme is further divided into three fields which have their own distinctive focus; faunas and floras, systematics and evolution, and ecological systematics. Project work is carried out by research and curation groups, with individual staff frequently contributing to a range of projects and, in many instances, to more than one of the programmes. Groups often attract and benefit from participation by externally funded researchers. This blend of people produces dynamic and flexible teams, well equipped to undertake primary research as well as offering an excellent environment for training.

Emphasis here is given to those programmes which have particular relevance to the issues of biosystematic resources, sustainable development and conservation. For each of the seven programmes, the general emphasis and direction of research is outlined and a few examples of collaborative projects are given.

The Fauna and Flora Programme concentrates on discovering, defining and documenting biodiversity, in both living and fossil groups. Areas of special emphasis are taxonomy and inventory. In the former, expertise is focused on the description of new taxa, the resolution of nomenclatural problems and the production of novel methods and guides for species identification, whilst inventory efforts are directed towards producing floras and faunas and describing patterns of species richness.

Faunas and floras are being prepared for several areas of the world, in both the tropics and temperate regions. World treatments of various major invertebrate fossil groups provide a picture of overall diversity through time. Studies in nomenclature, the naming of plants and animals, provide a firm basis for the application of scientific names used throughout biology. **Box 4.2** gives an example of one botanical project on tropical tree seedlings.

Box 4.3 Lake Baikal: a cooperative project in Central Asia between the Natural History Museum, the Royal Society and the Russian Academy of Sciences

Studies of the biodiversity of ancient lakes have provided many insights into the evolutionary processes by which new species originate. The most ancient of the world's lakes, and the deepest, is Lake Baikal in eastern Siberia and, since the late 1980's, groups of UK scientists in collaboration with Russian colleagues, have been involved in detailed scientific research on the unique fauna and flora of Baikal under a bilateral agreement between the Royal Society and the Russian Academy of Sciences.

What is unique about the diversity of life in Lake Baikal? Baikal is home to the world's only freshwater seal, to 29 species of sculpins (cottoid fishes), to over 250 species of gamarid amphipod shrimps, to over 200 species of oligochaete worms, to over 170 species of molluscs, to 150 species of ostracod crustaceans, to over 120 species of copepod crustaceans, to over 110 species of chironomid midges and to over 100 species of flatworms. The great majority of these species are endemics found only within Baikal. The lake also contains representatives of several smaller groups that are relatively rare in freshwater habitats, including an abundant species of polychaete worm, and several species of bryozoans and sponges. The presence of so many endemic assemblages of species and the dominance of intralacustrine (within the lake) speciation processes make Baikal an ideal natural laboratory for the study of the origin and maintenance of biodiversity.

This immense diversity of animal life is found primarily in the benthic habitats of the main lake. The great mass of water away from the bottom is home to a rather low diversity zooplanktonic community dominated by just two species of copepods and one species of amphipod. The benthic communities are much more varied; some are dominated by sediment which varies in amount with depth and from site to site around the lake, others are rocky, and yet others have their physical structure provided by 'forests' of erect sponges. These sponges are green due to the presence of symbiotic algae and have been utilized for food by particular amphipod and copepod species. Some copepods of the genus *Diacyclops*, for example, have highly modified mouthparts and clawed antennae which reflect this trophic specialization.

Paul Clark/NHM

The submersible craft used in the project approaching the mother research vessel on Lake Baikal.

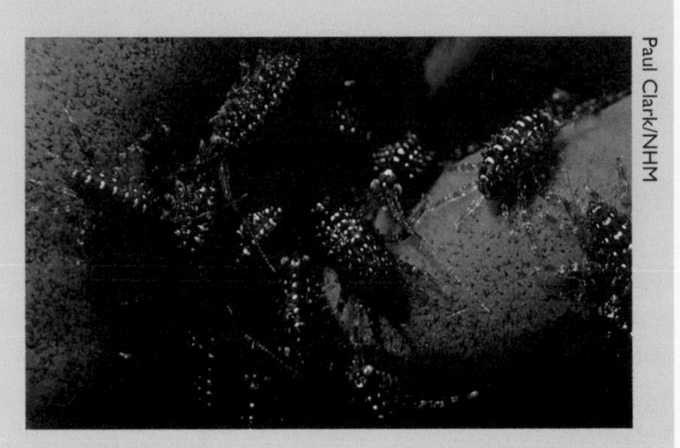

Paul Clark/NHM

An endemic amphipod associated with a dominant, also endemic, freshwater coral.

The opportunities for scientific collaboration with Russian scientists working on Baikal were opened up by the signing of a bilateral cooperation agreement between the Royal Society and the Russian Academy of Sciences. Joint projects have now been undertaken on topics such as the identity of the virus infecting the Baikal seal, eye structure and the evolution of visual pigments in the endemic cottoid fish, speciation in copepod crustaceans, the dynamics of carotenoid pigments in the food chain, the biology and ecology of unicellular planktonic algae, and the history of the lake as revealed by study of animal and plant remains in cores of bottom sediment. These studies have involved scientists from several British universities including the University of Ulster at Coleraine, as well as research institutes and the Natural History Museum. A multinational organization, the Baikal International Centre for Ecological Research (BICER) has now been established to coordinate international research effort on the lake and the UK, through the Royal Society, is a founder member of BICER.

(*Source*: G. Boxshall, *The Natural History Museum*).

C.H.C. Lyal/NHM

The NHM Belize Research Station.

The Systematics and Evolution Programme builds on descriptive studies by examining broad patterns of biodiversity, both past and present, and the causes of these patterns. Directions explored include higher classification and phylogeny estimation (often using DNA sequences), and integration of phylogeny with other data to reconstruct evolutionary history. Evolutionary work in the molecular laboratories of the Museum covers a wide range of projects, including studies of mollusc phylogeny and examination of the endosymbiotic origin of chloroplasts in algae. Computer programmes for the analysis of phylogenetic data in co-evolutionary studies are being developed at the Museum and techniques to establish conservation priorities are being used for a wide variety of organisms.

The Ecological Systematics Programme analyses the occurrence and distribution of organisms in space and in time to investigate the relationships between ecological processes and biodiversity. This work provides the basis for making predictions about how patterns of biodiversity will change in the future.

Patterns of community structure and trophic interactions in high diversity ecosystems such as coral reefs and tropical rain forests are being studied with the objective of identifying structuring processes. Changes to biodiversity following human intervention, such as timber extraction or road building, are being studied in tropical habitats and these studies will provide baseline data for the responsible management of biodiversity. Geographical Information Systems are being used to produce forest vegetation maps for Belize and to monitor changes in biodiversity.

The Environmental Quality Programme involves strategic taxonomic research focused on environmentally important and sensitive groups such as lichens, algae, invertebrates and protozoa, to enable their effective use in monitoring programmes. Methodologies involving the use of these organisms as biomonitors are being assessed, evaluated and developed.

Lichens have been used to monitor and assess damage from air pollution (acid rain), mining, fire in tropical ecosystems and ozone depletion. Monitoring of marine communities has been used to predict impact of coastal development in Britain. Microorganisms (fungi, plant and animal) and insects are currently being used in the evaluation of environmental contamination and its history. Freshwater quality is assessed and monitored using key groups of algae. Studies on toxic minerals and their effects are also a part of this programme.

The Living Resources Programme focuses on plants and animals of economic importance, thus providing a biosystematic basis for resource management and sustainable development. Cost effective and environmentally benign pest control in economically important crops and resources requires accurate identification of the pest organism and its differentiation from the myriad of related species that are not harmful.

Production of field manuals and identification systems for common pests of fish, horticulturally and agronomically important plants, tropical forest trees and stored products such as grains are an important component of work being carried out in this programme. The potentially useful relatives of presently economically important species are recognised and investigated and this is also a focus for research.

The Mineral Resources Programme examines the interactions between organisms and their physical environment as an important component in the maintenance of biodiversity. In this programme research is focused on the physical environment, especially on aspects of geology and earth history. The programme is involved with the oil and mineral industries and contributes to accurate targeting of exploration efforts.

Past climatic history is being studied using both diversity of microfossils and advanced techniques of isotope analysis. These studies will provide baseline data for comparison with present patterns of climatic change. Mechanisms of chemical movements and mineral growth in the solid earth are being investigated, as are the fundamental processes that produce ore deposits and particular groups of rocks. These processes and factors affecting the earth's surface ultimately influence all of life.

The Biomedical Sciences Programme is directed toward improving the health and well-being of humankind and domesticated animals. In vector-borne diseases precise identification of hosts and their parasites is essential for the implementation of cost-effective control programmes and for the prediction of outbreaks of these diseases.

The World Health Organisation and the Food and Agriculture Organisation recognise the Museum as a centre for the study of onchocerciasis, schistosomiasis and leishmaniasis, and screw worm and animal myiases, respectively. Research efforts are concentrated in tropical countries where vector-borne diseases are a major problem affecting both animal and human health.

The Human Origins Programme studies the evolution of mammals, primates and humans and their interactions with changing ecosystems. Research is broadly based encompassing mammalian biostratigraphy, animal migrations, and trends in climatic change, a background against which studies of patterns of human and primate microevolution and variation take place. Other projects investigate the origins of the human species and of other mammalian groups.

Science programmes and projects at The Natural History Museum build on the foundations of biosystematic research, experienced personnel, comprehensive and properly maintained collections, and information resources, to create opportunities for exciting, collaborative studies with a variety of research teams.

4.3 Royal Botanic Gardens, Kew

The Royal Botanic Gardens, Kew (RBG Kew) is one of the world's leading botanic gardens, with its combination of wide-ranging plant sciences, horticulture and education, all directed towards an understanding of the world's plants and their place in the global ecosystem and economy. The enduring excellence of its scientific and horticultural work over the past 235 years has given it a unique role.

RBG Kew is a corporate body employing over 550 people and having charitable status with a Board of Trustees sponsored by the Ministry of Agriculture, Fisheries and Food (MAFF), through which it is primarily funded. Its location and collections are described in Chapter Six.

The mission of the Royal Botanic Gardens, Kew is: *"To enable better management of the Earth's environment by increasing knowledge and understanding of the plant kingdom – the basis of life on earth."* These objectives are met by maintaining large global reference collections, by undertaking world-wide research into plants, fungi and ecosystems, by producing Floras, monographs and other research publications, by training botanists and horticulturists, including those from overseas countries, and by disseminating information widely both to the world's scientific community and to the public. RBG Kew places strong emphasis on collaborative research and on enabling developing countries to document and protect their plant resources.

Kew has become increasingly concerned in recent years that it must be ethically responsible for its activities (see Chapter Two), in such areas as intellectual property rights, ownership of biological resources, biological control and technology, and the transfer of skills to developing countries. Kew is committed to the equitable sharing of benefits arising from its research activities, disseminating the data resulting from research and applied conservation projects and pursuing an active policy of training. RBG Kew has adopted clear public positions on these issues and has encouraged other botanic gardens to do likewise.

As a primary international centre for the classification, identification and naming of plants and fungi, Kew's research programmes are designed to improve understanding of the plant and fungi kingdoms and maintain RBG Kew's position as one of the world's leading institutions for innovative basic and applied botanical and mycological research, conservation, teaching and the maintenance of biodiversity. They aim to contribute towards the documentation of a rapidly shrinking resource and answer questions about the rational utilisation and preservation of plant and fungal biodiversity.

Biodiversity in all its aspects is researched, from rainforest composition through to genome analysis. Scientific activity is centred on the Herbarium (see Chapter Six, Section 6.2.2), Jodrell Laboratory and the Living Collections Departments and covers five fronts: Systematics, Economic Botany and Ethnobotany, Phytogeography, Environmental Botany and Conservation, and Botanical Horticulture. Work is focused on families and groups of economic importance or particular biological significance, including Acanthaceae, Araceae, Asclepiadaceae, Cactaceae, Compositae, Cyperaceae, Euphorbiaceae, Gramineae, gymnosperms, Lamiales, Leguminosae, Malvales, Myrtales, Orchidaceae, Palmae, petaloid monocotyledons, Pteridophyta and Rubiaceae as well as the fungal divisions, Ascomycotina, Basidiomycotina and Zygomycotina. Increasingly, interest is centred on areas of important biodiversity, particularly in the tropics and sub-tropics, such as NE Brazil, Borneo, Irian Jaya, Madagascar and several African countries, especially East, West and South Central Africa.

The research programmes range widely: taxonomy and evolutionary studies; the preparation of plant inventories (Floras); the conservation and re-introduction of endangered plants; the search for new leads for biological control and drugs to treat diseases like AIDS, diabetes and cancer; improving the choice and quality of fuel-wood species for the developing world and the search for food, fodder, drought-resistant plants and wild seed stocks for future generations.

Royal Botanic Gardens, Kew

Visiting botanists working in the herbarium at Kew.

The major objectives of the unified Research and Dissemination Programme – for angiosperms, gymnosperms, pteridophytes and fungi – are to identify, define, name, describe and list designated taxa, to classify them in the most predictive system possible, to determine their inter-relationships and evolution, and to define their distribution, ecology, rarity and usefulness. Kew aims to achieve these objectives through interdisciplinary research on targeted taxa undertaken at the highest level of scientific excellence.

The collection of new material and the study of plants and fungi in the wild continue to be essential aspects of the work of the whole Gardens and, indeed, active collecting must continue with some urgency in response to the large-scale destruction of natural habitats which still continues. Expeditions are organised in full collaboration with host country institutions and the establishment of agreements and memoranda of understanding have led to a continuing improvement in the content and representation of the host country's and of Kew's collections. All material is obtained in accordance with all the statutory requirements of quarantine, CITES and the Biodiversity Convention. Acquisitions are expected to have full provenance data, to be incorporated into ordered and cross-referenced systems and to be made available to ensure their future security and maximise their use. For the past three years Kew has been developing capability in GIS related to databases of collected material, and is also investigating the use of remote-sensed data for vegetation analysis.

The Royal Botanic Gardens, Kew is only one of two UK Government funded institutions specifically charged with the *ex situ* conservation of wild plant diversity (see also RBG Edinburgh, below) and is the Government's Authority on plant conservation questions. For many years it has been a leader in the development of international conservation policy and a centre for advice and action on plant conservation. Increasingly, Kew is invited to help with the development of international, national and regional conservation strategies and its research programmes tackle the problems of environmental degradation and rehabilitation and species endangered by either trade or habitat destruction.

Box 4.4 Plantas do Nordeste and practical biodiversity: Local plants for local people

Plantas do Nordeste or 'Plants of the Northeast' (PNE) is a progressive multi-disciplinary research programme concerned with the identification and sustainable use of plant resources in the semi-arid northeast region of Brazil. As a collaborative venture it involves the Royal Botanic Gardens, Kew and over 20 institutions in Brazil ranging from universities and government agencies through to non-government organisations.

PNE is a practical programme with the expected results and products designed to meet local needs – at academic, government, NGO and community levels. Where possible, emphasis is on the latter, in a focused thrust towards passing information about sustainable management and use of plant resources to those able to act on it.

Royal Botanic Gardens, Kew

Children of Quatro Varas slums, acting out the benefits of medical plants.

The programme is concerned with resolving key plant resource issues in NE Brazil. A broad brush, practical approach is taken towards biodiversity studies. Identification of plants, quality collections and production of accurate data on plant species, their distribution and vegetation dynamics provides an essential basis for ongoing applied research and conservation considerations. The applied research aims to find improved management systems for the vegetation, develop new plant-based products and ways of increasing the use of beneficial plants.

Royal Botanic Gardens, Kew

Sustainable management of the 'caatinga' vegetation ensures year-round production of forage for small livestock.

Use of information technology in the development of biological data management capacity will provide an invaluable, accessible centralised resource. Emphasis is on information flow – passing on knowledge in a form that will inspire and motivate people to take action throughout North-east Brazil, which is where NGO involvement in PNE is crucial. Training at all levels forms a significant element in the dissemination efforts.

Present research covers forage plants medicinal plants, vegetation dynamics, identification and plant species diversity. New research fields identified include vegetation management for increased honey production and fuelwood cultivation for domestic and industrial use,

Strategic direction for PNE is the responsibility of the board of the Association 'PLANTAS DO NORDESTE' formed in July 1994. Funds for PNE are raised independently from a range of sources – individuals, companies, charities, national and international organisations.

PNE's uniqueness lies in its determined effort to bridge the communication gap between academia and local communities, passing the results of research on to those who can take action, thereby benefiting local people. It aims to provide a working model of how to put the requirements of the Convention of Biological Diversity into practice.

(*Source*: K. Pipe-Wolferstan, RBG, Kew)

RBG Kew has developed a comprehensive range of *ex situ* and *in situ* conservation techniques to support field programmes world-wide, more specifically by storage and cryopreservation of seeds, embryos and pollen, and the micropropagation and re-introduction of threatened and endangered species. Since 1984, Kew has undertaken conservation projects in over 35 countries throughout the world, including the identification of priority conservation areas in Brazil, India, Malawi, New Guinea, Sarawak and Tanzania. Projects have also been initiated to promote equitable and sustainable utilisation of plant resources in the arid lands of Brazil and in the rain forests of South-east Asia. Single-species rescue operations have been undertaken, particularly for island endemics such as the Saint Helena Olive and for endangered native British species such as *Cypripedium calceolus*. Within the Kew estates themselves, environmentally sympathetic management of selected areas is designed to contribute increasingly to the conservation of the indigenous flora and fauna.

As the UK CITES Scientific Authority, Kew advises the UK CITES Management Authority, the Department of the Environment, on all matters concerning the international trade in plants controlled by CITES. The Conservation Advisory & Policy Section (CAPS) carries out background research in areas such as international trade in wild bulbs and carnivorous plants and the timber trade and analyses trade in the European Union. A CITES Orchid Checklist funded by the CITES Secretariat, is currently being prepared by RBG Kew and coordinated by CAPS. RBG Kew is also investigating breeding systems in a number of genera and in some species nearing extinction, the genetic diversity of the Garden's collections relative to wild species and the phylogenetic relationships of endangered species. In several instances these studies may directly result in re-establishment of species in the wild.

RBG Kew is a member of the Tropical Forest Resource Group, a multi-disciplinary consortium of organisations that forms a powerful source of information and expertise for projects and policy developers in the areas of forest management and conservation, agroforestry and related land use. A key remit for RBG Kew is to develop an integrated biology research programme in consultation with universities, botanic gardens, governments, NGOs and international agencies. Kew is also a partner in the WWF/UNESCO/Kew 'People and Plants' initiative, which encourages the sustainable utilisation of non-crop plants by local people and has active ethnobotanical research and training programmes in both the Old and New World tropics.

RBG Kew is engaged in collaborative research programmes in over 40 countries with fieldwork in regions where the vegetation is particularly diverse or of special concern for conservation of genetic resources. These include Brunei, Malaysia, Indonesia, East Africa, Cameroon, Malawi, Madagascar, China and Brazil. A single collaborative project with the University of Sao Paulo has resulted in 115 scientific publications. Major projects are exemplified by:

- a collaborative programme with the Forest Department of Brunei Darussalam to complete a floristic inventory of the country;
- joint involvement in Cameroon in an ODA-sponsored Limbe Botanic Garden Development and Conservation project, Cameroon (see **Box 4.17**);
- a long-term collaborative programme in North-east Brazil under the *Plantas do Nordeste Project* (see **Box 4.4**), and;
- field programmes in tropical Africa and South-east Asia within the Survey of Economic Plants for Arid and Semi-arid Lands (SEPASAL) project (see Chapter Six, **Box 6.4**).

The Seed Conservation Section now holds the world's most diverse collection of wild-source seeds banked to international standards, particularly targeting the British flora and species useful to man from and and semi-arid lands.

The Jodrell Laboratory at Kew, founded in 1876, carries out fundamental experimental research in plant anatomy, biological interactions, cytogenetics, molecular systematics and seed conservation. The original laboratory was replaced by a larger building in 1965 and an extension was opened in 1994, increasing its size to 3000 sq. metres, thus more than doubling Kew's facilities for experimental research.

Focusing on microscopic, biochemical and genetic characteristics, this work is directed towards solving taxonomic questions, studying useful plants and plant products, addressing conservation problems and the application of all aspects of its research. The Biological Interactions Section, for example, considers bioactive plant chemicals and those that may prove of value in medicine and in determining evolutionary relationships between plant groups. It also studies plant-fungus interactions (e.g. orchid mycorrhizas) and plant-animal interactions (including biological control agents under glass at Kew) using an array of behavioural, electrophysiological and biochemical techniques. The Anatomy Section, has a reference collection of more than 70,000 microscope slides and an invaluable database index to anatomical literature. It conducts research in systematic plant anatomy and

Box 4.5 Detecting and conserving ancient genes

Any appraisal of the biodiversity of a given area must take account of the wider distribution of the floras and faunas involved. Certainly long distance dispersal can result in disjunct patterns of distribution, but dispersal alone cannot explain many of the patterns seen. Larger, generally tectonic explanations are required to elucidate the geological origin of those areas and the past relationships of the land masses and continents concerned.

Geologists can provide us with a wealth of data on this subject, but many areas of the world are simply unexplainable. These include the Caribbean Islands and many of the world's other oceanic islands. Even for land areas such as Australia, New Guinea and India there is a fair degree of speculation about many episodes in their past. The best sources of information about these past movements almost certainly reside in the plants and animals that were on those land areas and actually moved with them. It is only through improved phylogenetic studies can we hope to understand these processes.

Conservationists need to know how unusual and isolated are the many narrow endemics that come up as potential targets for conservation. Ideally, we would like all species to be conserved, but with the financial limitations that exist, choices must be made. Studies of the silversword alliance (Compositae) in the Hawaiian Islands have demonstrated that these plants are closely related to members of the Californian tarweed group, to which morphologically they appear highly dissimilar. Knowing that this relationship is likely, we are forced to design follow-up studies that will look at the factors responsible for the high degree of morphological re-patterning that took place within the Hawaiian archipelago. Furthermore, it allows us to prioritise species based on an understanding of their relative degree of genetic isolation; those without close extant relatives should be given greater emphasis than those that have closely related and widespread congeners.

Royal Botanic Gardens, Kew

Hawaiian Silversword/*Argyroxiphium kauense*.

Work being done at the Royal Botanic Gardens, Kew, on the phylogeny of a number of puzzling and often rare, plants using molecular techniques, is defining further the genetic uniqueness of these species, genera and families. The close relationship of the Californian tarweeds and the Hawaiian silverswords makes the silverswords less important than *Lactoris fernandeziana* (from the Juan Fernandez Island off the South American coast), which has no close relationship to any other group of extant angiosperms (its closest relative appears to be *Aristolochia*, but this is nonetheless a very distant relative). Knowing the degree of relatedness provides the framework to make rational choices about efforts to conserve species, both *ex situ* and *in situ*.

(*Source*: M. Chase, *Royal Botanic Gardens, Kew*.)

breeding systems, identifies plant fragments and artefacts and studies plant structures such as those of tree-roots and of tropical woods including those of fuel-wood value. Much of the work is of applied significance and has resulted, among others, in two major series of reference publications – *The Anatomy of Dicotyledons* and *The Anatomy of Monocotyledons*. The plant groups currently studied by the Cytogenetics Section include Solanaceae and Gramineae and some genera of petaloid monocotyledons. The Molecular Systematics Section, set up in 1990, has focused on enlarging the international DNA/RNA sequence database and using sequencing to provide evidence for the relationships of groups of plants including endangered species (see **Box 4.5**).

Kew's knowledge is shared by means of its many publications (at least 3,000 pages of high-quality scientific text each year) and through its educational programmes and international conferences. Its expertise and high curatorial standards are augmented by the numerous research visitors (from over 40 countries representing some 10,000 man-days per year) who work at Kew. Kew scientists publish the results of their research in the scientific journal *Kew Bulletin* as well as in many other journals. Kew produces, both in printed and CD-ROM form, the *Index Kewensis* (which, for more than a century, has listed all new names of seed plants), the *Kew Record of Taxonomic Literature*, *Curtis's Botanical Magazine* and major reference works such as *Vascular Plant Families and Genera* and *Authors of Plant Names*. Scientists at RBG Kew are directly responsible for the production of the *Flora of Tropical East Africa*, *Flora Zambeziaca*, *Flore des Mascareignes* and the *Flora of Ceylon* among others. They also contribute to over 45 definitive Floras including *Flora Malesiana*, *Flora Neotropica* and *Ascomycetes of Great Britain and Ireland*. An international initiative to produce a new world Flora with computerised data is currently being promoted by Kew (see Chapter Three, Section 3.3.3). Several of Kew's major projects are long-term (e.g. monographs, reference volumes and databases), and provide a degree of perspective and flexibility which facilitates exploratory thought and innovation in research and planning. RBG Kew builds on its traditional strengths while developing new skills at the cutting edge of science.

4.4 Royal Botanic Garden, Edinburgh

The Royal Botanic Garden Edinburgh (RBGE) is a key member of an international network of establishments within which research on systematic botany is carried out. Originally established in 1670 as a 'physic' garden for growing and studying medicinal plants, it has developed to become a distinctive focus for world-wide botanical research, education, conservation and amenity purposes, with individual strengths and skills in particular specialisms. With a staff of about 200, it now comprises four gardens throughout Scotland: Inverleith in Edinburgh, Younger in Argyll, Logan in Wigtownshire, and Dawyck in Peebleshire. The locations and collections of the RBGE are described in Chapter Six.

The Garden's Acquisitions Policy requires that the Living Collections are augmented by expeditions, and by collaboration, cooperation and seed exchange with other institutions in areas which match RBGE's research, education and conservation interests, and of which climatic and habitat conditions make their plants suited to cultivation in one or other of the gardens. The differences in soils, climates and other conditions between the four RBGE sites provide a greater diversity of habitats than that available to any other UK botanic garden. On completion of research projects, de-accessioning of plant material takes place; surplus material is then distributed for research and conservation work elsewhere.

The Royal Botanic Garden Edinburgh's mission is *"to explore and to explain the plant kingdom – past, present and future – and its importance to humanity. Its prime task is to pursue whole plant science, notably through systematic research on the diversity and relationships of plants, and their significance to the environment"*. The Royal Botanic Garden Edinburgh contributes, nationally and internationally, to activities, studies and policy development in the fields of biodiversity and conservation. RBGE collections and staff have particular strengths in Asian floras, and the cryptogams, for which they are able to play a role in the documentation and study of this neglected aspect of UK biodiversity. The specialists in the systematics, evolution and distribution of

Propagation at Edinburgh involves the careful nurturing of seeds and cuttings of a vast range of plants, many of which have never been in cultivation before.

flowering plants currently bring their skills to bear on a number of projects, using rapid surveys or detailed analysis as appropriate and contributing to major works such as the Floras of Bhutan, Arabia, and China. Contributions to handbooks and concise Floras in Asia and in Central and South America reflect the importance of RBGE scientists in international programmes for the study and management of plant resources. RBGE, as a founding member of the Edinburgh Centre for Tropical Forests (ECTF), provides expertise in plant taxonomy and environmental surveys to overseas projects for wealth creation through the sensitive management of sustainable forest resources (see **Box 4.11**). Conservation issues, identified by the Conservation Advisory Committee, underpin much of the Garden's scientific work, and some of its specific projects relate directly to environmental and species conservation and habitat monitoring, through the provision of survey reports on contract to agencies and decision makers.

RBGE's research is pursued both independently and in collaboration with other institutions. Major overseas biodiversity research programmes include: 'Biodiversity of the Cerrados' in Brazil; Flora of Arabia; Flora of Bhutan; Flora of China; Malaysian ectomycorrhizal studies with the Forest Research Institute of Malaysia, African mycorrhizal studies in Zambia and Cameroon; the International Conifer Conservation Programme; Fern Spore Bank Project; and Darwin Initiative projects in Mauritius and China.

Overseas contracts undertaken at RBGE include work in: Brazil (Biodiversity of the Cerrados Project, funded by ODA); Belize (funded by Oxford Forestry Institute/ODA); Malaysia (funded by ODA); China (funded by the Darwin Initiative); Molecular screening of rhododendrons (as part of the EC Molecular Genetics Screening Tools Project, funded by the EU); Indonesia (ECTF Indonesian Forest Management Project, funded by ODA).

Taxonomic research produces classifications, establishing differences, similarities and relationships for the world's plants. Taxonomic information forms the fundamental backbone of plant names and descriptions derived from known and identified specimens (the biological standards), which establish consistency and accuracy in other fundamental and applied plant sciences, horticulture, plant breeding, biotechnology, agriculture, forestry, medicine and other fields. RBGE has expertise in many areas of the world including: Arabia, southeast Asia, Belize, Brazil, Chile, China, the Himalaya (especially Bhutan), the Mediterranean (particularly Turkey, Portugal and Spain), Mexico, and the UK.

Families and plant groups currently researched at RBGE include: Berberidaceae, Coniferae, Cruciferae, Ericaceae, Iridaceae, Leguminosae, Rosaceae, Umbelliferae, Zingiberaceae, Bryophyta, Bacillarophyceae (diatoms), fungi and lichenised fungi, and Pteridophyta. The Institute also has expertise in palaeobotany, taxonomic computing and herbarium management.

Conservation collections used in research and for *in situ* and *ex situ* conservation of rare and endangered species are maintained and developed at RBGE. Full documentation and data recording makes efficient use of information technology to disseminate horticultural information through publication and electronic information exchange. Links with other institutions in other countries broaden and extend the representation of the world's plants available for study at RBGE.

4.5 Office of Science and Technology

The Office of Science and Technology was created in May 1992. A year later the first White Paper on science for 20 years was published, entitled *Realising our Potential: A Strategy for Science, Engineering and Technology* (Cm 2250, May 1993). The White Paper called for an enhanced contribution by science and engineering to wealth creation and improving the quality of life, whilst maintaining high quality strategic research. The Office of Science and Technology led to a refocusing and a restructuring of the UK Research Councils and allocated to them the £1.3 billion science budget for basic and strategic research. Of the seven research councils (See **Box 4.6**), two have particular relevance to biodiversity research, the Natural Environment Research Council (NERC) and the Biotechnology and Biological Sciences Research Council (BBSRC).

Box 4.6 UK Research Councils

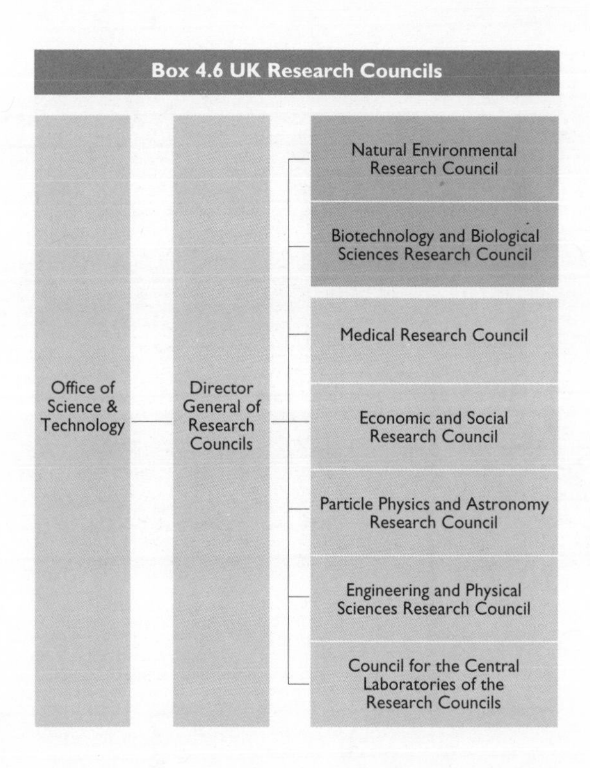

Box 4.7 NERC research centres, units and services

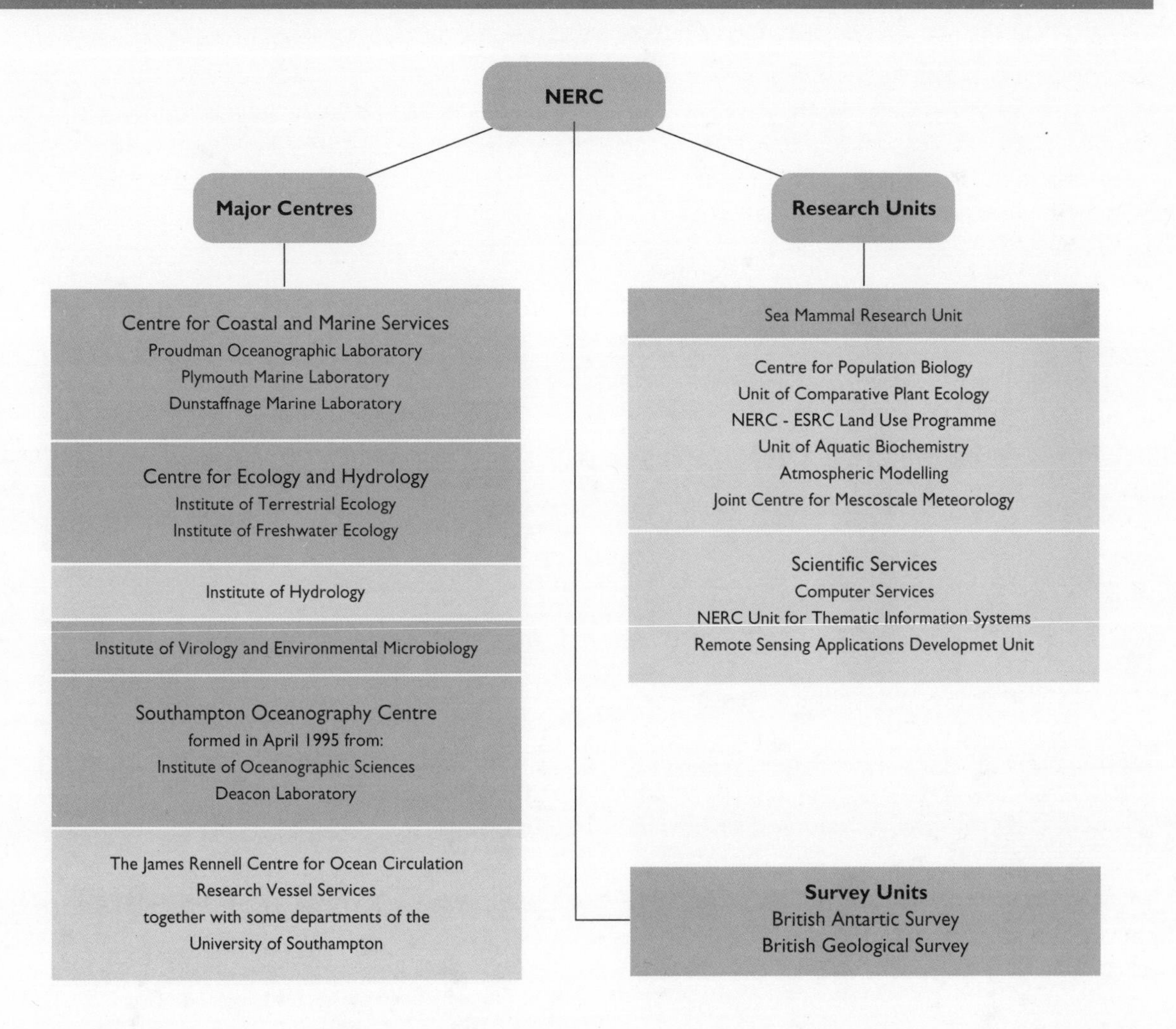

4.5.1 Natural Environment Research Council (NERC)

NERC is the UK's lead body for research, survey, long-term monitoring and related postgraduate training across the full breadth of environmental sciences.

NERC invests public money, 'The Science Budget', and seeks commissions from government departments and agencies, industry and international organisations to support environmental research in universities and its own Centres and Surveys (see **Box 4.7**).

NERC has identified six major environmental issues of concern for the UK on which it intends to focus its activities:

- management of land, water, and the coastal zone, the identification and exploitation of land-, freshwater-, and marine-based resources and their sustainability;
- understanding and prediction of biodiversity;
- waste management, bioremediation, and land restoration;
- pollution of air, land, sea and freshwater in relation to environmental and human health;
- environmental risks and hazards, including release of genetically modified organisms and improved prediction of extreme events; and
- global change, including prediction on a range of time and space scales.

NERC funding for research into the understanding and protection of biodiversity is to focus on: taxonomy, including evolution of biodiversity and phylogeny; measurement of spatial and temporal distribution of biota; improved understanding of the processes affecting distribution and abundance of natural populations; research to measure change in the UK and world biodiversity, understand the causes (both natural and anthropogenic) and effects of change on long and short timescales and underpin remedial measures; studies of the processes determining change and the consequences of variation in biodiversity and simulation and predictive modelling. The themes above, many of which have relevance to biodiversity, will be taken up by the NERC Core Programmes and new Thematic Programmes.

NERC funds best science at universities and in its own research establishments. About half of the Science Budget is invested in high quality

Box 4.8 NERC funding

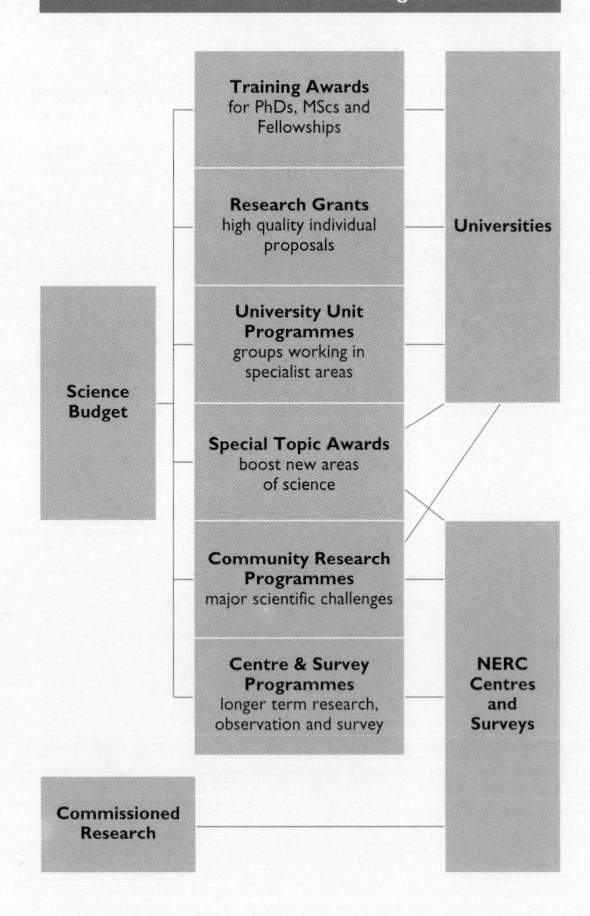

environmental science in UK universities. This is mainly in the form of research grants, fellowships and studentships to individuals, or contracts with NERC-funded specialist groups such as the Centre for Population Biology at Imperial College, London. (See **Box 4.8**).

NERC also has its own centres of excellence for studies of earth, water, land and air. Three new centres have recently been created. The Centre for Ecology and Hydrology, the Centre for Coastal and Marine Sciences, and the Southampton Oceanographic Centre, a joint facility between NERC and Southampton University. Their work is impartial and is frequently long-term to study natural cycles and slow, cumulative processes.

4.5.1.1 Centre for Ecology and Hydrology

The Centre for Ecology and Hydrology draws on the expertise of four NERC institutes, with a combined staff of some 700 researchers, to understand and manage terrestrial and freshwater environments within the UK and overseas in a better way. The component institutes are:

Institute of Freshwater Ecology

Seine netting for char, during studies on freshwater fish.

Institute of Freshwater Ecology (IFE)
IFE draws on 60 years experience into all aspects of freshwater research based on sound scientific knowledge of aquatic ecology, chemistry and the physics of lakes and rivers. Areas of expertise include:

- specialist research in aquatic ecology and chemistry;
- modelling behaviour of pollutants and their impacts on water bodies;
- control of algal blooms, water weeds and insects;
- remote sensing for survey and detection of change in water bodies;
- microorganisms in aquatic systems; and
- biology and management of freshwater fish.

IFE developed the RIVPACS system (River Invertebrates Prediction and Classification scheme) a key method for assessing water quality of a river from the micro-invertebrates present. IFE also is developing an Automated Network of Water Quality Monitoring Stations across European freshwater lakes, togther with a range of low maintenance packages for monitoring water quality and meteorological parameters. The data will be transmitted automatically by the international telephone network from the field sites to a control centre.

Institute of Hydrology
The Institute of Hydrology was established in the 1960s specifically to determine the effects of land-use changes on surface water resources. The Institute's 150 scientists worldwide work on:

- catchment hydrology related to land-use or climatic change including both water flow and quality;
- research in extreme hydrological events leading through to engineering applications;
- use of advanced instrumentation and computing in the collection, storage, analysis and dissemination of hydrological information; and
- process studies from local to global scales with emphasis on water and energy fluxes at the land/atmosphere interface.

Institute of Terrestrial Ecology (ITE)
ITE, established in 1973, aims to develop long-term, multi-disciplinary research and exploit new technologies to understand terrestrial ecosystems. ITE maintains six research stations within the UK. Part of its remit is to disseminate this research to decision-makers, particularly those responsible for environmental protection, conservation and the sustainable use of natural resources at national, regional, and global levels. Research is directed towards:

- composition, structure and processes of terrestrial ecosystems at an individual and species level;.
- interactions between atmospheric processes, terrestrial ecosystems, soil and water;
- modelling and predicting environmental trends due to natural and/or man-made changes;
- forest science; and
- terrestrial environmental microbiology.

The Institute has a range of facilities to support its research including remote sensing, aerial photo-interpretation and geographical information systems. The **Environmental Information Centre** (EIC) set up in 1989 and based at Monks Wood acts as a focus for data relating to terestrial ecology and the rural environment. The Centre is also custodian of a number of key national and European data sets, which have been developed to underpin research on species and ecosystems' responses to environmental change. EIC comprises three component units.

The Biological Databases Unit compiles and manages databases of spatial and temporal references on the biodiversity of the British Isles and databases of sites of importance for wildlife conservation in Europe.

The Remote Sensing Unit uses remote sensing for applications in ecology and land use, in particular natural resource management and land use planning.

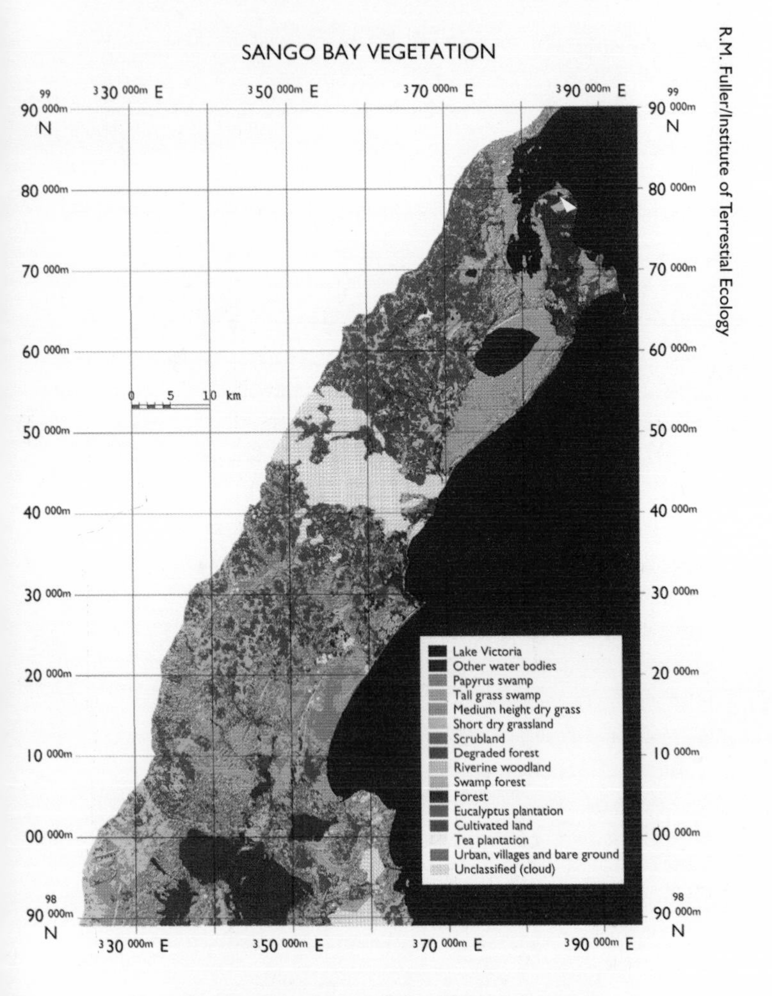

Land cover map of the Nabugabo area, Sango Bay, Uganda, produced by semi-automated computer-classification of Landsat satellite imagery.

The Geographical Information Systems Unit is developing and applying methods of spatial data analysis and modelling. The Unit also undertakes technical developments in the use of GIS in ecological research and the provision of technical support.

Outside the EIC there are other research programmes developing large spatial and/or temporal databases. The largest of these programes is conducted by the **Land Use Section** at Merlewood. This Section undertakes research which explains the relationship between the distribution of plants and animals and human activity in the UK. In addition to research on the ecology of land use, the **UK Environmental Change Network** (ECN) long-term monitoring programme is managed at Merlewood.

ITE is a partner in the Edinburgh Centre for Tropical Forests (see **Box 4.9**) where researchers are developing ways to regenerate forests for sustainable economic development and improve the quality and growth of timber, fruit and medicinal trees in tropical regions. Genetic improvement of indigenous trees, to maximise productivity and add value, involves their collection, selection and testing, propagation (with particular attention to low-cost methods), and planting to reduce soil degradation and minimise insect pest attack. Other work involves root symbiotic associations (mycorrhizas and rhizobin; see Volume Two, Chapter Three) for successful tree establishment and growth and includes examination of nitrogen-fixing trees on soil improvement in fallows of arid zone Africa. Molecular methods are now being used to target genetically-important tree populations for conservation and breeding.

***Institute of Virology and Environmental Microbiology* (IVEM)**

IVEM has its origins in the Pathology Unit of Oxford University formed in 1963. It has expanded and is now housed in a custom built facility in the heart of the University's science area. The Mission of the Institute is the study of viruses and other microbes in the natural environment, primarily in the following subjects:

- microbial relationships with vertebrates, invertebrates and plants;
- genetic microbes in the natural environment, including their replication, transmission vectors, host and ecology;
- behaviour of genetically modified organisms in the environment;
- risks and benefits; and
- identification, prevention and spread of diseases in species other than man.

IVEM, in conjunction with the Universidad Nacional Autonoma de Nicaragua and the European Union are developing novel biological control systems, targetting the insect pests of cotton. The development of Baculovirus 'pesticides' through advanced genetic engineering should greatly reduce the need for expensive and environmentally damaging chemical pesticides at local level.

4.5.1.2 Centre for Coastal and Marine Sciences (CCMS)

The CCMS with a combined staff of 230 researchers, brings together the environmental expertise and facilities of the Plymouth Marine

Box 4.9 Biodiversity and forest regeneration studies in West Africa by the Edinburgh Centre for Tropical Forests

The Edinburgh Centre for Tropical Forests is an association between the Schools of Forestry and Ecological Science (University of Edinburgh), the Institute of Terrestrial Ecology, LTS International, the Royal Botanic Garden, Edinburgh and the Forestry Commission. Since its foundation in 1990 ECTF has established itself as a leading research, training and consultancy organisation with an international repution in the sustainable management of tropical forests.

ECTF has been working on a number of projects concerned with reforestation in tropical countries. Sadly, the area reforested often tends to be small in relation to the area deforested. In West Africa, for example, the rate of deforestation during the 1980s was estimated to have been about 12,000 km^2 or 2%, annually. Against this, only 360 km^2 of plantation forests were established annually during the mid-1980s. These plantations, moreover, have had a poor record of maintenance and survival, and were usually established in areas of existing forest. A further problem is that reforestation may not lead to a full restoration of biodiversity. Plantations of exotic species, for example, are likely to contain less biodiversity than plantations of native West African species such as *Terminalia ivorensis*, *Terminalia superba* and *Triplochiton scleroxylon*.

Reforestation has many benefits, the most direct being the re-establishment of a crop of timber-yielding trees. Indeed underplanting in a thinned canopy provides an opportunity to increase the density of valuable timber tree species. In West Africa, for example, only 1-8 trees per hectare are currently extracted for timber from natural forest. Forest regeneration, however, can be done in contrasting ways: by the establishment of plantations after complete clearance of remaining forest cover, or by under-planting valuable tree species in partially cleared forest.

Different methods of rapid forest regeneration are likely to vary both in terms of their economic success and the levels of biodiversity they support. In order to quantify

Arthropod composition in the Mbalmayo Forest Reserve

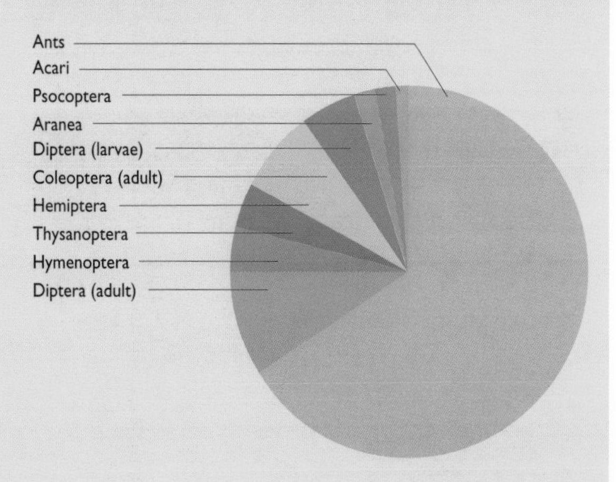

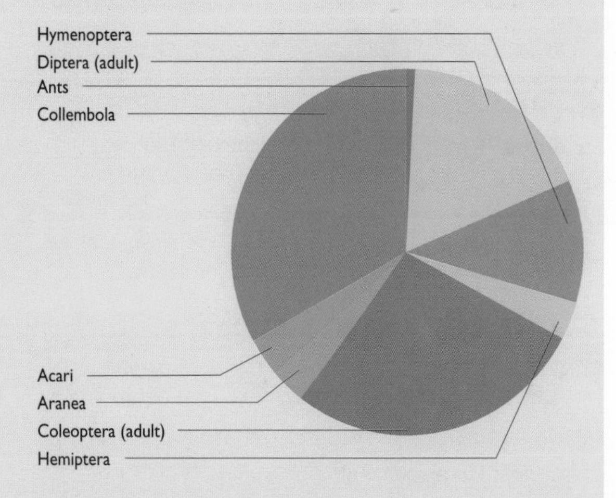

the economic and ecological consequences of different silvicultural methods for forest regeneration, the Forest Management and Regeneration Project (FMRP) was started in Southern Cameroon in 1991. This bilateral Cameroon (Office National de Développement des Forêts (ONADEF)) / UK (Overseas Development Administration) project also

Box 4.9 *Continued*

aims to develop a management plan for the Mbalmayo Forest Reserve, and to provide genetically improved planting stock of selected indigenous forest species. This project followed previous work by ONADEF and the Institute of Terrestrial Ecology (Edinburgh Research Station) in the Mbalmayo Forest Reserve in the late 1980s. They studied the vegetative propagation of indigenous hardwood trees, and the effects of different site preparation techniques on the physical and chemical properties of the soil, endomycorrhizal fungi, and tree physiology.

As part of the FMRP, ITE and the Natural History Museum set up a monitoring programme to quantify the impact of different methods of forest regeneration on arthropod abundance and diversity. This was done partly to investigate the impact of different approaches to regenerating logged forest on arthropod abundance and diversity *per se*, and partly to investigate the hypothesis that the risk of insect pest attack is minimised by adopting silvicultural methods which maximise biodiversity.

Malaise trapping, flight intercept trapping, and insecticide knockdown fogging were used to measure insect abundance and diversity in different forest types. Sampling was concentrated in regeneration treatments with contrasting amounts and diversity of vegetation: complete clearance plantation plots, where all trees were cleared prior to planting *Terminalia ivorensis* saplings; partial clearance plots, where approximately half the trees were removed prior to planting; and uncleared forest plots. Flight intercept traps caught mostly Collembola, Coleoptera and Diptera, but ants dominated the canopy fogging samples. On average, 63% of the arthropod individuals sampled were ants. A total of 96 ant species were recorded from the forest canopy in the Mbalmayo Forest Reserve. The dominant species included some common West African species such as *Crematogaster striatula* and *Oecophylla longinoda*, but the most abundant ant species recorded was a hitherto unknown species of *Technomyrmex*.

Canopy fogging showed that the abundance of several insect groups, such as bees and wasps, tended to be greater in partial clearance than in the complete clearance plantation plots. Moreover, preliminary work has shown that insect diversity also tends to be greatest in partial clearance plantation. To what extent this results in a reduction in insect pest attack cannot be determined from a short-term study. However, this study has already shown that the methods used in forest regeneration programmes can have a profound influence on biodiversity.

(*Source:* A.D. *Watt*, ITE *Edinburgh Research Station* & N.E. *Stork*, NHM, *London*.)

Laboratory, the Oceanographic Laboratory at Bidston, Merseyside and the Dunstaffnage Marine Laboratory at Oban.

The Plymouth Marine Laboratory studies the physics, chemistry and biology of marine and estuarine ecosystems, and the impact of human activities on systems both nationally and globally. PML works closely with its Scottish counterparts at Dunstaffnage Marine Laboratory, maintaining an active programme studying the diversity and ecology of marine invertebrates and algae, and and representing UK marine biodiversity research on the international front. **Box 4.10** refers to one international programme initiated by the PML.

4.5.1.3 Southampton Oceanography Centre

The Southampton Oceanography Centre is the national focus for all aspects of research, training, undergraduate and postgraduate teaching, technology and support services in marine sciences. Established on a purpose-built site, the Centre was formed from the combined expertise of NERC's Institute of Oceanographic Sciences Deacon Laboratory, James Rennell Centre for Ocean Circulation and Research Vessel Services, together with the University of Southampton's Departments of Geology and Oceanography and its underwater acoustics group.

Specialist facilities include a research aquarium, temperature controlled areas, pressure

Box 4.10 The biodiversity of coastal marine organisms: Research by the Plymouth Marine Laboratory

In a programme spanning 1994 – 1998, the PML is describing patterns of biodiversity of coastal marine organisms on both regional and global scales. It is acting as the co-ordinator of an international programme of research by members of the International Association of Meiobenthologists (IAM) who are developing and retrieving artificial substrate units (ASU) at the locations shown on the map below.

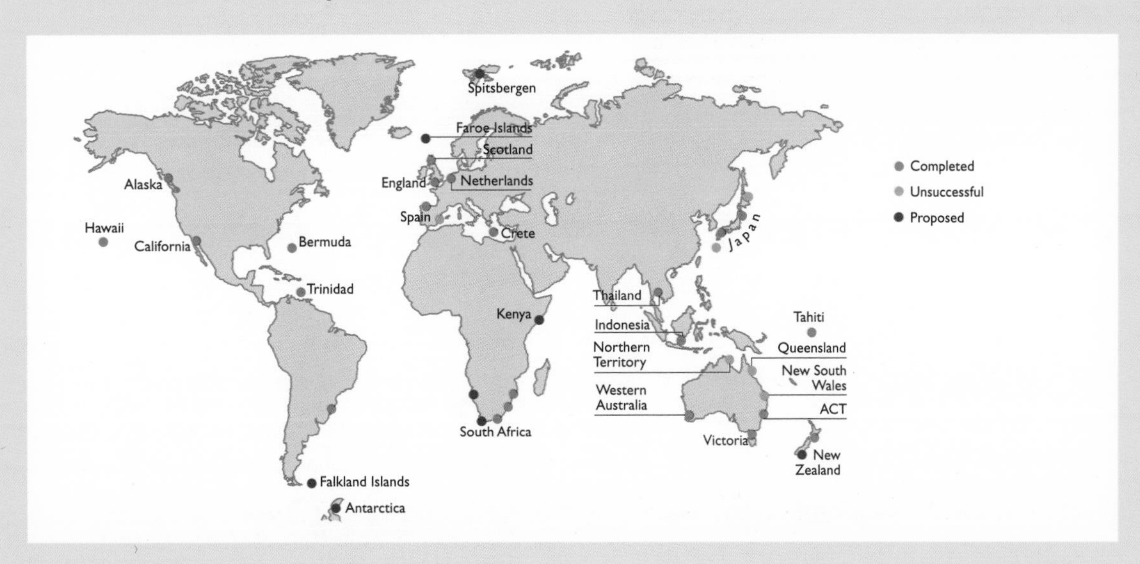

Fig. 1 Locations of sites for artificial substrate units

ASU's are provided as a pack which come with full instructions regarding deployment, retrieval and initial sample processing.

The ASU (see photograph) consists of 4 plastic mesh pads attached to a steel piton which is hammered firmly into rock substrate at a standard water depth (12-15m), with five replicate units at a number of different sites at each location. These should be left in place for five months in temperate and tropical waters and up to one year in polar regions where the rate of colonisation is much slower.

PML aims to identify much of the macro-fauna (amphipods and polychaetes) and meio-fauna (nematodes and copepods) whilst other taxa will be distributed to specialist taxonomists elsewhere in the UK and abroad.

An artificial substrate unit (ASU) ready for staking in place.

(*Source*: R.M. *Warwick* & M. *Gee*, *Plymouth Marine Laboratory*.)

calibration, test tanks and a wide range of different laboratory areas. There are 200 metres of quay for the sole use of the Centre. The National Oceanographic library is being housed on the site.

Institute of Oceanographic Sciences Deacon Laboratory is the UK leader in deep ocean research. It is the point of contact for advice based on research in the fields of physical and chemical oceanography, marine geology and geophysics, and deep ocean biology.

A relatively small part of the resources of this institute is used on marine biodiversity studies but biologists there play a significant, and international, role in research into the benthic

P. Bucktrout/British Antarctic Survey

Signy Research Station with R.R.S. Bransfield anchored behind.

flora and fauna. See Chapter Six for mention of deep sea biological collections.

The Research Vessels Services operates the only UK fleet of vessels equipped specifically for deep sea oceanograpic study and research. These include the RS Charles Darwin, RS Discovery and RRS Challenger.

4.5.1.4 British Antarctic Survey (BAS)

NERC's polar research activities cover both the Arctic and the Antarctic. However, the majority of NERC's polar science research is at present carried out in the Antarctic by the British Antarctic Survey. Research is carried out throughout the year from five permanently staffed research stations, including the field station, Bird Island, in the British Antarctic Territory, South Georgia and South Sandwich Islands. (See **Box 4.11**).

The Antarctic continent and its surrounding islands comprise a remarkable and unique range of terrestrial and aquatic ecosystems. The limited biodiversity ensures that community structure is unusually simple whilst the hierarchy of interacting stresses – low temperatures, dehydration, osmotic stress, increasing ultraviolet radiation, limited growing period – makes colonisation, establishment and survival difficult for organisms.

Two divisions of BAS are concerned with biodiversity studies. The Terrestrial and Freshwater Life Sciences Division conducts research over 20 degrees of latitude – from South Georgia to the southern point of the Antarctic Peninsula. This constitutes a gradient of increasing climatic severity and decreasing biodiversity, ideal in many ways for studying key processes which determine the survival of populations and the establishment, development and organisation of simple natural communities.

The Marine Life Sciences division undertakes primary research directed at improving and understanding the structure and dynamics of the Southern Ocean ecosystem. The Southern Ocean has a long history of uncontrolled exploitation, first of fur seals and then whales, and several nations are now exploiting intermediate levels of the food web (krill, squid and fish). Until there is a quantitative understanding of the major energy pathways and the principal interactions between key components of the system, it will be impossible to use the living resources wisely and conserve the ecosystem. The overall research is divided into three unequally sized

Box 4.11 British Antarctic Survey research stations

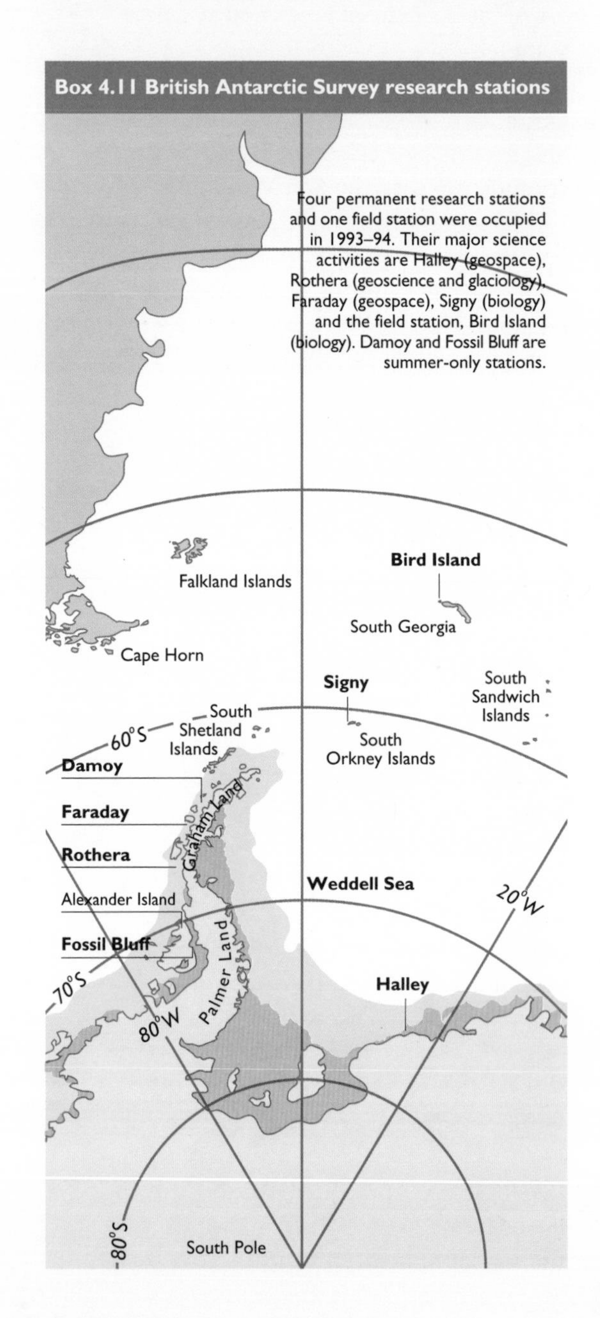

programmes, Pelagic Ecosystem Studies, Higher Predator Studies and Nearshore Marine Biology. BAS makes a significant contribution to the SCAR (Scientific Committee on Antarctic Research) programme on Coastal and Shelf Ecology of the Antarctic Sea-ice Zone (CSEASIZ), and to the work of the Commission for the Conservation of Antarctic Living Resources (CCAMLR). The latter body is responsible for the management of all living resources within the Southern Ocean with the exception of seals and whales which are covered by separate conventions.

A number of taxonomic studies are underway concentrating on the lichens and mosses, and BAS is working with the Environmental Research Centre at University College London on the production of definitive diatom guides for Antarctic freshwater systems. This will complement previously published work on the marine diatom flora of the Southern Ocean. Other biodiversity studies include breeding and ecological work on birds and work, in conjunction with the Sea Mammal Research Unit (see below) on seals. Taxonomic research has also provided the foundation for several BAS monitoring projects, some of which have been running for over 25 years.

4.5.1.5 Sea Mammal Research Unit

The Sea Mammal Research Unit, based at St Andrew's University undertakes strategic research on sea mammals, with special focus on their physiology, migration and behaviour, and on the implication of these for population processes. It is developing novel methods of tracking and monitoring animals at sea to analyse and interpret this information. Open ocean studies of marine mammals include those on elephant seals in South Georgia, white whales in the Canadian High Arctic, and the development of a satellite telemetry and visualisation system. Population biology projects include those on pilot whales, harbour porpoises in the N.E. Atlantic and dolphins and manatees in the Amazon.

4.5.2 Biotechnology and Biological Sciences Research Council (BBSRC)

BBSRC is one of the six UK research councils. It was formed in 1994 by the incorporation of the former Agricultural and Food Research Council with biotechnology and biological sciences programmes of the former Science and Engineering Research Council.

Box 4.12 Research Institutes and Interdisciplinary Research Centres sponsored by BBSRC

- Advanced Centre for Biochemical Engineering
- Babraham Institute
- Centre for Genome Research
- Institute for Animal Health
- Institute of Arable Crops Research
- Institute of Food Research
- Institute of Grassland and Environmental Research
- John Innes Centre
- Oxford Centre for Molecular Science
- Roslin Institute
- Silsoe Research Institute
- Sussex Centre for Neurocience

BBSRC's core scientific programmes in biology and related areas in chemistry, physics and engineering, comprise both basic and strategic research. More applied research is undertaken with, and full funding by, industrial and Government partners.

BBSRC supports research in eight research institutes and in four interdisciplinary research centres, as well as groups and units in universities (see **Box 4.12**).

The International Agricultural Development Unit, based in Harpenden, fosters links between our research community and agencies involved in agricultural development.

4.5.2.1 Babraham Institute

The Babraham Institute, in Cambridge, carries out a programme of research to advance understanding of the molecules, genes, cells and systemic processes involved in growth, nutrition, reproduction, lactation, health and welfare in the whole animal.

Its mission includes undertaking fundamental and strategic research which will provide the underpinning science for biologically-based industries (biotechnological, biomedical, agricultural, food, health-care and pharmaceutical), governments and agencies, and provide the training and basic knowledge necessary for advancement in relevant areas of biology.

4.5.2.2 Institute for Animal Health

The Institute for Animal Health has laboratories at Compton, Pirbright and Edinburgh.

Its scientific aims include the development of disease control measures which will increase the efficiency of agriculture and which will protect and enhance the environment, and maintain the integrity of the food chain. It also

Box 4.13 Some recent projects related to biodiversity studies and clients of IACR outside Europe

Tropical weeds

- resistance of cereals and cowpea to the parasitic weed, *Striga*, in sub-Saharan Africa (ODA/EU);
- combating propanil resistance in *Echinichloa colona* in Central America (ODA).

Plant parasitic nematodes

- microbial control of root-knot nematodes on vegetable crops in subsistence agriculture (EU/ODA);
- diagnosis of potato cyst nematodes in South America (EU, The British Council);
- burrowing nematode, *Radopholus similis* on tea, coffee, banana in various countries (ODA).

Biometrics

- experimental design and data handling for studies on perennial and annual crops, agroforestry and aquaculture in Cameroon, Colombia, Zanzibar, and Nigeria (ODA);
- assessment of research quality and data handling needs (ODA).

Tropical plant diseases

- lethal yellowling diseases of coconuts in Africa and the Caribbean (EU);
- *Pseudomonas solanacearum* (Bacterial wilt) diagnosis and control in Cuba and Brazil (ODA);
- viruses of black pepper in Sri Lanka (Sri Lankan Government);

Institute of Arable Crop Research

Black pepper cultivation is important in Sri Lanka where virus diseases can cause up to 80% yield loss.

- fungally-transmitted viruses of small grain cereals in China (EU).

Insect pests

- diagnostic kits for identification of insect pathogenic fungi (ODA);
- characterisation of insecticide resistance in whitefly (*Bemisia tabaci*) and its field management in Pakistan (ODA) and Israel (agrochemical industry).

Plant biotechnology

- improvement of cassava tuber quality for Brazil (ODA);
- improvement of durum wheat for Ethiopia (EU).

Contact: *Stephen James*, IACR, *Rothamsted*, *Harpenden*, *Herts* AL5 2JQ.
Tel: +44 (0)1582 763133
Fax: +44 (0)1582 469688

carries out research on new diseases and in this context relates, in a minor way, to biodiversity research.

4.5.2.3 Institute of Arable Crops Research (IACR)

IACR has laboratories at Harpenden, Bristol and Bury St Edmunds. Its mission is to undertake high quality, basic, strategic and applied research, with relevant postgraduate training, in biological and related sciences, integrating these to optimise existing and novel crop production systems and their interactions with the environment.

Its skills range from traditional strengths in statistics, agronomy, soil science, plant physiology and biochemistry, nematology, entomology and all aspects of crop protection, through to newer areas such as molecular biology and genetics (see **Box 4.13**).

Rothamsted International was set up in 1993 to raise funds and award Research Fellowships to scientists from overseas wishing to develop their research at IACR-Rothamsted in an innovative multidisciplinary environment.

4.5.2.4 Institute of Grassland and Environmental Research (IGER)

IGER has its headquaters in Aberystwyth and outstations at Trawsgoed, Dyfed, North Wyke, Devon, and Trecastle, Powys. It carries out research to improve the efficiency, potential and sustainability of grassland-related agriculture. Its programmes embrace plant genetics, forage and cereal breeding, cellular and molecular biology, environmental plant science and ecology (including thermal and drought tolerance, animal research and land use. Its experienced professional staff and extensive experimental facilities make IGER the major centre for independent research into grassland and the environment in the UK.

It maintains gene banks of oats, grasses and legumes and a culture collection of the nodule bacterium, *Rhizobium*. (See Chapter Six, Box 6.9).

4.5.2.5 John Innes Centre (JIC)

The JIC is in Norwich. Its mission is to contribute to enhancing knowledge, the quality of life and the economic well-being of societies worldwide by:

- conducting high quality basic and strategic research and postgraduate training relating to the understanding and exploitation of selected plants and microorganisms;
- providing knowledge, technology and trained scientists to meet the needs of users and beneficiaries that include the worldwide scientific research community, the agriculture, bioprocessing, food, healthcare, pharmaceutical, chemical and other biotechnological-related industries, governments, the European Union and the general public; and
- providing advice, disseminating knowledge and promoting public understanding in important areas of biological sciences.

Its research in plant breeding (see **Box 4.14** for example) is of international reknown and it maintains (with part support by MAFF) gene banks in cereals, peas and field beans (see Chapter Six, Box 6.9). Its research programme also includes plant propagation by tissue culture, biological nitrogen fixation, herbicide resistance, pathogen variation and host resistance, photosynthesis research, plant environmental stress (including drought tolerance/resistance), seed development and plant virus research.

4.5.2.6 Roslin Institute

The Roslin Institute in Edinburgh, carries out basic and strategic research relevant to farm animal production. It is the major UK centre for the mapping, analysis, and use of the genomes of livestock, with excellent laboratory and farm facilities for research on all livestock species. The Roslin Insitute currently has European-centred genome mapping and biodiversity research programmes on pigs, cattle and poultry.

4.6 Ministry of Agriculture, Fisheries and Food (MAFF)

The Ministry of Agriculture, Fisheries and Food is responsible for Government policies relating to the safety and quality of food in the UK; for the protection and enhancement of the countryside and marine environment,

Box 4.14 Plant breeding research at the John Innes Centre

M. Ambrose/John Innes Centre

Fig. 1 *Brassica creta* growing on the island of Limnos, Greece.

M. Ambrose/John Innes Centre

Fig. 2 Chromosome painting using fluorochrome labelled genomic probes on an octaploid oat (carrying ABCD genomes). AB and D genomes fluoresce yellow while the C genome regions fluoresce green.

The study of wild relatives of cultivated species features in a number of different John Innes programmes; an example is the work on wild *Brassica* (cabbages) species collected from the Mediterranean region (Fig. 1) and used in order to increase the genetic variation available to breeders in *Brassica napus*.

Chromosome studies

The introgression of characters from one species has occurred often in the wild and in the domestication of many crop plants. Important genes or traits found in wild relatives of wheat, oilseed rape and peas, particularly for quality characters, disease resistance and abiotic stress are being introduced to produce lines for UK and international breeders. Techniques such as 'chromosome painting' have been developed to locate and display the introduced regions. Through differential staining the translocated regions can be visualised (Fig. 2).

Comparative mapping

Biochemical and molecular markers are having a great impact on the methodologies of plant breeding. JIC laboratories were the first in Europe to develop these for mapping in plants and the first in the world for cereals. A major objective remains to produce comprehensive genetic maps of important cereals.

(*Source*: M. *Ambrose*, *John Innes Centre*)

for flood defence and other rural issues. MAFF is the licencing authority for veterinary medicines and the registration authority for pesticides. It commissions extensive research to assist in the formulation and assessment of policy and to underpin applied research and development work by industry. In the context of this Manual, much of MAFF's research and monitoring data is relevant to the protection and enhancement of biodiversity. Main areas of research include: pests and pesticides, wildlife conservation, farm woodlands, climate change, soil and land protection and set-aside.

The Ministry assists in funding for the conservation and utilisation of Plant Genetic Reources to ensure that these resources are available as basis for fundamental botanical research, and from which to develop new crop varieties to meet future changes in climate, and product requirements; thereby, enabling the UK to meet specific obligations relating to international gene banks. MAFF supports three major collections: the Vegetable Gene Bank at Horticulture Research International, the National Fruit Collection at Brogdale and the Pea Bank at the John Innes Institute in Norwich. (See

Chapter Six, Box 6.9 and discussion under Section 6.3.3.)

MAFF is the Ministry that sponsors the Royal Botanic Gardens, Kew (see Section 4.3: now a Non-Departmental Public Body), and maintains that institute's link with central government.

MAFF also has responsibility for farm animal genetic resources. It conducts a programme of research, acts as a co-ordinator for UK activities and represents UK interest in international fora.

4.6.1 The Directorate of Fisheries Research

The Directorate of Fisheries Research is a division of the Ministry of Agriculture, Fisheries and Food with main laboratories at Lowestoft and smaller laboratories at Burnham-on-Crouch, Conwy and Weymouth. Its main remit is fisheries management and aquatic environment protection. Advice is given to Ministers in relation to UK and EC fish conservation and management policies. Over 40 stocks of fish and shellfish are regularly monitored. These are mostly marine and include salmonid fish and cultivated species. Research covers all aspects of ecology, behaviour, physiology and resource management, including the development of theoretical models. A new Fish Diseases Laboratory has recently been opened at Weymouth.

4.7 Scottish Office

The Scottish Office channels its support for agricultural, biological, aquatic, environmental and engineering research principally through ten organisations known as the Scottish System. This includes the Royal Botanic Garden (discussed in 4.4, above) and six agricultural research institutes of which the Hannah, Moredun and Rowett Research Institutes are predominantly researching into animal physiology, biochemistry and pathology; they and the Scottish Agricultural Statistics Service are not discussed further here. The three institutes mentioned below contribute to UK biodiversity research to a greater or lesser extent.

4.7.1 Scottish Crop Research Institute (SCRI)

The Scottish Crop Research Institute is a major international centre for research on agricultural, horticultural and industrial crops, and on the underlying processes common to all plants. It aims to increase knowledge of the basic biological sciences (and so has a bearing on biodiversity); to improve crop quality and utilization by the application of conventional and molecular genetical techniques and novel agronomic practices; and to develop environmentally benign methods of protecting crops from depredations by pests, pathogens and weeds. A broad multidisciplinary approach to research is a special strength of the institute, and the range of skills available from fundamental studies on genetics and physiology, through agronomy and pathology to glasshouse and field trials is considered to be unique within the UK research service.

SCRI works closely with a number of international agricultural research centres which are members of CGIAR, the Consultative Group on International Agricultural Research (**Box 4.15**) set up to eradicate human hunger and poverty through research.

SCRI works closely with the Centro Internacional de la Papa (CIP), Peru, on potato research and both institutes house potato gene banks (see **Box 4.16**). In a study for CIP on drought tolerance in potatoes, considerable diversity in rooting types was discovered at SCRI and the relative importance of the balance between root and shoot characters for the water economy of the plant was highlighted. Further collaborative work has included research on resistance to potato leaf roll virus.

However SCRI's contribution to developing countries has not been limited to collaborative projects with the CGIAR units. Studies on viruses of tomato, cassava, cotton and okra have also been done in association with national agricultural research groups and the French aid organisation, ORSTOM. The sensitive 'ELISA' and PCR-based virus detection methods developed at SCRI were used in a screening system for cassava clones supplied by CIAT to IITA, via Scotland to prevent the inadvertent introduction of South American cassava viruses to Africa. In addition, molecular genetic studies involving several countries

Box 4.15 Members of the Consultative Group on International Agricultural Research (CGIAR)

Acronym	Name	Founded	Headquarters
CIAT	Centro International de Agricultura Tropical	1967	Colombia
CIFOR	Centre for International Forestry Research	1993	Indonesia
CIMMYT	Centro Internacional de Mejoramiento de Maiz y Trigo	1966	Mexico
CIP	Centro Internacional de la Papa	1970	Peru
ICARDA	International Centre for Agricultural Research in the Dry Areas	1975	Syria
ICLARM	International Centre for Living Aquatic Resources Management	1977	Philippines
ICRAF	International Centre for Research in Agroforestry	1977	Kenya
ICRISAT	International Crops Research Institute for the Semi-Arid Tropics	1972	India
IFPRI	International Food Policy Research Institute	1975	USA
IIMI	International Irrigation Management Institute	1984	Sri Lanka
IITA	International Institute of Tropical Agriculture	1967	Nigeria
ILCA	International Livestock Centre for Africa	1974	Ethiopia
ILRAD	International Laboratory for Research on Animal Diseases	1973	Kenya
INIBAP	International Network for the Improvement of Banana and Plantain	1984	France
IPGRI	International Plant Genetic Resource Institute	1974	Italy
IRRI	International Rice Research Institute	1960	Philippines
ISNAR	International Service for National Agricultural Research	1979	The Netherlands
WARDA	West Africa Rice Development Association	1970	Ivory Coast

supported by ODA and European Union funding, are currently in progress on cocoa, tea, coffee and selected tropical trees.

4.7.2 *Macaulay Land Use Research Institute (MLURI)*

The MLURI's remit is research in the context of rural land use and resource management, with the object of assessing the environmental, economic and social impacts of agriculture and related uses, and the changes resulting from policy and management, climate and pollution. Its work on the Scottish environment places the latter in a European context.

4.7.3 *Other agricultural institutes*

The Scottish Agricultural Colleges in Edinburgh, Perth and Ayr are basically teaching institutes with an element of research. Those in Edinburgh and Perth maintain living collections of the cereal barley, and of fibre flax and oil-bearing linseen. (See Chapter Six, Box 6.9).

The Scottish Agricultural Science Agency, basically a coordinating department for agricultural services in Scotland, maintains a field-pea gene bank at East Craigs.

Box 4.16 The utilization of potato gene resources in the UK

E.K. Balls/J.G. Hawkes

J.G. Hawkes and E.K. Balls travelled in 1939 to the Andes in search of *Solanum*. They collected many interesting varieties from local farmers, as here on the altiplano near Eucalyptus, Peru.

The conscious and unconscious selection of desirable forms has been a feature of crop improvement since the birth of agriculture. The removal of individuals not possessing desirable attributes means that the selection process also leads to the gradual loss of genetic diversity. Continual cycles of selection over many centuries has left most major food crops with a tiny fraction of the genetic variability found originally in their progenitor species. Whilst agronomically useful characteristics have been improved by this process, genes not subject to selection have diminished in number. Without doubt, resistance to pests and diseases comprise the most significant category of genes affected in this way. Their loss has rendered many crops extremely vulnerable to attack following exposure to new pathogens or pathotypes.

One of the best known and most tragic examples of the consequences of such genetic erosion were the famous Irish potato famines of 1845-7. Susceptibility of the potato to the late blight disease *Phytophthora infestans* resulted in complete failure of the crop in Ireland and many parts of mainland Europe following a widespread epidemic of the fungus. Screening of the tuber-bearing *Solanum* species from Mexico and Southern America were both unsuccessful. Genes from S. *edinense* which contain some resistance to infection were later introduced into breeding material and eventually released commercially in the form of the German cultivar 'Sandnudel' in 1934.

Attempts were made to broaden the range of variation available to the breeders and many expeditions took place. Probably the largest of these was the British Empire Potato collecting Expedition of 1939, sponsored by the Imperial Bureaux and led by E.K. Balls and J.G. Hawkes. In excess of 1400 accessions of primarily cultivated potatoes were collected. These, together with material donated from other sources, formed the basis of the Empire Potato Collection, later renamed the Commonwealth Potato Collection (CPC) now developed and maintained at the Scottish Plant Breeding Research Institute, Fife.

Scottish Crop Research Institute

Development of new potato varieties in SCRI, Fife; progeny test site of 20,000 clones.

This substantial gene bank was used to combat another major pest, the soil-borne Potato Cyst Nematode (PCN) which devastated the potato crop in the UK and mainland Europe in the early 1950s. Resistance to this pest was identified in five accessions in the bank, apparently conferred by a single major dominant gene (later designated the H gene) that is inherited tetrasomically. These properties allowed the resistance to be transferred rapidly into S. *tuberosum* and by the mid 1960s the first cultivars containing the gene became available on the UK national list.

An excellent review of the potato's history by Prof. J.G. Hawkes *The Potato: Evolution, Biodiversity and Genetic Resources* was published by the Smithsonian Institution Press. Washington, DC, USA, in 1990.

(*Source*: M. *Wilkinson, Scottish Crop Research Institute.*)

4.7.4 Fisheries Research Services

The Fisheries Research Services provides government with expert advice and technological information, much of it policy-oriented on

- Scotland's existing and potential fisheries for marine, migratory and freshwater species and their management and enhancement;
- existing and potential aquaculture;
- the living resources on which these fisheries and aquaculture are based, and their availability and quality, including aspects which may give rise to public health concerns;
- the capture and farming methods used by these industries; and
- the aquatic environment and wildlife and their protection from man's activities, including waste disposal.

FRS also performs certain statutory and regulatory duties, and represents Scottish, and where appropriate, UK interests at national and international meetings.

4.8 Forestry Commission Research Division

The Forestry Commission is a Government Department reporting to the Secretary of State for Scotland, the Minister of Agriculture, Fisheries and Food, and the Secretary of State for Wales. Although its main remit is to advise on and implement the UK Government forestry policy which encourages the preservation and enhancement of biodiversity in Britain's forests, its staff have also provided technical and scientific support in tropical forestry regions. One of its strengths, which it is willing to share, is in developing effective use of financial resources for the best implementation of the Forest Principles enshrined in the Rio Declarations.

The Forestry Commission established an Overseas Consultancy Service in 1994, through which expertise is available on a wide range of forestry-related matters including training, forestry policy and forest management. In addition, the two Forestry Research Stations at Alice Holt in Hampshire and at Roslin, near Edinburgh, are able to provide help through contract research advice and partnerships.

4.9 Foreign & Commonwealth Office Overseas Development Administration

The Overseas Development Administration (ODA) is the part of the Foreign and Commonwealth Office responsible for managing Britain's official aid programme. The overall purpose of the ODA is to promote economic and social development in other countries and the welfare of their people. Within this objective the ODA commissions and sponsors research on topics relevant to those geographical regions designated as the primary targets of the aid programme and of benefit to the poorest people in those countries.

In 1994/95 the cost of ODA's aid programme for developing countries was £2,072 million of which some £120 million was spent in support of research and development. The ODA's support for research is organised in five main programmes covering renewable natural resources, engineering and related sectors (water and santitation, energy efficiency and geoscience, urabanisation and transport), health and population, economic and social, and education. The Renewable Natural Resources Programme includes those on integrated pest management, forestry and agroforestry, plant sciences (including support for the Centre for Arid Zone Studies), fisheries (including fish genetics) and environment conservation and policy. In addition to these programmes, research is carried out as part of the UK's bilateral aid to particular countries.

Many of the challenges facing developing countries are of a complexity and geographical range which requires the application of research resources on an international scale. The ODA contributes to a number of international centres and programmes undertaking research relevant to developing countries.

The ODA assesses how each of its development projects might affect the environment and has produced a *Manual of Environmental Appraisal* for the guidance of all staff.

Britain supports the conservation of biodiversity in developing countries through research and funding at the global, international, regional and national levels, and participates in the negotiations on the international Convention on Biological Diversity. Actions taken by developing countries under the Convention, and which have global benefits, are funded through the Global Environment Facility (GEF). British aid also supports conservation of the world's genetic material in gene banks, botanic gardens and zoos, and has provided funding for the Commonwealth Science Council's Biodiversity Programme. See **Box 4.17** for an example.

Most of ODA's research programmes on renewable natural resources are managed by scientific institutes in the UK, and in particular by the Natural Resources Institute. These institutions in turn sub-contract a proportion of the work to other centres of expertise. For example the Forestry and Agroforestry programme is managed for the ODA by the Oxford Forestry Institute, the Plant Sciences programme by the Centre for Arid Zone Studies at the University College of North Wales, Bangor, and Aquaculture programme by the Institute of Aquaculture at the University of Stirling.

4.9.1 Natural Resources Institute (NRI)

NRI is an Executive Agency of the ODA staffed by over 300 natural and social scientists concerned with promoting the sustainable use of renewable natural resources in the developing world. The Institute carries out a wide variety of research and development activities in the renewable natural resource sector of developing countries for ODA and other donor organisations. These activities generally have a development focus but much of the work concerns sustainability issues in the sense of maintaining the quantity and range of biological resources ('natural capital') for use by future generations.

NRI's activities relate to all stages of the project cycle, including identification, assessment, planning, implementation, monitoring, and evaluation, and cover research, management, consultancy and advisory duties. Work is conducted in both natural and managed (human influenced) ecosystems, with a concentration on traditional agricultural systems. Geographical experience is wide, and work has been conducted in a range of ecosystems in many parts of the developing world, including tropical forests, wetlands, mountain and coastal r egions, and savannas (with locations in Africa, Asia, Central and South America, and the Pacific islands). The central concern of the majority of activities is sustainable development of local farming and livelihood systems, and in this context, biodiversity and wider environmental issues play an important part. NRI also has considerable expertise in the management and conservation of natural ecosystems and protected areas where biodiversity issues are of primary concern. NRI also has major management responsibilities for the implementation of ODA's Renewable Natural Resources Research Strategy and several of these programmes impinge on biodiversity issues.

A unique feature of NRI is its broad disciplinary range which includes biological sciences, with an emphasis on ecology and the environment, agricultural, livestock, fisheries and forestry sectors, and social sciences, including agricultural economists and sociologists. NRI staff specialise in natural resource issues and problems, and working in multi-disciplinary teams which include socio-economists. The Institute is noted for its integrative skills, and the use of interdisciplinary and systems approaches; a particular interest is the interlinkages between biological, human and farming systems. NRI's expertise principally lies in the scientific and practical problems of systems and taxa which are heavily utilised, often unsustainably, as a result of population growth and agricultural intensification. Research tends to be strategic or applied, and field rather than laboratory based. Contrasting examples of NRI research activities are given below, covering natural and managed systems and policy development.

Three projects are currently supported in whole or in part under the Darwin Initiative:
(1) Darwin Onion Project, a project to collect and multiply seed of onion landraces from West Africa, and carry out trials to evaluate useful characteristics; (2) invertebrate biodiversity of the Mkomazi Game Reserve, Tanzania, a project to provide baseline information on the diversity of fauna and flora to be used to assess the effects of future management strategies; and (3) fungal resistance in wild species of chickpeas, involving studies of the diversity of resistance mechanisms, in India and Syria.

Box 4.17 Mount Cameroon: a biodiversity conservation joint project between the ODA and the Cameroon Forestry Department

Mount Cameroon, at 4,095m is the highest mountain in West Africa and is unusual in having natural vegetation from sea-level to the sub-alpine peak. Rainfall at the foot, Cape Debundscha, is at 10-15m per year (one of the highest levels in the world), but other parts of the mountain receive less than 2m per year. Lowland rainforest, including freshwater swamp and mangrove at sea-level, extends to c. 800m, merging into montane forest which gives way to grassland at 1,900-2,400m. Unlike E. African mountains, no ericaceous zone is present. Species diversity decreases with altitude, until, at the summit, only six plant species are present.

Mount Cameroon has been known as the site for about 150 endemic plant species, about half are strictly endemic to the mountain the rest occurring nearby, e.g. on Fernando Po (Thomas & Cheek 1992; Cheek & Hepper, 1994). This number is expected to increase considerably when inventory work is more advanced.

The Mount Cameroon Project

The Limbe Botanic Garden and Forest Genetic Resources Conservation Project began in 1988, staffed and supported by the Government of Cameroon Forestry Dept. and the UK ODA, and advised by the Royal Botanic Gardens, Kew. A new, more broadly based, three year project has been approved by ODA (Aug. 1994) and the Government of Cameroon which will operate alongside long-term GTZ (German Overseas Aid) support for the conservation work on Mt Cameroon.

M. Cheek/RBG, Kew

The upper slopes of Mt Cameroon, a habitat rich in endemics.

Inventory method

The first inventory was conducted in a small, far from pristine patch of forest in the eastem foothills in 1992. Mabeta-Moliwe consists of about 36 km^2 lowland forest, with freshwater swamp forest, mangrove and littoral forest. It consists of a steep N-S ridge, c. 400m high, and a series of E-W ridges and streams.

For the purpose of a timber inventory 12 25x25m plots were intensively investigated, every species being recorded and about 1,000 voucher specimens being taken. Up to 150 species per plot were noted. In addition, c. 1,300 fertile specimens were made and from these c. 20 new taxa were detected of which 10 have been described and illustrated so far. About 900 species of vascular plant are now known from the forest.

The three succeeding inventories were carried out in the Etinde forest (c. 300 km sq.) on the main massif, and the Onge and Mokoko forests (c. 90 km sq.) in the western foothills. A total of c. 9,800 specimens have now been gathered.

Data collecting and mapping

A specimen data-base of all species likely to occur in S.W. Cameroon was set up in 1992. The Kew herbarium was trawled for all specimens from the area in the Rubiaceae in 1993 and the plan is to add more families in this manner if labour becomes available. Since mid 1993, all collections have been entered using the BRAHMS utility RDE (Rapid Data Entry).

A mapping function for BRAHMS is being developed at Oxford Forestry Institute by William Hawthorne. When this is ready it should be possible to map instantly, on screen, all known localities for any plant species around the mountain for which there is data. The hope is that with this database and the associated specimens, the populations of rare plant species on the mountain can be identified, monitored and managed more effectively for long term conservation.

(*Source*: M. *Cheek and* S. *Cable*, RBG, *Kew*.)

NRI's wide-ranging research and development activities in the field of biodiversity may be characterised in the following ways.

- Understanding and characterising diversity, including developing knowledge of what biodiversity exists and what are its present and potential functions. This has been a special feature of NRI's work in integrated pest management (IPM) and post-harvest storage. Emphasis has been on economically useful plants and those pests, pathogens and weeds which cause serious economic damage.

- Monitoring biodiversity through natural resource survey and mapping of terrestrial and aquatic ecosystems. Short and long term monitoring has been carried out of forest, grassland, wildlife, wetland and coastal ecosytems in over 60 countries. NRI has been closely involved in the development of inexpensive satellite-based systems for monitoring vegetative, climatic and hydrological conditions and change.

- Conserving biodiversity, including the development of strategies to conserve the diversity of landraces and wild relatives of crops in a framework of traditional farming systems (see **Box 4.18**). NRI is also involved in demarcation, planning and management of forest reserves, and wildlife conservation in these areas.

- Managing biodiversity through the management of projects and programmes concerned with understanding, conserving and utilising biodiversity of valuable natural resources. NRI has experience of managing projects linking scientific and practical issues and requiring prioritisation procedures for conservation. The Institute also has considerable collective experience of assessing the impact of projects and other interventions on biodiversity and the environment.

- Utilisation of biodiversity, taking account of commercial and local subsistence demands as well as the conflicting pressures for immediate exploitation and development and the need for long term sustainability (see **Box 4.19**). Biological resources are widely utilised by local people and NRI has expertise in the analysis of household incentives and strategies, property rights, stakeholder analysis and market failure.

Box 4.18 Bean diversity in traditional farming systems

NRI has an extensive programme of studies on useful biodiversity in traditional farming systems and natural ecosystems. For example, a project in Tanzania is looking at diversity in traditional *Phaseolus* bean mixtures. This project aims to document and describe diversity in traditional bean mixtures by assessing the resistance in the individual landrace components of these mixtures to major bean pathogens. Useful resistance sources have already been identified, and these will contribute directly to the improvement of global bean production. A planned extension to this project will be to look at how farmers assess bean biodiversity, and how farmer practices have contributed to the generation and maintenance of this diversity over time. Related studies are in progress on disease resistance in traditional sweet potato mixtures in Southeast Asia and traditional sorghum culture in South and West Africa.

D.M. Tererson/Natural Resources Institute

Bean mixtures brought into the market at Mbeya, in the southern highlands of Tanzania.

(*Source*: K. *Shawe*, NRI.)

- Policies for biodiversity which provide a rational framework and appropriate incentives for biodiversity conservation and exploitation in different situations. This is a growing area which demands developing policies and incentives for balancing the interests of multiple users, usages, scales and time perspectives.

Box 4.19 NRI Rosewood study

Rosewood oil is produced in Brazil for use in perfumery by the destructive harvesting of whole trees, and similarly destructive extraction procedures, in unmanaged ecosystems. Until recently, the exploitation of wild resources has been unrestricted and natural regeneration has been unable to keep pace. The depletion of natural populations, the imposition of harvesting restrictions, and the advent of substitute sources have led to a detailed examination of opportunities for improvement. NRI activities have included surveys of tree distribution, the establishment of nurseries and collection of germplasm, investigation into prospects for coppicing and oil production from leaves, and detailed examination of market opportunities. Revitalisation of the industry and protection of biodiversity are dependent upon the development of low-cost propagation and production techniques and raising oil yield and quality. Preliminary results suggest that leaf oil is of similar composition to wood oil but, to minimise production and transportation costs, must be grown on plantations or in farm forestry systems rather than collected from scattered sites in the wild. Work on natural variability suggests there is scope for breeding selected material for improved oil quality and yield.

(*Source*: R. *Grimble*, NRI.)

4.10 CAB International

CAB International (CABI) is an international, intergovernmental organisation with its headquarters in the UK, established in 1928. It is owned by its member governments, which currently number 37. The organisation is largely self-supporting financially, through the sale of its products and services, and other sources of income.

CABI is dedicated to improving human welfare worldwide through the dissemination, application and generation of scientific knowledge in support of sustainable development. The emphasis is on agriculture, forestry, human health and the management of natural resources, with particular attention to the needs of developing countries.

CAB International has been active in providing an international service for biodiversity for more than 80 years, at and from its four scientific Institutes, from its field stations in different countries, and through the products from its research information database CAB ABSTRACTS.

Scientific and technical information services are provided, derived principally from bibliographic databases built by CABI staff from the world's relevant published literature, with assistance from free-lance workers and contracted organizations.

CABI's main database, CAB ABSTRACTS (see Box 7.5 in Chapter Seven), contains abstracts of over 3 million items of world literature related to the mission of CABI (increasing by c. 150,000 p.a.).

Outputs from these databases are distributed in printed form, and in electronic form on diskette, CD-ROM (compact disc) and magnetic tape. On-line access is provided by commercial database vendors. These products and services are used by research scientists, teachers and students in higher education, development planners, health workers, and other professionals. Copies of source documents are available through CABI's Document Delivery Service.

CABI also publishes books, primary journals, newsletters and training manuals.

Research and development is undertaken in innovative information systems. Training in information management is offered, especially for developing countries. Advice and practical assistance are given in the design and implementation of science-based information systems, and in the use of modern information technology. Where necessary, partnerships with development assistance agencies are sought to help developing countries acquire the products and services they need.

During the last eight decades CABI has established four scientific institutes, each now recognized as a centre of excellence. Three are Biosystematic institutes, Entomology, Mycology and Parasitology, conducting taxonomic research and providing diagnostic identification services for organisms of agricultural and economic importance. The fourth institute is the International Institute of Biological Control, established in 1929 to provide advice, information and research support on this economic and environmentally important science.

CABI provides a series of training courses within its scientific specialities in developing countries and in the UK, often working closely with universities and other educational institutions. CABI has taken the lead in an ambitious international cooperative initiative, BioNET-International. This global network of biosystematists of arthropods, nematodes and microorganisms depends on the creation of sub-regional LOOPs (locally organised and operated partnerships; see Chapter Three, Section 3.3.5). CABI has also undertaken advanced biosystematics training under the Darwin Initiative for more than 20 specialists, who spent 12-month periods at CABI Institutes as Darwin Fellows. CABI scientific institutes have links with universities worldwide, and with UK universities particular affiliations exist with the University of Reading, University of London (Imperial College, the London School of Hygiene and Tropical Medicine, Wye College, Royal Holloway College, and the Royal Veterinary College), University of Oxford (Oxford Forestry Institute), University of Wales (Cardiff), the University of Kent and the University of Surrey.

4.10.1 International Institute of Biological Control (IIBC)

IIBC advocates the use of naturally occurring and self-perpetuating natural enemies in controlling pests in crop ecosystems and conservation areas, so as to reduce the destructive impact of invasive alien species or chemical pesticides on the indigenous biodiversity and the environment. IIBC operates through seven bases around the world (see **Box 4.20**), most of them linked to leading national research institutes. The international staff of IIBC are involved in over 80 separate projects each year ranging from small-scale local farmer training field schools to multi-million dollar international programmes.

IIBC undertakes a wide range of research. The Leverhulme Unit for Population Biology and Control, a joint IIBC-Imperial College initiative, undertakes fundamental research in experimental ecology and population modelling to improve the practice of biological control.

Box 4.20 International Institute of Biological Control field stations

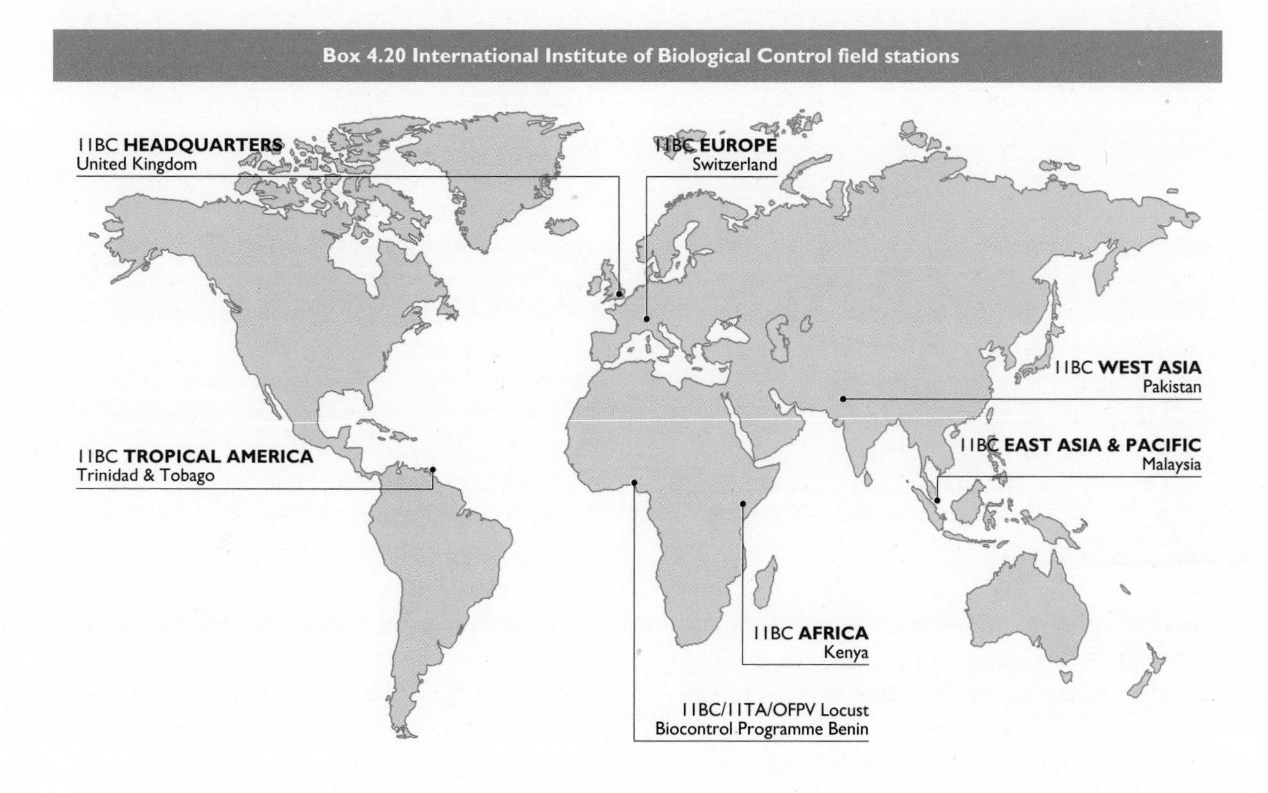

This research complements the practical training courses run by IIBC around the world for farmers, extension scientists and researchers. These include participatory training on biological control in rice and cotton cropping systems in Asia, in plantation crops in Latin America and in forestry systems in Africa. One long-running course is directed at national quarantine officers, to improve their knowledge of safety aspects of biological control and increase their participation in the introduction of exotic natural enemies. The most recently established training course is the two-week course on evaluating pesticide effects on natural enemies, designed for plant protection specialists and pesticide registration authorities.

Customized, illustrated manuals are prepared for training courses, as well as special self-training packages. Recently, IIBC has begun work with national programmes and FAO on the development of training materials for farmers, to help them discover and appreciate the role of natural enemies in their crops.

In addition to training courses, IIBC cooperates with universities in Europe, Africa, Asia and Latin America to support degree-level training in biological control, and organizes bench training in specialized techniques for scientists and technicians sponsored by their own institutions or through training grants.

4.10.2 International Mycological Institute (IMI)

IMI provides specialist services to mycology in support of agriculture, industry, the environment and public health, in accordance with the policies of CAB International and with the emphasis on developing countries. It provides a worldwide service for the identification of fungi (including lichens, slime moulds and yeasts) and plant pathogenic bacteria. The provision of an authoritative and rapid identification service is a key aim of IMI's team of specialists.

These services are backed by dried and living collections of these microorganisms (see Chapter Six, Section 6.2.4).

IMI's team of specialists (the largest number of mycologists at any single centre in the world), who have a wide range of experience in different countries, place the Institute in a unique position to undertake inventorying and survey work in support of biodiversity studies, plant disease surveys, ecological investigations, and obtaining novel or unusual taxa. IMI also participates in planning, inventorying and monitoring strategies, and in the preparation of checklists and schemes.

IMI has Official Correspondents in 38 countries; a network under active expansion. These Official Correspondents act as national focal points for information on IMI and its services.

Information products including books, serial publications, mycological papers and electronic products are produced by the Institute through CABI.

IMI offers a wide range of courses and tuition opportunities for individuals. Since 1985, it has trained 900 people from 96 countries. Topics for tuition include specialist courses on the systematic identification of particular groups of fungi and courses on a range of fungi of relevance to agriculture, the environment, industry and health. They are summarised in Box 5.26 in Chapter Five.

4.10.3 International Institute of Parasitology (IIP)

IIP provides identification, consultancies, research and information services to agricultural, biological and medical scientists on parasites of animals and man and on nematode parasites of plants and invertebrates. The Institute is the FAO Collaborating Centre on Helminthology, Taxonomy and Informatics and has collaborative links with the Institute of Arable Crops Research (IACR), Rothamsted Experimental Station, the Royal Veterinary College and Imperial College of Science, Technology & Medicine of London University and Reading University.

The IIP scientists have international experience with plant parasitic nematodes in the tropics and subtropics. They are actively involved in surveys and field trials, supporting local and regional research programmes in developing countries. Staff advise on sustainable methods of crop protection in both food and cash crops. Institute-based research projects complement in-country activities. The Institute provides an internationally recognized, authoritative biosystematics service on plant parasitic nematodes of economic importance, and other soil-dwelling nematodes with potential as bioindicators of habitat degradation. Staff produce monographs on nematode systematics, publish descriptions of species new to science and train numerous visiting workers.

Institute scientists also provide a world service for the authoritative identification of nematodes, cestodes, digeneans, monogeneans and acanthocephalans parasitic in animals and man. Supported by reference collections and an excellent literature resource, the helminthologists publish identification aids in the form of keys, monographs, checklists and descriptions. Collaborative research on biosystematics, biodiversity and veterinary parasitology is undertaken with other organizations.

IIP provides a molecular based service developed for the identification and characterization of entomopathogenic nematode species. Members of the Unit have experience in the use of polymerase chain reaction (PCR) to identify single nematodes to species level. Molecular characterization is also being done on parasitic nematodes of both plants and animals. IIP is actively involved in the isolation and identifications of entomopathogenic nematodes, particularly in the tropics, and developing them for insect control. A large reference collection of living and preserved specimens is maintained to provide the basis for an authoritative identification service based on DNA techniques.

The Institute runs six-week training courses on the identification of helminth parasites and plant parasitic nematodes of economic importance in alternate years. Staff contribute to several MSc courses and the Institute offers training facilities for PhD students and specialized study visits.

4.10.4 International Institute of Entomology (IIE)

IIE is a centre of excellence in research, training and provision of information on insects and mites. It conducts core research as well as research for national and international organizations and donors. The Institute's headquarters, training facilities and ecological staff are located at 56 Queen's Gate, London, opposite the Natural History Museum, where the biosystematic staff are based.

The Institute's research activities focus on agricultural entomology, biodiversity and environmental change in both natural and managed ecosystems. Research on biodiversity employs the combined skills of taxonomists and ecologists and focuses on biodiversity mechanisms, measurement and monitoring and on regional characterization.

The development and use of new methods for dissemination of taxonomic data is an important activity of IIE's staff.

IIE offers a wide range of training courses on the taxonomy and biology of insects and mites, particularly those of economic importance. Courses are either held at the London headquarters or in conjunction with Universities and Museums around the world. Specialist one-to-one training is a popular means of using the expertise of individual staff members, and PhD students working in conjunction with UK universities are welcome.

Services provided by IIE include an identification service for insects and mites and the provision of distribution maps of insects. The Bulletin of Entomological Research is edited and managed by staff of IIE.

4.11 Bibliography and references

Much of the information contained in this chapter is based on the brochures and annual reports of the research institutes described above. They have not been listed here; many are substantial documents in their own right. Increasingly these institutes are also using electronic networks to publically make available information about their services and research projects. Enquiries should be made direct to the institutes; addresses of all are given in Appendix Two.

BBSRC (1994). *Biotechnology and Biological Sciences Research Council handbook* 1994/95. BBSRC, Swindon.

Cabinet Office, (1994). *Multi-departmental scrutiny of Public Sector Establishments*. HMSO, London.

Cabinet Office, Office of Public Service and Science, and Office of Science and Technology. (1995). *Forward look of Government-funded science, engineering and technology*. Vol. 2. *Research Councils and Departments: Forward look Statements*. HMSO, London.

Cheek, M. and Hepper, F.N. (1994). Progress on the Mount Cameroon Rainforest Genetic Resources Project. *Nigerian Field Society* (UK) *Occ. Paper*, No 1: 15-19.

DoE (1994). *Biodiversity: The UK Action Plan*. HMSO, London.

DoE (1995). *Biodiversity: The UK Steering Group Report*. *Vol 1. Meeting the Rio Challenge, Vol 2. Action Plan*. HMSO, London.

Hawkes, J.G. (1990). *The potato: evolution, biodiversity and genetic resources*. Smithsonian Institution, Washington, DC.

HMSO (May, 1993). *Realising our potential: A strategy for science, engineering and technology*. (CM2250). HMSO, London.

Scottish Office (1993). *Policy for Science and Technology*. Agriculture and Fisheries Department, The Scottish Office, Edinburgh.

Sheail, J. (1992). *Natural Environment Research Council: a history*. NERC. Swindon.

Thomas, D. and Cheek, M. (1992). *Vegetation and plant species on the south side of Mt Cameroon in the proposed Etinde reserve*. Cameroon Govt./ODA/Royal Botanic Gardens, Kew.

Winter, Michael (1995). *Networks of Knowledge: research, advice and environmentally friendly agriculture in the UK*. Countryside and Commuity Research Unit/Gloucester College of Higher Education, Cheltenham.

BIODIVERSITY RESEARCH AND TRAINING IN UK UNIVERSITIES AND LEARNED SOCIETIES

5.1 **Introduction**
5.2 **Universities and colleges**
5.2.1 University courses and specialised biodiversity training
5.2.2 Research in universities
5.3 **Societies and other organisations with a biodiversity interest**
5.3.1 The multidisciplinary and specialist Learned Societies
5.3.2 National specialist botanical Societies
5.3.3 Botanical and zoological garden organisations.
5.3.4 Ornithological Societies
5.3.5 Other Societies and Trusts specifically studying and/or conserving animals
5.3.6 Societies set up to study particular ecosystems
5.3.7 Organisations with an international brief
5.4 **Training in museums, botanical gardens and zoos**
5.5 **UK Government promotion of biodiversity**
5.5.1 Direct Government grant-aid
5.5.2 The Department of the Environment and the Darwin Initiative for the Survival of Species
5.6 **Funding for training**
5.6.1 General information
5.6.2 Applying for funding
5.6.3 Awards for postgraduate study and research
5.6.4 Awards for other training
5.7 **References and bibliography**

Chapter authorship

This chapter has been drafted and edited by David Rae *with additional material contributed by* R. Atkinson, D. Bridson, M. Claridge, R. Cubey, G. Douglas, M. Fitton, E. Fryd, A. Gammell, M. Gosling, G. Halliday, J. Holloway, C. Jermy, S. Jury, D. Long, N. Maxted, S. Mosedale, W. Knight, R. Ormond, P. Raines, J. Ratter, N. Riddiford, J. Rodwell, E. Rogers, M. Sands, B. Schrire, S. Sutton, L. Thompson, E. Watt, S. Winser *and* P Wyse Jackson.

5.1 Introduction

There can be little doubt that the UK, in world terms, makes a leading contribution to the study of biodiversity. This does not simply reflect the total number of universities, museums, botanic gardens and allied institutions around the country. It is equally a measure of the vast array of expertise (both amateur and professional) in learned societies and other non-government organisations concerned with natural history and environmental conservation, which exists on a scale unequalled around the world. A great reserve of knowledge is spread throughout these organisations and high-quality research is carried out in Britain and abroad, and a wide range of training (from local excursions and workshops to post-graduate degrees and diplomas) is part of their function. It is this established base of resources and expertise which provide the country with the capability to play a major role in addressing the present environmental challenges.

The World of Learning (Europa, annual) gives brief details of the UK's universities, learned societies, professional associations, research institutes, libraries and museums. This chapter will review the range of work in the universities, the scope of their specialised training for biodiversity assessment, and the role played by the learned societies and other NGOs. *The Directory of University Expertise and Facilities: Biosciences* (Horne & Zeitlyn, 1995) reviews biosciences research, facilities and training in 30 UK universities in what will become a biennial service.

5.2 Universities and colleges

There are over 200 universities and colleges within the UK serving all sectors of academia. Details can be found in the *Commonwealth Universities' Yearbook* (ACU, 1995a). These include, not only those thought of as the older and 'established' universities found in most cities, but also those former polytechnics with more technically applied courses now given full university status and funding. Others, whose contribution to biodiversity studies must not be overlooked, are the colleges or universities of technology such as those at North East Surrey, Dundee (the Dundee Institute of Technology), Farnborough and Loughborough; and the Colleges of Higher Education such as those at Bedford, Winchester (King Alfred's), Southampton (LSU) and Cheltenham & Gloucester, the latter, for instance, with interests in South Georgia and the Falklands Islands. Medical Schools attached to many of the larger universities also contribute to our understanding of microorganisms as a result of their pathological research programmes.

5.2.1 University courses and specialised biodiversity training

Most universities provide a complete range of courses organised from a number of faculties and departments. The University of Edinburgh, for instance, which is by no means the largest UK university, runs more than 300 undergraduate courses from well over 100 teaching and research departments. Other specialised colleges run only a few courses and research in specific areas, such as Askham Bryan College and Harper Adams Agricultural College, both of which specialise in the land-based industries. The Universities and Colleges Admissions Service *Handbook* (UCAS, 1995) gives details of available courses in all disciplines.

Biodiversity is an interdisciplinary subject requiring an appreciation of many organisms, at habitat, species and genetic levels. This means that the expertise of the ecologist, taxonomist, and cytogeneticist are equally important and they must all have an input into biodiversity training. However, one of the many positive aspects of the concept of biodiversity is that it has the capacity to unite divergent ends of the biological sciences. Many other disciplines are important for the study of the function of biological organisms in the environment ranging from geology, agronomy and marine biology to mycology and microbiology. Many find that it is the interdisciplinary nature of biodiversity that makes it so interesting. The range of courses and qualifications offered at this educational level includes Higher National Diploma (HND), BA and BSc and, at postgraduate level, MSc, MPhil or, PhD and D. Phil. HND, BA and BSc courses usually last for three years, or four at honours level, while MSc and MPhil may be a one or two year taught course, or may be done by research, PhDs

Box 5.1 Plant conservation training in countries rich in biodiversity: the University of Birmingham Intiative

The continuing challenge for the world's agricultural, botanical and conservation scientists is twofold: to feed an ever increasing population and to halt the rate of plant extinctions. The World Bank has estimated that, currently, 500 million people in Africa, Asia and Latin America suffer from chronic undernourishment while IUCN and WWF estimate that up to 60,000 plant species may become extinct by the middle of the next century. There is, therefore, an urgent requirement to provide scientists throughout the world with the skills to conserve and utilise their native plant biodiversity effectively.

N. Maxted/Birmingham University

Collecting wild legume species in a field in Azerbaijan.

University of Birmingham Off-site Short Courses

Aims and Objectives

To provide managers, researchers and technicians with the practical and theoretical skills they require to conserve and utilise their native flora for the benefit of all humankind.

With the realisation that not all those who require such training are able to travel to the UK, the School of Biological Sciences, University of Birmingham, has established a range of training courses in Plant Conservation and Genetics 'offsite', in regions of the world rich in biodiversity, many of which are in developing countries. The objective of these courses is to provide intensive in-service, refresher or introductory training.

In recent years courses have been run in Australia, Mexico, Morocco, Namibia, Syria, Sri Lanka, Spain, Turkey, Russia, the United States of America, and Zambia. Where appropriate, international specialists are invited to contribute and staff from the following institutes have contributed: the Royal Botanic Gardens, Kew, World Conservation Monitoring Centre, Oxford Forestry Institute, Institute of Grassland and Environmental Research, commercial companies and international NGOs involved in plant conservation.

The topics taught 'off-site' have included:

- Conservation data management and analysis;
- Plant exploration and in situ conservation strategies;
- Plant identification and systematics (para-taxonomy);
- Seed conservation and gene-bank management;
- Biotechnology and in vitro conservation;
- Plant tissue culture of 'new' species;
- Development & utilisation of molecular markers in plant breeding;
- Quantitative genetics and plant breeding.

However, it should be stressed that individual courses and curricula are tailored to meet the specific requirements of national or regional conservation programmes and developmental strategies, which permits a highly flexible approach.

Programme Director: Dr Nigel Maxted, School of Biological Sciences, University of Birmingham.

usually take three years and are based on original research though there may be a set number of compulsory taught classes.

Universities have the great advantage of having large numbers of skilled staff from different disciplines in one institution. They deal with the large bulk of higher education and have the facilities to deal with large numbers of students. For these reasons, universities are responsible for the greatest amount of biodiversity training and are developing specific courses on biodiversity for particular countries (see below). Increasingly, however, specialised training at the postgraduate level may be undertaken in (and funded by) research institutes (see Section 5.2).

At least 75 universities and colleges offer some type of biodiversity training, involving almost 10,000 students. There are no undergraduate courses specifically in biodiversity, nor is it desirable that there should be. At this level it is far better that students should do a general course in for example, biology or ecology in order to obtain a general appreciation of the subject and become familiar with some of its components. Thereafter, at postgraduate level it is appropriate to focus on biodiversity. Nonetheless, the concept of biodiversity is taught within general undergraduate courses and a recent survey has shown that "significant amounts" of biodiversity are taught within at least 38 different undergraduate course subjects in a total of at least 75 universities and will be included in a further 23 course subjects planned for the near future. It should be noted that many of these courses (e.g. Biology, Zoology and Environmental Science) are run at several universities. See UCAS (1995).

It is at postgraduate level that a student can really start to benefit from training in biodiversity, building on his or her existing knowledge of the natural world, gained at undergraduate level. In the UK at least 36 universities offer postgraduate courses that include training in biodiversity. At least 775 students attend such courses each year. Many of the courses are tailored to give overseas students, especially those from developing countries, a firm grounding in biodiversity assessment and conservation. These training programmes are not a direct outcome of the recent worldwide interest in biodiversity but benefit from many years of experience gained from such teaching. The University of Birmingham, for instance, has provided master's degrees, diplomas and certificate courses for over 550 students from 90 countries over the past 25 years. In recent years courses such as the Botanical Diversity: Classification, Conservation and Management MPhil., have been run 'off-site' (in the country where the training is needed) directed towards habitats within a specific biome. (see **Box 5.1**). Other specialist courses are on specific organisms, e.g. on insect systematics and indentification as with the Applied Insect Taxonomy Diploma run at the University College of Wales, Cardiff, and the parasitic Hymenoptera course run at Sheffield University in conjunction with The Natural History Museum (see **Box 5.2**), or the Biodiversity of Fungi course run at London University. Many courses, such as those just mentioned, are run jointly with specialist staff from other research institutes being involved. Other courses are specific to environmental matters, conservation or pollution such as the Conservation Biology course run at University College London, the Environmental Monitoring Diploma/MSc run at the University of Bradford, the Global Biodiversity Monitoring & Conservation MSc that is given at the University of Hull jointly with WCMC, or the Ecotoxicology and Pollution Studies MSc run at the University of Luton over 18 months.

A few universities are capitalising on unique expertise built up over a number of years from UK studies and are developing training courses for a specialist international market. For example, the Unit of Vegetation Science, at Lancaster University, whilst formulating the UK National Vegetation Classification has over 12 years' experience in providing training to help agencies and individuals understand the description and assessment of vegetation resources. Besides courses at MSc level, a variety of general and specialised short-courses and workshops are organised at Lancaster and elsewhere, with commissioned training for customers who wish to concentrate on sites and problems of their own choosing, e.g. the U.S. Nature Conservancy and of Tiraha University in Albania, the latter supported by the UK Know-How Fund.

Those universities which have built up expertise, and often research units, to study the biodiversity of particular ecosystems (e.g. coral reefs, savanna grassland) are increasingly involving students, national parks staff and other researchers from the host countries concerned, in some cases in the form of planned training programmes.The UK Government, as a response to the 'Rio Summit', has increased funding for such initiatives through the programme known as the **Darwin Initiative for the Survival of Species,** and a wide range of training initiatives especially for developing world members has been launched. This aspect is discussed further under Section 5.5 and in

Box 5.2 Insect systematics courses: Cardiff and London

University College Wales, Cardiff and the National Museum of Wales: Applied insect taxonomy course

Insects are the major single cause of damage or loss of yield in agriculture, and through their role as vectors are an important threat to the health of humans and their livestock. Crucial to any form of pest control is an accurate knowledge of the identity of the species involved.

M. Claridge/University of Wales, Cardiff

Planthoppper, *Nilapavata lugens*, a major pest of irrigated rice in Asia.

M. Claridge/University of Wales, Cardiff

Student studying museum collections at Cardiff.

The University College Wales, Cardiff provides a one-year diploma course, designed primarily for people from tropical and developing countries where there is a serious shortage of expertise in identification of insects. The training includes a thorough grounding in theoretical subjects such as insect systematics, biological species concepts and speciation.

The Diploma is taught jointly by staff of the University and the National Museum of Wales, making full use of the extensive collections in that institute. Good collection management, invaluable in establishing programmes in developing countries, is an integral part of the training.

A vital part of the diploma is that where the student tackles an original taxonomic project under the Course Director's supervision. Should students wish to continue their research at Cardiff, they have the opportunity to undertake a full-scale research programme leading to the award of an M.Phil or Ph.D.

This course has attracted people from 35 countries, the students coming from several institutions in each country and returning to positions of resposibility

Course Director: *Prof. Michael Claridge*, School of Pure and Applied Biology, UCWC.

Natural History Museum, Imperial College, University of London: Taxonomy and biology of parasitic Hymenoptera

Parasitic wasps are very numerous, diverse, of great ecological significance and economically important. Their correct identification is vital for workers in many fields but, because of their diversity and, often, their small size, they are regarded as inaccessible except to experts. However, recent advances in wasp systematics, encapsulated in various identification manuals, mean that most entomologists and ecologists can learn to identify the major groups.

The first *Taxonomy and Biology of Parasitic Hymenoptera* course took place in September 1989 and it now runs annually at Easter. Organised jointly by the Natural History Museum's Department of Entomology and Imperial College, the intensive seven day course makes use of both the Museum's collections and the university's teaching and residential facilities. Course tutors come from the national museums in Edinburgh and Leiden as well as from Imperial College and the NHM. Interest in the course is worldwide; the 26 students, from postgraduate ecologists to agricultural entomologists, in 1993 represented 13 countries.

Participants receive an overview of the parasitic Hymenoptera and an introduction to their biology and systematics. Practical instruction is given in the handling and identification of all of the groups, and in topics such as collecting and rearing. Participants are also able to meet and seek advice on their own particular problems from leaders in Hymenoptera taxonomy. The course therefore covers the needs of both graduate students and applied researchers involved in biological control and in the agrochemical industry.

Course Director: *Dr Mike Fitton, Natural History Museum, London.*

Box 5.27. An example is seen in **Box 5.8** where the Tropical Marine Research Unit of York University is sharing its knowledge through a one-week training course in reef fish identification and survey methods given to four rangers at the Ras Mohammed National Park, Egypt.

Box 5.3 lists current and proposed postgraduate courses which include training in biodiversity and related subjects. Many of the courses combine thorough academic training with a strong element of field study. Indeed many of the MSc-type courses consist of 6–8 months of compulsory lectures culminating, if successfully examined, in a Diploma, followed by 4–6 months of field investigation and research leading to the award of the full MSc.

5.2.2 Research in universities

Biodiversity research in universities is dependent on the interests and experience of the lecturing staff. In many cases it would probably be true to say that university staff are more concerned with devising methods of describing, recording and monitoring biodiversity while a great deal of the practical application is done by other institutions (e.g. the major museums and herbaria, and NERC research units; see Chapter Four). A good example of this is seen in botany where it is the staff of botanic gardens and museums who are regularly involved in field research and compiling plant inventories and identification manuals. It is the universities which have led the way in developing taxonomic methodology in areas such as cladistics, cytogenetics, chemotaxonomy and ultrastructure. The reason for this is partly an historical one, the consequences of which are discussed in a forward-thinking Report, *Evolution and Biodiversity – The New Taxonomy*, drawn up by The Natural Environment Research Council (Krebs, 1992). Over the last fifty years there has been a move in systematic biology (i.e. taxonomy) research from morphology and anatomy being the most significant elements, to a biochemical and molecular level approach. This has resulted in university biological schools replacing retiring systematists/taxonomists by those with 'molecular' skills and little understanding of biodiversity, a state which has had a significant knock-on effect and led to a dearth of taxonomists. It is further reflected in the diverting of already scarce research funds towards molecular studies. This subject, and the national (and indeed the international) lack of those able to identify and classify biological organisms, has been the basis of various reports (ABRC, 1977; HoL, 1992; Claridge & Ingrouille, 1992).

Universities and colleges tend to have many faculties dealing with a broad spectrum of disciplines. In biological institutes research tends to be more focused, often in a restricted area of expertise. An exception to this is The Natural History Museum which has the potential and policy of promoting interdisciplinary projects. The interests of an individual within a particular university faculty may influence the reputation of that faculty within the lifetime of the individual, or even the lifetime of his/her students.

The table in **Box 5.4** illustrates the range of the biodiversity-related research taking place in a selection of UK universities. Biodiversity here has been interpreted in its broadest sense. Thus the topics investigated, which have been taken from questionnaires returned by the department concerned, and personal contacts, show some topics to be highly pertinent to biodiversity as discussed in this manual, while others relate to the Convention of Biological Diversity in a more tangential way (e.g. studies in ecology that relate to sustainable use). Of the projects listed, very roughly half are related to the species level of biodiversity while about a quarter each relate to the genetic level and habitat level. Projects that study biodiversity in the UK alone are omitted.

Over the past 150 years, many universities have built up preserved collections of plants and animals to demonstrate biodiversity to students. These collections formed the basis of research programmes and were later expanded as a result. One such an example is shown in **Box 5.5** in which longterm collecting expeditions are providing material for an account of the flora of Eastern Greenland. From the early 1960s, there has been an increasing practice to take students on overseas field trips (initially to European areas, latterly further abroad), particularly to sample other floras. This has often led to cooperation with host institutes abroad and to the initiation of longer-term biodiversity studies. An example is given in **Box 5.6**. Studies in animal biodiversity tend to develop wider ecological programmes including those on population sizes, species

Box 5.3 Current and proposed postgraduate courses which include training in biodiversity

(adapted from the Directory of Environmental Courses 1992–1993 Baines & James, 1994). Cert = Certificate; Dip = Diploma; Pg = Postgraduate: Note: PhD degrees in biodiversity subjects can be taken at most universities where appropriate supervision is available.

UNIVERSITY	COURSE TITLE	LEVEL OF COURSE
Aberdeen University	Ecology Environmental remote sensing Environmenal science Forestry Marine and fisheries science	MSc, Pg Dip MSc, Pg Dip MSc, Pg Dip MSc, MPhil, Pg Dip MSc
Anglia University, Cambridge	Environmental assessment	MSc
Aston University, Birmingham	Management and safety technology	MSc
Bath University	Development studies Environmental science, policy and planning	MSc/ Diploma MSc/ Diploma
Belfast: see Queens University		
Birmingham University	Conservation and utilisation of plant genetic resources Botanical diversity classfication, conservation and management Plant breeding and crop improvement Applied genetics	MSc MPhil MPhil MSc
Bournemouth University	Coastal zone management	MSc, Pg Dip
Bradford University	Business strategy and environmental management Environmental monitoring	MSc, Pg Dip MSc, Pg Dip
Brighton University	Environmental impact assessment	MSc
Bristol University	Ecology and management of the natural environment	MSc, Diploma, Certificate
Brunel University	Environmental science with legislation and management	MSc
Cambridge University	Environment and development	MPhil
Central England University, Birmingham	Environmental protection, control and monitoring Environmental management	Pg Dip MA, Pg Dip
Cheltenham and Gloucester College of Higher Education	Environmental policy and management	Diploma, MA, Certificate
Chester College	Environmental biology	MSc/ Diploma
Christ Church College, Canterbury	Fresh water biodiversity	MPhil, Diploma
Cranfield University, Silsoe	Environmental diagnostics Agroforestry Range management Soil conservation Land resource management Land resource planning Environmental water management Environmental monitoring	MSc, Pg Dip MSc MSc MSc MSc Pg Dip MSc MSc
Durham University	Ecology Environmental management practice	MSc, Diploma MA, Advanced Dip
East Anglia University, Norwich	Agriculture, environment and development Environmental science	MSc MSc, MPhil, Pg Dip
Edinburgh University	Reproductive biology Resource management, forestry, agriculture and ecology Environmental health Environmental chemistry Rural science Biodiversity and taxonomy of plants (in conjunction with Royal Botanic Garden Edinburgh) Ecology and resource management	MSc, Diploma MSc, Diploma MSc, Diploma MSc, Diploma MSc, Diploma MSc, Diploma MSc
Farnborough College of Technology	Environmental management	MSc, Pg Dip, Pg Cert
Glamorgan University, Pontypridd	Environmental conservation management Quality environmental management	MSc MSc
Glasgow University	Evolution in taxonomy: Principles and practices Applied entomology	10 day course MSc
Greenwich University	Environmental risk assessment Environmental research	MSc MSc
Heriott-Watt University, Edinburgh	Marine resource management	MSc, Diploma
Hull University	Environmental policy and management Estuarine and coastal science and management Global biodiversity: monitoring and conservation	Diploma MSc, Diploma MSc, Cert, Diploma

Box 5.3 *continued*

UNIVERSITY	COURSE TITLE	LEVEL OF COURSE
Imperial College, College of Science Technology and Medicine, London	Environmental analysis and assessment Environmental analysis, radionuclide metrology	MSc, Diploma MPhil, Diploma
Institute of Earth Studies, Dyfed, South Wales	Protected landscape management	MSc
Kent University, Canterbury	Ecology Conservation biology Environmental law and conservation	MSc, MPhil MSc, MPhil MSc
Kingston University	Earth science and the environment	MSc
Lancaster University	European environmental policy and regulation Environmental and ecological sciences	MSc MSc, Pg Dip
Leeds University	Ecology and evolutionary biology	MSc
Leicester University	Natural resource management	MSc, Diploma
Liverpool University	Behavioural and evolutionary ecology Environmental assessment Recent environmental change Environmental and evolutionary biology	MSc MSc MSc MSc
London Guildhall University	Quaternary environmental change	MSc, Pg Dip
London (East) University	Environmental sciences	MSc
London School of Economics and Political Science	Environmental management	Pg Dip
London University, Wye College	Landscape ecology, design and management Applied environmental science Environmental management Rural resources and environmental policy	MSc MSc MSc/ Diploma MSc
Manchester University	Plants, microbes and environmental biology	MSc/ Diploma
Manchester Institute of Science and Technology (UMIST)	Pollution and environmental control	MSc, Dip Tech Sci
Manchester Metropolitan University	Behavioural ecology Environmental and heritage interpretation Conservation biology Environmental and geographical sciences Countryside management	MSc, Pg Dip, Pg Cert MA, Pg Dip MSc, Pg Dip MPhil MSc, Pg Dip, Pg Cert
Middlesex University	Conservation Policy	MA, Diploma
Moray House Institute Heriot-Watt University	Marine resource development and protection	MSc, Diploma
Napier University, Edinburgh	Biology of water resource management	MSc, Pg Dip
Northumbria University, Newcastle Upon Tyne	Environmental monitoring and control	MSc
Newcastle Upon Tyne University	Tropical agricultural and environmental science Tropical coastal management	MSc, Diploma MSc, Diploma
Normal College, Bangor	Countryside management	Pg Dip
Nottingham University	Environmental planning for developing countries	MA, Diploma
Nottingham Trent University	Heritage studies	MA, Diploma
Oxford University	Evolution, taxonomy and ecology Forestry Forestry and its relation to land use	MSc MSc MSc
Oxford Brooks University	Environmental assessment and management	MSc, Diploma
Paisley University	Environment management	MSc, Pg Dip
Plymouth University	Applied marine studies	MSc
Portsmouth University	Coastal and marine resource management Environmental change	MSc, Pg Dip, Pg Cert MSc, Pg Dip, Pg Cert
Queens University, Belfast	Environmental and evolutionary biology Applied environmental sciences	MSc MSc

Box 5.3 *continued*		
UNIVERSITY	COURSE TITLE	LEVEL OF COURSE
Reading University	Forestry extension	MSc
	Wildlife management and control	MSc, Diploma
	Land management	MSc, MPhil, Adv Dip
	Pure and applied plant and fungi taxonomy (with International Mycological Institute)	MSc
	Evaluation of plant ecosystems	4 Month Course
	Plant taxonomy for conservation	3 Month Course
	Plant taxonomy methods	MSc, Diploma
	Environmental resourses	Pg Dip
	Environmental protection	MPhil
	Botanical diversity: Its classification	MPhil
	Conservation management	MSc
	Molecular and immunolocal parasitology	MSc
	Vegetation survey and assessment	MSc
Robert Gordon University, Aberdeen	Ecological design	MSc, Diploma
South Bank University, London	Environmental monitoring and assessment	MSc, Pg Dip
St Andrews University	Land resources and land utilisation	MSc, Diploma, M Phil, M Litt
Stirling University	Environmental science	MSc
	Environmental management	MSc, Diploma
	Aquaculture	MSc
Strathclyde University, Glasgow	Environmental studies	MSc, MEnvS
	Environmental management	MSc, Diploma
	Ecology and conservation	MSc
	Plant biology	MSc
Strathclyde University Jordanhill College of Education	Environmental studies	Certificate
Sunderland University	Environmental management	MSc, Diploma
Surrey University	Environmental management	MSc
Sussex University	Environment, development and policy	MA
	Rural development	MA
Ulster University, Coleraine	Environmental management	MSc, Diploma
Wales University College, Aberystwth	Environmental auditing	MS, Diploma
	Environmental rehabilition	Diploma
	Environmental impact assessment	MSc, Diploma
Wales University College, Bangor	Ecology	MSc, Diploma
	Environmental forestry	MSc, Diploma
	Forestry	Diploma
	Rural resource management	MSc, Diploma
	Marine and environmental protection	MSc
Wales University College, Cardiff	Applied insect taxonomy	Diploma
Wales University College, Swansea	Environmental biology	MSc, Diploma
	Biogeography	MPhil
Warwick University, Coventry	Ecosystems analysis and resource management	MSc
West of England University, Bristol	Environmental policy and management	MA, Pg Dip
	Environmental biotechnology	Pg Dip
Wolverhampton University	Environmental science	MSc, Pg Dip, Pg Cert
Worcester University	Environmental issues	MSc, Pg Dip
York University	Environmental economics and environmental management	MSc, Diploma

interaction, territories and food requirements. This usually relates to plant-animal interrelationships and through that to plant species diversity. It does not always follow that high animal diversity is found in areas of primary forest with a high plant diversity. **Box 5.7** is an example.

With the increased interest in sustainable use of living resources, university research into speciation, breeding patterns and ecological niches is being targeted at those groups with potential for food. Fish is a group in which a knowledge of what species are to be found and in what quantity, is very relevant to community needs. Researchers in UK universities are working on marine fish in the North Sea in conjunction with MAFF; and also in parts of Africa where habitats are threatened, populations are rising and all resources are

Box 5.4 Some examples of research relevant to biodiversity at UK universities	
UNIVERSITY AND DEPT/FACULTY	RESEARCH AREAS
Aberdeen University Dept of Biological Sciences	Studies on endangered species of snails in Hawaii. Tropical forest diversity and regeneration in West Africa and SE Asia. Reproductive biology of mediterranean plants. Transmission of waterborne diseases. Biodiversity of soil microbial populations.
Anglia Polytechnic University Biology Dept	Invertebrate studies in Egypt. Cellular slime moulds (Myxomycota) and other fungi.
Bangor: see University of Wales	
Bradford University Dept of Environmental Science	Ecotypic variation in *Capsella bursa-pastoris* (Brassicaceae: Angiospermae). Biodiversity and ecosystems in response to management of upland moorland.
Bristol University Dept of Biological Science	Diversity in genus Bacillus – taxnomic considerations of bacteria. Avian diversity in Kenya and Uganda.
Cambridge University Dept of Plant Sciences	Primate ecology and distribution in Malesia, Africa and S America. Molecular Taxonomy of Tea (*Camellia*).
Cardiff: see University of Wales	
Dundee University Dept of Biological Sciences	Variation in *Rhizobia*. Molecular diversity in *Chamaecrister* (Leguminosae). Pan-tropical Leguminosae. Parasitic hymenoptera Algae. Boreal and pan-tropical chitons in marine communities. Neotropical invertebrates.
Durham University Dept of Biological Sciences	European plants. Insects from Europe and Antarctica. Worldwide algae. Mammals of Latin America and Europe.
East Anglia University School of Biological Sciences	Ecosystem management in Nepal. Morphological and molecular variation in and between species of insects, fishes, trees, coastal plants. Studies on Lagomorphs (rabbits etc). World-wide aphid host preference studies. Speciation and phylogeny of fishes and fish parasites.
Edinburgh University Division of Biological Sciences	Genetics of parasitic protozoa, malarial drug resistance, species differences, HIV-1 and variation in plant pathogenic fungi. Conservation of genetic resources. *Drosophila* evolution. Ecology of mammalian herbivores. Impact of global change. Environmental physiology of plants and vegetation. Atmospheric pollution. Tropical and community foresty. Organisation and management of seed programmes in developing countries.
Essex University Dept of Biology	Studies on European diatoms. Characteristics of sulphate reducing bacteria in saltmarshes. Studies on aphids, soil microorganisms. Studies on European and Antarctic bacteria.
Exeter University Division of Biological Sciences	Helminth communities in Mexico stetichlids and in Irish and Australian fish. Insect hormonal analogues in plants. Ant biology. Ecophysiology of tropical plants. Bryophyte species limits. Isoenzyme variation in apomictic *Sorbus*. *Carex* leaf-surface diversity. Adaptions and polymorphisms within plant populations.
Glasgow University Faculty of Sciences	Molecular systematics in a wide range of organisms. Taxonomy of Angiosperms in Europe, oceanic islands, Argentina, Egypt, India. Bryophyte studies in Azores, Europe. Taxonomic studies of amphibia, corals, marine invertebrates. Prostista biology, including Amoebae.
Greenwich University Dept of Biological and Chemical Sciences	Studies on European Ostracoda (Crustacae).
Keele University Biological Sciences Dept.	Studies on European thrips (Thysanoptera: Insecta).
Kent University Durrell Institute of Conservation Ecology	Study of amphibian diversity and ecology. Sustainable development in National Parks (India, Indonesia, Zimbabwe), *Adansonia* in Madagascar. Application of GIS to biodiversity studies in Tanzania.
Kings College, London	Studies in tropical fish ecology, especially E. Africa.
Lancaster University Institute of Environmental and Biosciences	European vegetation classification and mapping: Arctic flora studies.
Leeds University Dept of Pure and Applied Biology	Biosystematics of Toona spp (Meliaceae: Angiospermae). Molecular phylogenetics of fungi. Lichen species distribution. Determinants of species diversity and boundaries. Genetic variation within species especially in Streptomycetes. Speciation in European and SE Asian grasshoppers, plant-hoppers and *Drosophila*. Fig tree/animal associations.
Leicester University School of Biological Sciences	Plankton and fresh water invertebrates in Africa. Insect pest studies in Europe, Africa and SE Asia. Higher plant studies in Europe and N America.
Liverpool University Dept Environmental and Evolutionary Biology	Systematics of Simulidae (East Africa), Bryozoa, marine and freshwater ciliates, marine Crustacea, fish parasitic flatworms (Europe & E Africa). Aquatic plant-algal interactions, salt-tolerant varieties of angiosperms, evolution to resistance to herbicides and heavy metals. Fish, incl scallop and finfish ecology and marine ecological genetics; seaweeds.
Liverpool John Moore's University School of Biological and Earth Sciences	Molecular and classical taxonomy of Echinodermata, Polychaetae, Nemertines, Homoptera: Psylloidea, Cyclids (Lake Tanganyika). Distribution and ecology of Monochid nematodes. Speciation in Amazon parrots. Conservation action plans for world's Galliforms and Sichua Hill Partridge. Ascomycete sytematics; molecular taxnonomy of *Salix* and conifers.
Luton University Faculty of Applied Science	Botanical studies in Borneo, Hawaii and Europe. Zoological studies in West Indies and Europe.
Manchester University School of Biological Sciences	Evolution of Mollusc fauna of Madeira. Mammal fauna of Ethiopia. Studies of arthropod cuticles. Effects of global change on Arctic ecosystems. Taxonomy of Planarians.

Box 5.4 *continued*	
UNIVERSITY AND DEPT/FACULTY	RESEARCH AREAS
Middlesex University Environmental Science and Engineering	Butterflies, Scarabidae (Coleoptera: insecta) and the ants of Colombia. Aquatic macro-invertebrates of Northern Europe.
Nottingham University Faculty of Environmental Studies	Plant genetic manipulation. Arctic and sub-arctic lichens. Biodiversity generation, conservation using micropropagation and cryopreservation. Insect and plant studies in Egypt. Ecology and diversity in tropical urban ecosystems, and peatlands. Ecology, biogeography and conservation of forest island areas. Helminth and protozoan parasites of rodents and birds in Portugal. Southern Sinai Syphidae biodiversity. Physiology and reproductive behaviour of crickets and cockroaches.
Oxford University Oxford Forestry Institute	Island floras, taxonomy and conservation biology. Molecular biology of tropical trees. Taxonomy, nomenclature and flora of tropical trees, especially Meliaceae. Population and genetic variation. Exploration and evaluation of African accacias. Reproductive biology of tropical trees.
Queen Mary and Westfield College University of London School of Biological Sciences	Algae, fish and invertebrates of Indian Ocean. Studies in Pteridophyta and Bryophytes; European grasshoppers; European and Indo Pacific barnacles. Worldwide Chiroptera. European peat bogs. Chrionomids of Europe and Japan.
Reading University School of Plant Science	Taxnomic investigation of at least 13 groups or families. Systematics of at least 22 groups or families, genetic diversity of at least 17 groups or families. Flora studies of Morocco, Azores, European Garden Flora; bryophtes of Malawi. Conservation of rare plant species. Landscape pattern and predicting biodiversity. Malaysian logging impact. Soil ecology and biodiversity. Biochemical variation in plant compounds. Biological studies on seed storage and gene banks.
Royal Holloway and Bedford New College University of London	Dinoflagellates of the N Atlantic. Taxonomy of water beetles and biology of bruchid beetles; insect-plant interactions. Bird biodiversity and ecology, especially in relation to song. Ecology of shingle banks.
St Andrews University Plant Sciences Dept	Biodiversity in the Arctic and insect biodiversity. Brazilian Cerado Biodiversity.
Stirling Univeristy Institute of Aquaculture	Gene banking of aquatic organisms. Population and molecular genetics of fish and aquatic invertebrates.
Strathclyde University Faculty of Science	Studies in Basidiomycota and Fungi 'Imperfecti': biodiversity and chemistry.
Sussex University School of Biological Science	Amphibia, insects especially butterflies, orchids of Europe. Indian rice and chickpea studies. Vertebrates of Africa. Arctic plants.
University College London	Primate studies, genetics of small mammals. Freshwater microbiology.
University College of North Wales (Bangor) Faculty of Science and Engineering	Ecology of crops, rangelands, birds, tropical forests, grasslands, shore and marine and upland invertebrates. Biology of aquatic microbes, protozoa, and clonal animals. Marine invertebrates and microalgae studies. Studies in European bryophyta. Crop production in semi-arid areas.
University of Wales (Cardiff) College School of Biological Sciences	Taxonomic and ecological studies on nsects in particular Mediterranean, Arabian Gulf and SE Asia. Pest species of insects, especially leaf-hoppers.
University of Wales (Swansea) College School of Biological Sciences	Genetic diversity of marine invertebrates, tropical fishes and aquatic habitats generally. *Limnothrissa miodon* – central East Africa ecology. Studies on Nile Perch of Lake Victoria. Genetic diversity databases.
Westminster University Dept of Biological Sciences	Freshwater fish of Europe. Marine phytoplankton North Sea and N Atlantic. Tropical invertebrate studies.
York University School of Biology (Dept of Applied Biology)	Mammals and invertebrates of Europe, and of N Hemisphere. Invertebrates and marine vertebrates of tropics coral systems. Ecology and distribution of microorganisms and their biodiversity.

dwindling. Other priorities in biodiversity research are endangered habitats or ecosystems, and in particular, coral reefs (**Box 5.8**) and tropical forests (**Box 5.7)**

Other collections emphasise variation within a family or a genus and have been, or may be in the future, the basis for taxonomic research. In the case of plants this could involve the growing of live material, and thus universities have developed botanic gardens as well as museums. It is not surprising therefore, to find that both the first and second botanic gardens to be established in Britain were at Oxford and Edinburgh universities respectively (see Chapter Four). With the increase in research into biodiversity, and especially with the advent of techniques for analysing DNA (see Chapter One) dried specimens of both plants (herbaria) and animals (skins) have an added value, and researchers are looking increasingly to larger, national collections for such taxonomic studies. Universities are developing academic links with the major institutes, e.g. Reading University with RBG Kew and the NHM, London University with the International Mycological Institute and NHM, and Edinburgh with its Royal Botanic Garden. The latter association has a strong historical base when, for nearly two centuries, the Chair of Botany in the University was contiguous with that of the Regius Keeper of the RBG.

There has been a significant increase in biodiversity studies since UNCED and the acceptance of the Biodiversity Convention, as

Box 5.5 Floristic studies in East Greenland: University of Lancaster

Our knowledge of the East Greenland higher plants prior to 1939 is based very largely on the activities of Danish and Norwegian scientific expeditions. Some of these were summer expeditions operating during the brief and intermittent periods of ice-free off-shore waters and always at the whim of the strong east Greenland polar current. Notable among these were the the Denmark Expedition (1911–1912) based on Scoresby Sound (70–71°N), the Thule Expeditions to south-east Greenland (1930–1933) and the three-year expedition (1931–1934) based on Clavering Island (74°N). The botanists on these expeditions confined their summer expeditions almost entirely to the relatively rich ground along the fjords and the outer coast. A notable exception was the overwintering British expedition (1936–1937) led by the geologist L. Wager based on the Kangerlussuaq Fjord (68°N) during which important collections were made from inland nunataks at altitudes of up to 2000m.

With the establishment of air-strips during and immediately after the second world war it became possible for expeditions to utilise the whole of the summer relatively unconstrained by the vagaries of the ice.

South-east Greenland

Most of the post-war expeditions to south-east Greenland were primarily climbing expeditions. They rarely included members with botanical expertise but following prior briefing and armed with a plant press, collections were made from about 100 sites over the range 500–2400m alt. Although individually certainly not comprehensive, the sheer number of collections makes it likely that few species were overlooked. Analysis of the material has provided a remarkably complete set of altitudinal data as well as a detailed picture of the flora of a remote area which otherwise would have remained largely unvisited by professional botanists. One particularly interesting and unsuspected feature is the way several high-arctic species, for example *Melandrium affine*, which are absent from coastal regions south of Scoresby Sound, penetrate far to the south on the high inlands nunataks.

The central fjord region

During the last thirty years there have been at least 20 expeditions primarily of a botanical nature to the central fjord region lying between Scoresby Sound and Dove Bay (77°N). As in SE Greenland the majority were British. Seven were led by Dr Geoffrey Halliday of the University of Leicester (latterly Lancaster) but practically all collected data for incorporation in his projected Flora of the region.This data has been used to produce species distribution maps.

Most expeditions were highly mobile, moving by dingy or foot far from base and penetrating far inland. In this respect they differed markedly from the pre-war boat expeditions which concentrated their efforts on a selected number of coastal sites. The mobility of some expeditions was vastly enhanced by prior air-drops of strategically placed supplies by the UK RAF.

Over the entire period of these investigations close liaison has been maintained with the Greenland Botanical Survey at Copenhagen University and herbarium material is deposited in their Greenland herbarium and in the Arctic herbarium at Lancaster University.

(*Source:* G.Halliday, *Institute of Biological and Environmental Sciences, University of Lancaster.*)

G. Halliday/University of Lancaster

Small, but species rich sedge moss mire above Nordbugt, Northeast Land.

Box 5.6 Moroccan flora project: University of Reading

Morocco has a rich flora estimated to comprise over 5,000 species of which 20% are endemic, many with very restricted distributions. Visible devastation of habitats is occurring through overgrazing and the extensive cutting of timber for local needs. There is no complete Flora for the country and the only checklist is now over 55 years old. To remedy this unfortunate situation, the Commission of the European Communities has sponsored a collaborative research project, involving the Department of Botany of the University of Reading, Spanish botanists in Seville and Barcelona and the Institut Agronomique et Veterinaire Hassan II, Rabat. which aims to provide a synonymic checklist with identification keys, an atlas of plant distribution and a red-data book. The main area covered is that north of the Rabat-Oujda road, together with Jbel Tazzeka and its surrounding National Park, since this has a flora more closely-related to the Rif Mountains to the north than to the rest of the Middle Atlas.

S. Jury/University of Reading

The conifer, *Tetraclinis articulata* near Cap Mazari, Morocco. Picture of hard cones to supplement herbarium material.

Moroccan botanists have delineated twenty floristic regions. Habitats are numerous and include coastal ones, lowland irrigated fields, intensively cultivated and with spectacular weed floras; arid deserts, limestone gorges, oak forests, exposed schistose and granite mountains, and cedar forests.

Although records are being taken from the earlier published works and the previous herbarium collections made by botanists such as Sennen and Font Quer in the old Spanish Morocco, very extensive field work is currently being undertaken throughout the year using modern GPS and portable computer systems.

The team in Rabat have joined in all expeditions and provided Land Rovers, drivers and extra technical support. Use of off-road Government vehicles is enabling easy access to the mountains and freedom of movement in the difficult territories of *Cannabis* cultivation.

Not only have numerous new distribution records and several new species already been found, but the observations made and material collected have allowed several research projects to be undertaken in both Rabat and Reading on economically important genera such as *Rosmarinus*, *Lavandula*, *Thymus* and *Convolvulus*. This collaborative research is also aimed at helping with the development of a herbarium, library and research facility in Rabat together with the training of Moroccan staff and students, both in Morocco and on exchange visits to the University of Reading.

(*Source*: S.L. *Jury, Plant Sciences, Univ. of Reading*.)

new sources of finance for such studies become available. One of the UK Government's contributions to this has been through awards given under the Darwin Initiative (see Section 5.5) which have allowed universities and other organisations to develop and elaborate programmes.

The effect of the House of Lords Report mentioned above (HoL, 1992) has also improved the academic climate in favour of taxonomy and biodiversity studies. One outcome has been the creation of the UK Systematics Forum which is compiling a directory of those working in the UK on the taxonomy of plant, fungi, protista and animals, a database that was launched on the Internet late in 1995. There are compelling reasons for systematists to develop better communications: more than any other group of scientists, the systematics community has a major contribution to make towards improving our knowledge of the species on Earth and the phylogenetic relationships between them

Box 5.7 Mammal biodiversity in Central Africa: Edinburgh University and the International Centre for Medical Research, Franceville, Gabon

A programme to conserve mammal biodiversity in Central Africa by **Edinburgh University** was begun in 1989 as part of a long-term collaboration between the University and the **International Centre for Medical Research, Franceville, Gabon**. Gabon's lowland forest has a high diversity of plants and animals. In the Lope Reserve, the biomass of elephants, primates and ungulates is the highest ever recorded for a tropical forest site. Low intensity logging (1-2 trees/ha) does not seem to affect the population density of elephants or lowland gorillas, but chimpanzee density decreases rapidly and remains low for years. Lee White's work in Gabon could provide a useful model for sustainable logging, because large populations of diverse mammals are coexisting with timber extraction. This work has also suggested that a distinct forest type, which is associated with a stage of savanna recolonisation, is the key to supporting such a large biomass of mammals. It therefore seems that, in order to conserve mammal biodiversity in Central Africa, it may be necessary to conserve large areas of this forest type, known as Marantaceae forest, and hitherto regarded as a product of forest degradation.

(*Source*: E. *Rogers*, *University of Edinburgh*.)

(Blackmore, 1995). An ESF European Network in Systematic Biology has been set up to develop communication links, and this, it is hoped, will promote an increased awareness of the need for taxonomic research, which will itself catalyse the university biodiversity research programmes.

5.3 Societies and other organisations with a biodiversity interest

The greatest strength of learned societies is in the expertise of their membership. The function of most learned societies is to bring together those of like mind and ability into an organisation for mutual benefit and the advancement of that subject. To this end most learned societies have a scattered membership throughout the UK and overseas and arrange conferences and seminars at which members can read papers and debate current issues. The ensuing conference proceedings often constitute the most up-to-date texts on a particular subject. Many societies also publish regular journals and reviews. Larger learned societies may award grants and commission research. Learned societies are often regarded as the 'official organ' of a profession or subject and may be called upon by Government or other official bodies to give expert evidence or offer opinions.

There are learned societies serving all sectors of disciplines, from history, geography and archaeology to cultural institutions, medicine and the natural sciences. In terms of the study of biodiversity there are more than 30 relevant learned societies.

Some societies, such as, The Royal Society (founded in 1660) and the Royal Society of Edinburgh (founded in 1783), serve all sectors of science, not just the biological sciences, but are selective in accepting members, who must be of very high scientific merit. Others are specific in aims and have small memberships but can still have activities and publications of a high standard. The aims, numbers of members, details of publications and secretary, are given for the more important Societies in **Box 5.9**. Those with significant biodiversity programmes are mentioned below.

5.3.1 The multidisciplinary and specialist Learned Societies

5.3.1.1 Linnean Society of London

The Linnean Society of London is the oldest biological society in the UK. Founded in 1788 by Sir James Edward Smith, it is the owner of the Linnean Collection of fish, mollusca and plants (see Box 6.2 in Chapter Six). It promotes studies in biodiversity through its meetings and symposia. The Linnean Society as part of its 200th anniversary celebrations sponsored a multidisciplinary expedition with the Royal Geographical Society (see below) to Kimberley, northern Western Australia. (See **Box 5.10**).

Box 5.8 Reef research in Egypt and the western Indian Ocean

For over a decade, the Tropical Marine Research Unit, at the University of York, has had an ongoing programme on the biodiversity and conservation of coral reefs in the Red Sea. Since about 1990 work has concentrated on the Ras Mohammed National Park, Southern Sinai, Egypt, where research has been undertaken in cooperation with staff of the National Park and with the support of the Egyptian Environmental Affairs Agency. The principle studies have been

a) investigation of the abundance and distribution of reef fish of seven different families (Pomacentridae, Labridae, Chaetodontidae, Pomacanthidae, Lutjanidae, Serranidae and Haemulidae);

b) continuation of a more detailed study of the ecology and behavioural ecology of one specific family, the Chaetodontidae (butterfly fishes) which, because it includes species which are obligate corallivores, have been proposed as indicator species for reef status and health;

c) investigation of the rates of damage to corals by SCUBA divers and of the effects of environmental awareness programmes in reducing these rates of damage;

d) monitoring of reef sites within the Ras Mohammed National Park subject to different degrees of impact by tourism and visitor use; this work has included development of new direct visual and photographic methods for estimating coral cover.

Recently, studies of the ecology and zoogeography of butterfly fishes (Chaetodontidae) have been taken further with support from the Darwin Initiative, in a programme to compare species richness and abundance in the Red Sea and western Indian Ocean. Within the western Indian Ocean species richness increases with decreasing latitude to reach values considerably greater than those in the Red Sea, where the lower species number has been related to the Red Sea's relatively recent geological origin. A high proportion of Red Sea species (about 50%) is endemic to the Red Sea / Gulf of Aden region – by contrast most western Indian Ocean species are widely distributed through the Indo-Pacific region. In the Red Sea, both numbers of species and abundance are greatest in the central area, decreasing both to north and south.

C. auriga – Threadfin

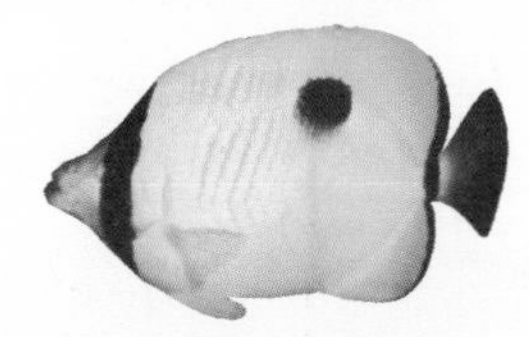

C. unimaculatus – Teardrop

C. falcula – Saddleback

C. madagascariensis – Madagascar

C. fasciatus – Red Sea Racoon

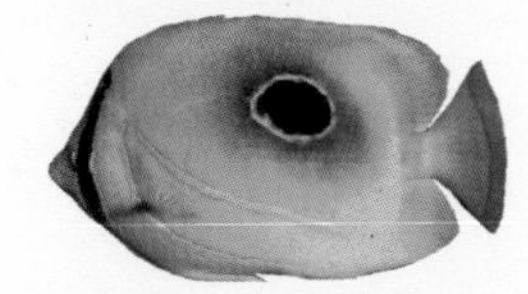

C. bennetti – Bennetts

Selection of Butterfly fish images prepared from computer database to aid identification of the genus *Chaetodontis*.

Programme Director Rupert Ormond,
Tropical Marine Research Unit, University of York.

[Note: please label colour bars by hand to confirm correct use of colour (colours 1, 2 and 3?)]

Box 5.9 Major Learned Societies and other organisations involved in biodiversity assessment based in the UK

Note: International organisations with similar aims with headquarters in Great Britain are placed at the end.

Overseas biodiversity high — Overseas biodiversity medium — Overseas biodiversity low

SOCIETY	ADDRESS	MEMBERSHIP	SCIENTIFIC JOURNAL	NEWSLETTER/ BULLETIN
Amateur Entomologists' Society	5 Oakfield Plaistow Billinghurst RH14 0QD	2,050		Bulletin
Biology Curators Group	Steve Thompson Scunthorpe Museum Oswald Road Scunthorpe DN15 7BD	c. 300	The Biology Curator	
Botanical Society of the British Isles	Mrs Mary Briggs 9 Arun Prospect Pulborough West Sussex RH20 1AL	2,700	Watsonia	BSBI News
Botanical Society of Scotland	c/o Royal Botanic Garden Edinburgh EH3 5LR	570	Botanical Journal of Scotland	BSS News
British Arachnological Society	MJ Roberts Burns Farm Cornhill AB45 2DL	600	Bulletin of the British Arachnological Society	Newsletter
British Bryological Society	Dr ME Newton c/o Dept of Botany Liverpool Museum William Brown Street Liverpool L3 8EN	585	Journal of Bryology	Bulletin of BBS
British Chelonian Group	Mrs Diana Desmond c/o Dr R Avery School of Biological Sciences University of Bristol Bristol BS8 1UG	1,600	Testudo	BCG Newsletter
British Dragonfly Society	Jill Silsby Hadun Avenue Purley Surrey CR8 4AG	1,300	Journal British Dragonfly Society	Newsletter
British Ecological Society	Dr Hazel Norman 26 Blades Court Putney London SW15 2NU	4,700	Journal of Ecology; Journal of Animal Ecology; Journal of Applied Ecology; Functional Ecology	Bulletin
British Entomological & Natural History Society	30 Penton Road Staines Middlesex TW18 2LD	700	Proceedings & Transactions	
British Herpetological Society	c/o Zoological Society of London Regents' Park London NW1 4RY	1,050	Herpetological Journal	Bulletin
British Lichen Society	c/o Department of Botany The Natural History Museum Cromwell Road London SW7 5BD	1,342	The Lichenologist	Bulletin
British Mycological Society	Dr ST Moss School of Biological Sciences University of Portsmouth King Henry I Street Portsmouth PO1 2DY	2,150	The Mycologist; Mycological Research	Newsletter
British Ornithologists Union	c/o Natural History Museum Lakeman Street Tring HP23 6AP	1,800	The Ibis	
British Phycological Society	c/o Department of Botany The Natural History Museum Cromwell Road London SW7 5BD	620	European Journal of Phycology	Newsletter
British Pteridological Society	c/o Department of Botany The Natural History Museum Cromwell Road London SW7 5BD	760	Fern Gazette; The Pteridologist	Bulletin
British Trust for Ornithology	The Nunnery Thetford Norfolk IP24 2PU	10,000	Bird Study	BTO News

Box 5.9 *continued*

SOCIETY	ADDRESS	MEMBERSHIP	SCIENTIFIC JOURNAL	NEWSLETTER/ BULLETIN
Conchological Society of Great Britain & Ireland	School of Geography & Geology Cheltenham & Gloucester College Shaftesbury Hall St George's Place Cheltenham GL52 3AP	780	Journal of Conchology	Newsletter
Federation of Zoological Gardens of Gt. Britain and Ireland	Zoological Gardens Regents Park London NW1 4RY	60 corporate members		Zoo Federation News
Freshwater Biological Association	The Ferry House Ambleside Cumbria LA22 0LP	2,000	Scientific Studies	Freshwater Forum
Genetical Society	Prof JPW Young Dept of Biology University of York York YO1 5DD	1,600	Heredity	Genes & Development
Henry Doubleday Research Association	Ryton Organic Garden Ryton-on-Dunsmore Coventry CV8 3LG	17,000		Leaflet; Vegetable Finder
Linnean Society of London	Burlington House Piccadilly London W1V 0LQ	2,400	Botanical Journal of the Linnean Society; Biological Journal of the Linnean Society	The Linnean
Malacological Society of London	c/o Dr GBJ Dussart Canterbury Christ Church College North Holmes Road Canterbury CT1 1QU	300	Journal of Molluscan Studies	
Mammal Society	Unit 15, Cloisters House Cloisters Business Centre 8 Battersea Park Road London SW8 4BC	1,500	Mammal Review	Mammal News
Marine Biological Association of the UK	The Laboratory Citadel Hill Plymouth PL1 2PB	1,350	Journal of the Marine Biological Association	MBA News
National Council for the Conservation of Plants and Gardens	The Pines Wisley Garden Woking, Surrey GU23 6QB	8,500	Plant Heritage	Newsletter from local groups
North of England Zoological Society	Chester Zoo Upton-by-Chester CH2 1LH	5,000		Chester Zoo Life
Plantlife	c/o The Natural History Museum Cromwell Road London SW7 5BD	6,000		Plantlife
PlantNet	c/o Timothy Walker University of Oxford Botanic Garden Rose Lane, Oxford OX1 4AX	34 corporate members		Newsletter
Primate Society of Great Britain	c/o Institute of Biology 20–22 Queensbury Place London SW7 2DZ	570		Primate Eye
Rare Breeds Survival Trust	National Agriculture Centre Stoneleigh Warwickshire CV8 2LG	10,000-		The Ark
Royal Entomological Society	41 Queens Gate London SW7 5HR	2,025	The Entomologist; Ecological Entomology; Systematic Entomology	Newsletter
Royal Geographical Society	1 Kensington Gore London SW7 2AR	13,000	The Geographical Journal The Geographical Magazine	News
Royal Horticultural Society	80 Vincent Square London SW1P 2PE	170,000	New Plantsman	
Royal Scottish Geographical Society	Graham Hills Building 40 George Street Glasgow G1 1QE	2,100	Scottish Geographical Magazine	Geogscot
Royal Society	6 Carlton House Terrace London SW1Y 5AG	1,200	Philosophical Transactions of the Royal Society; Proceedings of the Royal Society	Bulletin of the Royal Society; Notes & Records
Royal Society of Edinburgh	22, 24 George Street Edinburgh EH2 2PQ	1,070	Proceedings of the Royal Society of Edinburgh; Transactions of the Royal Society of Edinburgh	

Box 5.9 continued

SOCIETY	ADDRESS	MEMBERSHIP	SCIENTIFIC JOURNAL	NEWSLETTER/ BULLETIN
Royal Society for the Protection of Birds	The Lodge Sandy, Bedfordshire SG19 2DL	885,000	Birds	Bird Life
Royal Zoological Society of Scotland	Scottish National Zoological Park Edinburgh EH12 6TS	13,000	Annual Report	Arkfile
Scientific Exploration Society	Expedition Base Motcombe, Nr Shaftsbury Dorset SP7 9PB	480		Sesame
Scottish Association for Marine Science	PO Box 3 Oban, Argyll PA34 4AD	420	Annual Report	Newsletter
Scottish Ornithologists Club	21 Regent Terrace Edinburgh EH7 5BT	2,700	Scottish Birds	Scottish Bird News
Society for General Microbiology	Marlborough House Basingstoke Road Spencers Wood Reading RG7 1AE	5,000	Microbiology; Journal of General Virology	
Society for Environmental Exploration	77 Leonard Street London EC2A 4QS	600	[Project Reprints]	
Systematics Association	GP Larwood Department of Geological Sciences University of Durham	600	Symposia Reports	
United Kingdom Federation for Culture Collections	c/o Dept. Biological Sciences Heriot Watt University Edinburgh EH14 4AS	11 corporate plus 33 sustaining members		Newsletter
Zoological Society of Glasgow and West of Scotland	Calderpark Zoological Gardens Uddington, Glasgow G71 7RZ	500		Zoolife
Zoological Society of London	Regent's Park London NW1 4RY	20,000	Journal of Zoology; International Zoo Year Book	
International organisations based in the United Kingdom				
Association of South-East Asian Studies	c/o Centre for SE Asian Studies The University of Hull Hull HU6 7RX	>120		ASEASUK News
BirdLife International	Wellbrook Court Girton Road Cambridge CB3 0NA	na		World Birdwatch
Botanic Gardens Conservation International	Descanso House 199 Kew Road Richmond, Surrey TW9 3BW	400	Botanic Gardens Conservation News	Roots
Coral Cay Conservation	154 Clapham Park Road London SW4 7DE	na		Newsletter
Earthwatch Europe	Belsyre Court 57 Woodstock Road Oxford OX2 6HU	6,000		Earthwatch Magazine
Fauna and Flora International	Great Eastern House Tenison Road Cambridge CB1 2DT	5,000	Oryx	Fauna and Flora News
Jersey Wildlife Preservation Trust	Les Augres Manor Trinity, Jersey Channel Islands JE3 5BF	10,500	Dodo	On the Edge
Raleigh International	27 Parsons Green Lane London SW6 4HZ	na		Field Research News
Tropical Biological Association	c/o School of Biological Sciences University of Bristol Woodland Road Bristol BS8 1UG	na		Newsletter
Wetlands International	Slimbridge Gloucestershire GL2 7BX	47 member countries		IWRB News
World Pheasant Association	PO Box 5 Lower Basildon Reading RG8 9FP	2,000	[Project Reprints]	
World Wide Land Conservation Trust	PO Box 99 Saxmundham, Suffolk IP17 2LB	na		WWLCT News

Box 5.10 Kimberley Research Project, Western Australia 1988

M.J.S. Sands/RBG, Kew

Napier Range near Windjana Gorge, an exhumed, deeply disected Devonian reef, with tree savannah in foreground.

M.J.S. Sands/RBG, Kew

Entomologists collecting in riverine vegetation, in Bell Gorge.

In northern Western Australia lies the Kimberley region, a wilderness area three times the size of England. The landscape is weathered and rugged, formed of a vast upland of ancient rocks bordered on the south western side by an exhumed Devonian limestone reef stretching for hundreds of kilometres. Lying within the tropics, the Kimberley region experiences high temperatures and very marked wet and dry seasons, with vegetation ranging from Pindan woodland and tree savannah to vine thickets and riverine forest.

Working in this impressive region, a major multidisciplinary project, **The Kimberley Research Project**, ran for 5 months in 1988, involving some forty Australian and British scientists. It was jointly organised by the **Royal Geographical Society** and the **Linnean Society of London** with the cooperation of the Western Australian Government, and timed to take place in Australia's Bicentennial year.

The field programmes began in March, towards the end of the wet season, and continued until the end of July. Throughout this period, while a team of ten geomorphologists conducted a varied programme of research, the life sciences team undertook a wide range of biological and ecological studies involving extensive collecting of floral and faunal specimens.

Botanical projects included specific work on water plants, grasses, Tiliaceae and the distinctive Boab tree, *Adansonia gregorii*, a characteristic feature of the Kimberley. Several thousands of flowering plant and fern specimens and, for the first time in this region, hundreds of fungi and lichen specimens were collected. Bush foods, including both plants and animals, were studied in collaboration with the local aboriginal Bunaba people of the Junjuwa community and a survey of rare and uncommon mammals was conducted. Other **zoological projects** included extensive insect collecting associated with a specialised study of bees as well as studies of the biology and territorial behaviour of dragonflies and damselflies, aquatic faunas especially crustaceans associated with permanent springs, snail population studies and the acoustic and foraging behaviour of several bat species. The Kimberley Project has yielded many new records for the region as well as several new taxa and research is still continuing based on data and specimens collected during the field period.

In several instances the biologists had a direct and important interface with the work of the geomorphological team. For example, research was undertaken on the possible contribution of mosses in the formation of tufa, a survey of the vegetation occurring on 'gilgai' (a distinctive micro-relief feature) was conducted and consideration was given to the influence of the region's geomorphological history on snail distribution and morphology in the limestone ranges. The biological programme was notable for the important and extensive involvement of several cryptogamic botanists, and of particular significance in a region having received little or no previous attention from specialists in any cryptogamic discipline.

(*Source*: *M.J.S. Sands, Project Deputy Leader, Royal Botanic Gardens, Kew.*)

Box 5.11 Project Wallace: a study of primary rain forest organised by the Royal Entomological Society

Project Wallace was launched to celebrate the 150th anniversary of the Royal Entomological Society of London and was based in the primary rain forest of the catchment of the Dumoga and Bone rivers in North Sulawesi. The area had been recently designated a National Park by the Indonesian Government, and the programme was developed with the full cooperation of their Institute of Sciences (LIPI). Over 100 international based scientists were involved in the following five programmes extending over a 12 month period, 1 Jan – 31 Dec, 1985.

1. Agricultural entomology: to investigate and evaluate the deleterious effects of insect pests on agricultural and forestry production in the Dumoga Valley and to publish guides to the identification of such pests for local use.

2 Medical entomology: to investigate and evaluate the deleterious effects of insects as vectors of disease to man and domestic animals in the Dumoga Valley.

3. Forest regeneration: to establish to what extent insects modify the regeneration of the tropical rain forest and to estimate their value as indicators of successive stages in this regeneration.

4 Insect diversity and conservation: to prepare an inventory of the insects present in the Dumoga-Bone National Park and to carry out studies on the rain forest ecosystem for comparison with other parts of Sulawesi and S.E. Asia as a basis for efficient site management.

5. Geophysical survey: to undertake such geophysical surveying and weather recording in the Dumoga-Bone National Park as may be required for purposes of assisting site management.

J. H. Martin/NHM

Visiting scientists collecting agricultural pests near Dumoga Bone National Park Field Station, Indonesia.

The programme was designed to provide field training for 50 Indonesian students in entomology and for some, a period of study in UK and elsewhere. Well over 100 papers have been published and a symposium held in 1988 brought together much research of significance (see Knight & Holloway, 1990). Collections of insects made will form a substantial addition to the reference collection in the Bogor Zoolological Museum for future reference by Indonesian entomologists. The Base Camp is now a permanent field station for use by Indonesian and international researchers.

(*Source*: W.J. *Knight* & J.D. Holloway, *Natural History Museum & International Institute of* Entomology, London)

Box 5.12 International field research projects mounted by the RGS 1967–1997

ECOSYSTEM	LOCATION	YEARS	NO. OF PARTICIPANTS
Tropical forests	Mato Grosso, Brazil	1967–69	69
	Gunong Mulu National Park, Malaysia	1977–78	145
	Maraca Island, Brazil	1987–88	202
	Batu Apoi Forest Reserve, Brunei	1991–92	121
Savanna	Kora National Reserve, Kenya	1983	52
	Mkomaizi Game Reserve, Tanzania	1993–97	50+
Mountains	Karakoram, Pakistan	1980	73
	Middle Hills, Nepal	1991–94	12
Drylands	South Turkana, Kenya	1968–70	34
	Wahiba Sands, Oman	1985–86	40
	Kimberley, Australia	1988	35
	Badia, Jordan	1992–97	60+

Box 5.13 Royal Geographical Society's field research programmes

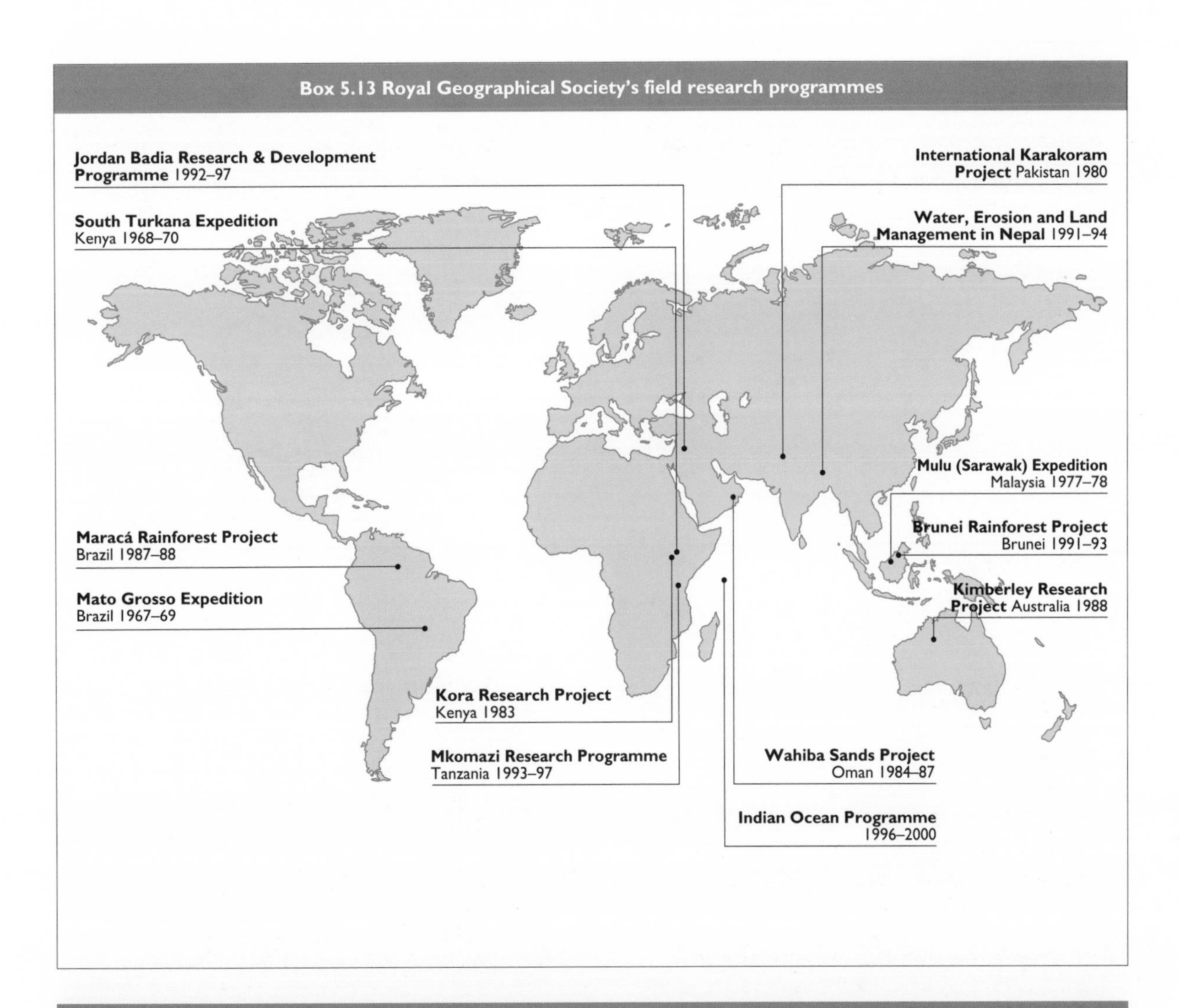

Box 5.14 Mkomazi Research Programme 1993–1997 a joint project between the Royal Geographical Society, University of Oxford and the Tanzanian Department of Wildlife.

The Mkomazi Game Reserve in north east Tanzania is an area of particular biological interest as it contains a high level of local physical and biotic diversity. The Reserve lies on the Kenya border adjacent to the Tsavo West National Park, between Kilimanjaro and the coast, covering an area of 3,600km^2 (3°50'–4°25'S and 37°35'–38°45'E).

Altitude varies between 630m and 1,594m with the land sloping in a northern and easterly direction. The surface is broken by a series of hills, which largely represent outliers of the Pare and Usambara massifs. These have a profound effect on the local climate in terms of the amount of rainfall (or lack of it), and strong seasonal winds. The annual rainfall of 300–900mm is split fairly evenly between the 'long rains' (March to mid-May) and the 'short rains' (late October to December).

The Reserve provides a wet season refuge for many of the larger herbivorous mammals from Tsavo West National Park. However, the higher land near the Reserve boundaries to the south and west has, unlike many other semi-arid savannah conservation areas, a large human population. The Pare and Sambara peoples are agriculturalists who have traditionally farmed the slopes of the surrounding mountains.

A. Jackson/RGS

Box 5.14 *continued*

The major habitat types are largely determined by altitude, rainfall and soil type. Over 70% of the total land area is occupied by a range of habitats broadly classified as *Acacia-Commiphora* scrub or bush. Travelling from west to east within the Reserve it is clear that the percentage of *Acacia* canopy cover decreases from figures as high as 80% to little more than 20%. This is attributed to soil type, decreasing altitude and rainfall, but may also be a result of the pressure from increasing human population and their livestock living around the periphery and encroaching into the Reserve.

A further cause of change is the recurrence of fire, either generated by pastoralists, agriculturalists or naturally. From present studies it is clear that the effect of fire has been severe. In many areas, where the natural habitat appears to be open grassland, the presence of scattered mature *Commiphora* trees indicate an original habitat of fairly dense *Acacia-Commiphora* woodland, as still exists to the east of the Ndea Hills.

The result of burning is a progressive reduction in tree cover, eventually resulting in loss of the upper soil 'seed bank' and establishment of open grassland.

The Programme headquarters is at Ibaya Camp, comprising a series of huts providing accommodation for up to 25 people.

A further and important facility to be built at the Ibaya camp is a dedicated Conservation Centre, for use by visitors and scientists alike. This will provide a laboratory and a meeting room, ensuring the camp and the Reserve remain a focus for research and an invaluable source of local environmental information.

> *"This three year study is more than just producing a detailed inventory of animals and plants. It is also about bridging a geographical gap between the conservators of nature on the one hand and the needs of the guardians of the regions on the other. To do this we need a new generation of Tanzanians committed to finding these solutions and I believe there is a plan to involve a good mix of young and old field scientists."*
>
> Hon. Juma Hamad Omari, *Minister for Tourism, Natural Resources and Environment, Dar es Salaam.* 28th July 1994

5.3.1.2 Royal Entomological Society

A leading Society in entomology of international status, presently with no major biodiversity programme. In 1984–85 it organised a major collecting expedition to Dumoga-Bone in northern Sulawesi. See **Box 5.11**.

5.3.1.3 Royal Geographical Society

The RGS was founded in 1830 to promote geography. Throughout its existence, the Society has encouraged exploration and supported scientific expeditions to all parts of the globe. In planning such projects, it is currently RGS policy to bring together a mix of earth, life and social science disciplines to provide accurate information about a region, ecosystem or habitat and the processes taking place there. Such interdisciplinary information, shared with decision-makers, contributes to planning for sustainable development and re-inforces the integration of policies for land-use and beneficial conservation of natural resources.

Overseas research programmes. In the past three decades, the RGS has mounted 13 major international field research projects (expeditions) (see **Boxes 5.12, 5.13 & 5.14**) in partnership with host-country government and research institutes. These multidisciplinary teams involved over 700 scientists, many of them

Box 5.15 The RGS Expedition Advisory Centre

The EAC provides advice, information and training to anyone embarking on fieldwork overseas. It is the leading such centre in the world. Each year it assists more than 500 teams, the majority of which are university-based undergraduates or recent graduates. Many of these young scientists are inspired by the annual *Planning a Small Expedition* Seminar, run by the EAC, or at specialist workshops organised around the country.

The EAC publishes a range of training manuals on every aspect of field research and expedition logistics. The Expedition Field Techniques series, is a low cost suite of publications aimed at encouraging undergraduates to adopted appropriate methodologies. Titles currently include: *Ethno-biology, Ecotourism, People-Oriented Research , Simple Surveying, Fishes, Reptiles & Amphibians, Small Mammals (excluding bats)* and are often published to coincide with workshops on similar topics. During 1995, these workshops also included bird and habitat survey techniques, and wildlife sound-recording.

Contact: Shane Winser, Information Officer, Royal Geographical Society

Box 5.16 Location of RGS supported expeditions: Europe, the Arctic and North America, 1995

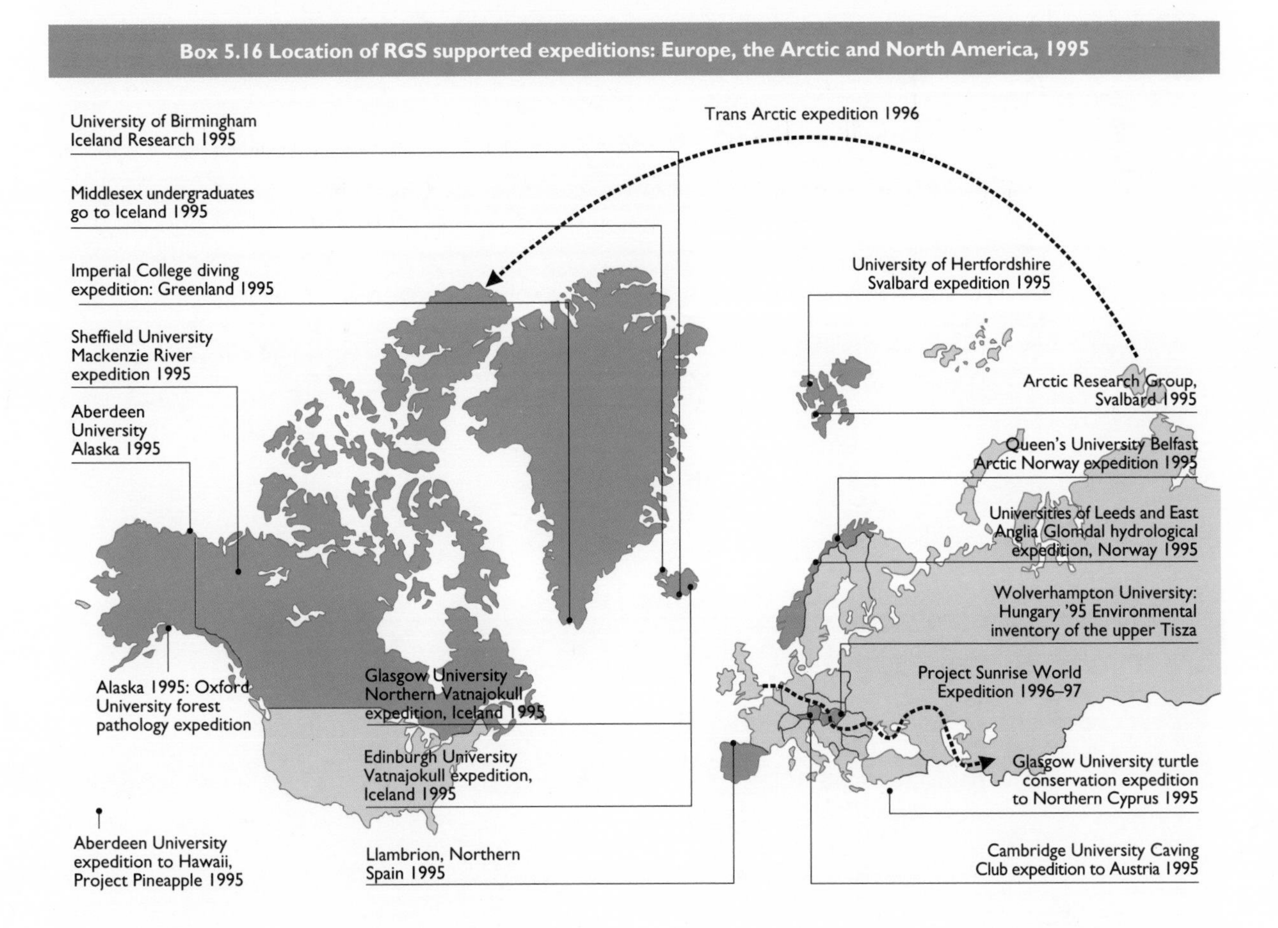

collecting data relevant to land-use change and biodiversity studies in a wide variety of habitats.

Expertise. Through the organisation of these research programmes the RGS has built up expertise for establishing field centres for research in remote, little-disturbed ecosystems. It is the policy of the Society to secure geographical data from a wide variety of sources, incorporating traditional knowledge held by local communities on the one hand and making full use of new computer and satellite technologies on the other, thus creating a multi-disciplinary approach to conservation and development priorities.

The RGS only works at the invitation of host governments and in the closest possible co-

Box 5.17 Location of RGS supported expeditions: Africa, 1995

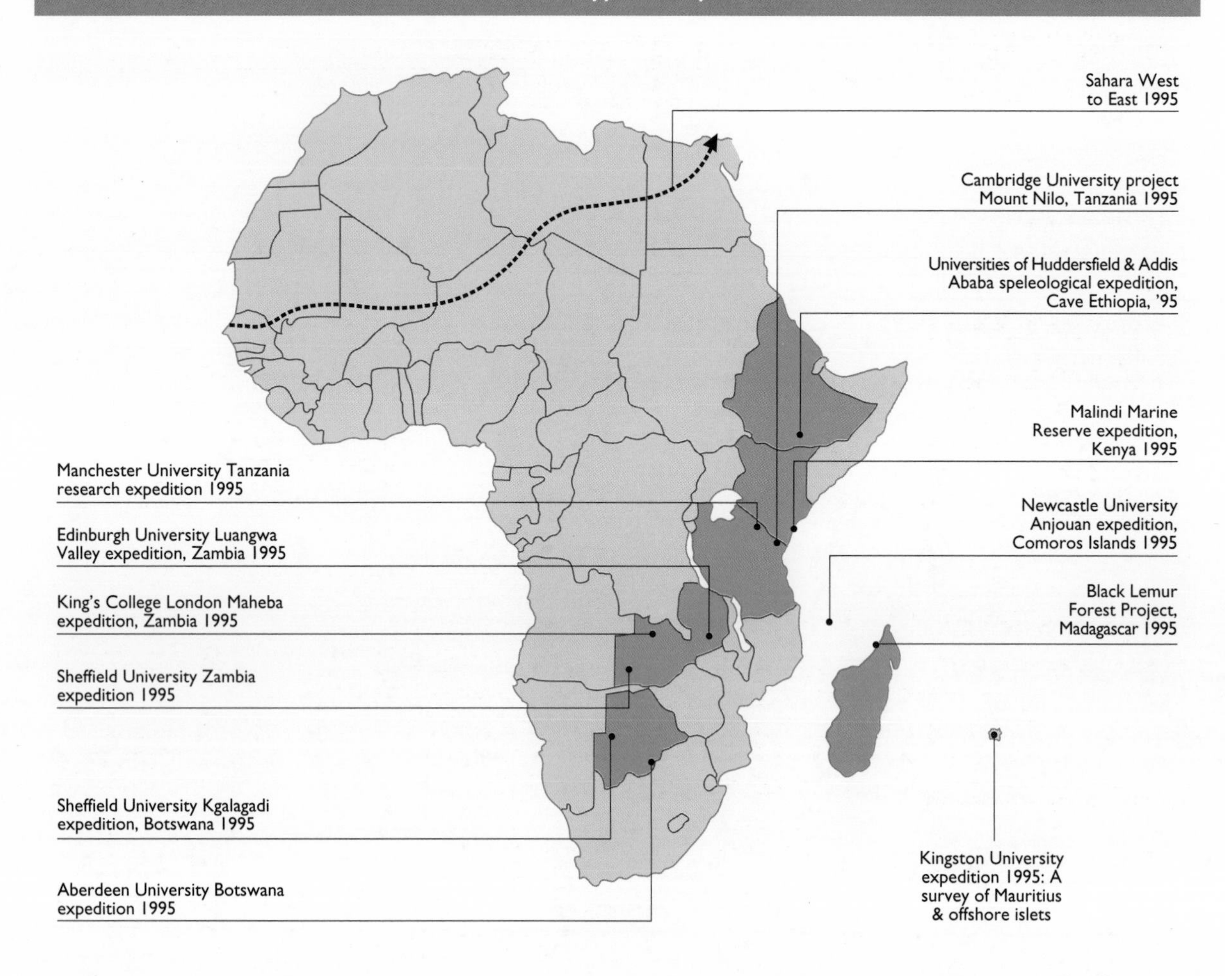

operation with local scientists and educational establishments. Findings are presented first in the host country, and as a result the RGS has become respected as an apolitical research and discussion forum. Results of these programmes reach a wide audience through, books, teacher resource packs, magazines, television films and national lecture programmes.

Education and training. In recent years, the RGS major research programmes have all included NERC-funded postgraduates, research students, and scientists from host country institutes working alongside UK researchers, thus providing valuable field experience and transfer of technical and local knowledge. The main training arm of the RGS is the Expedition Advisory Centre (**Box 5.15**).

In addition to help and advice, the RGS awards grants to research projects overseas. Following a rigourous screening procedure, approximately 70 groups are grant-aided each year from the Society's own funds and from other donations to the RGS from both charitable and commercial sources. In 1995, over £40,000 was awarded in grants to expeditions, the great majority of which are undergraduate groups doing their first fieldwork overseas in association with local students and local NGOs. **Boxes 5.16–19** show the range and number of these supported expeditions.

To increase its commitment to co-ordinated global research, the Society recently established a 15-month consultancy to determine the role of the RGS in long-term research and monitoring. To complement current work by climatologists and oceanographers, the Global Observatories Programme (GOP), (see Chapter Three, Section 3.5.1 and Box 3.10) has already been in discussed with both the International

Box 5.18 Location of RGS supported expeditions: Australasia and Asia, 1995

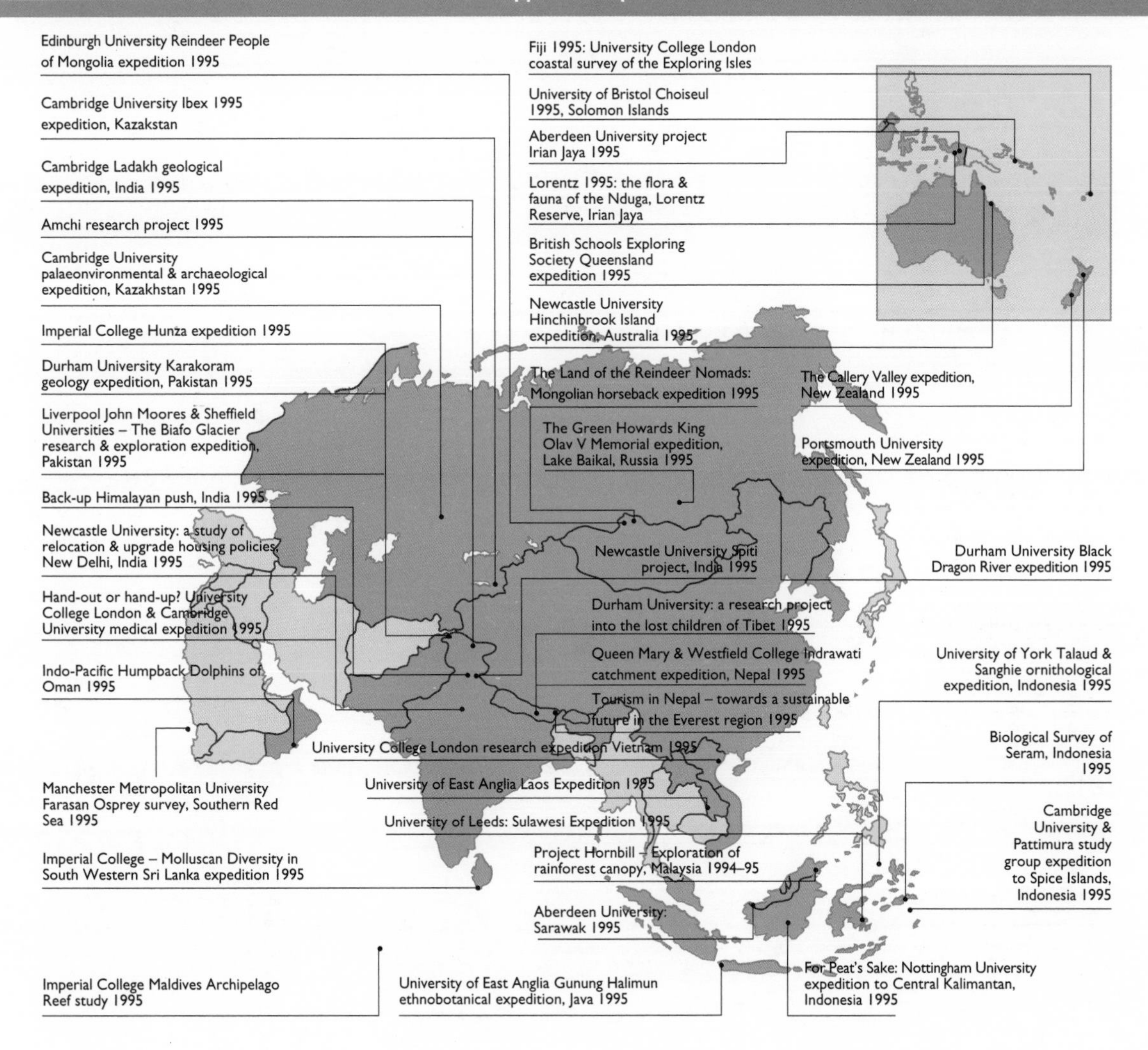

Council for Scientific Unions and the United Nations networks concerned with global change, especially the Global Terrestrial Observing System (GTOS) group. It is proposed that the RGS, in conjunction with a number of partners in the European Union, will establish a small number of permanent 'field universities' in key biomes of the world for long-term research, monitoring, education and training.

5.3.1.4 Royal Society

The Royal Society is an independent academy promoting the natural and applied sciences, founded in 1660 and incorporated by Royal Charters. Among its many activities to promote and advance science, the Society makes grants to individual scientists to support them in their research and provides funds to enable UK-based scientists to visit other laboratories or to host overseas co-workers in their own laboratories. The Society participates in two major programmes which contribute to biodiversity assessment, one on Lake Baikal, Siberia, and one in Danum Valley in Sabah. (See Box 4.3 in Chapter Four and **5.20** below).

5.3.1.5 Royal Society of Edinburgh

The Royal Society of Edinburgh was established in 1783 under a Royal Charter granted by George III for the "Advancement of Learning and Useful Knowledge"; it now represents all branches of learning. The principal function of the society is the promotion of research and scholarship, through meetings and symposia, the publication of learned journals, and the award of research fellowships, scholarships and prizes. The society provides an important forum for interdisciplinary activity and this provides a balance to the need for specialisation. The

Box 5.19 Location of RGS supported expeditions: Central and South America, 1995

Society also administers various bequests and research fellowship schemes for post doctoral research. These are designed to promote research and scholarship at various centres of learning in Scotland.

5.3.1.6 Royal Zoological Society of Edinburgh

The Society promotes, through the presentation of the Society's living collections, the conservation of animal species and wild places by captive breeding, environmental education and scientific research.

5.3.1.7 Systematics Association

The SA was founded in 1937 to study systematics in relation to biology and evolution. It regularly holds Symposia on taxonomy and systematics and publishes the proceedings as a significant contribution to biodiversity studies.

5.3.1.8 Zoological Society of London

The Zoological Society of London is a charitable trust supported by grants, visitors to the zoos and by donations. The mission of the Society is to conserve animal species and their habitats by conducting scientific research, by caring for and breeding endangered species, by fostering public interest and by participating in conservation action world-wide. See Zuckerman (1976) for an account of the history of the Society. The Society has five divisions: the London Zoo, Whipsnade Wild Animal Park, the Institute of Zoology (see **Box 5.21**), the Conservation and Consultancy Division and the Learned Society.

A number of projects at the Institute of Zoology are undertaken in collaboration with other divisions of the Society, especially where this involves captive breeding of threatened species (e.g *Partula*) within one of the two zoological collections, and reintroduction programmes by the Conservation and Consultancy Division.

Box 5.20 Biodiversity research within the SE Asia Rainforest Research Programme of the Royal Society at Danum Valley Field Centre, Sabah, Malaysia.

The S.E. Asia Rainforest Research Programme was set up by the Royal Society of the UK 10 years ago under the title 'Patterns and Processes of Regeneration in Tropical Forest'. As from March 1995 this Programme was succeeded by one entitled 'Restoration of Rainforest' which will run until the end of 1999. The main site for fieldwork is Danum Valley Field Centre, situated at the boundary of a 38K ha area of undisturbed lowland dipterocarp forest and an area ten times as large of selectively logged forest. DVFC is situated in south-eastern Sabah (the north-eastern part of Borneo) 70km inland from the town of Lahad Datu. Some research is done elsewhere, notably on the islands of Krakatau in Indonesia.

Biodiversity is not the principal aim of either of the existing or future Programmes, which encompass everything from meteorology through hydrology to patterns of tree mortality, gap dynamics, fruit and seed dispersal, seedling physiology and the responses of birds, mammals and insects to forest disturbance. However it is a measure of the fundamental place of biodiversity knowledge in environmental studies that much biodiversity work has been done, and is planned in the future.

Even the hydrologists become involved. Major erosion events are often associated with the breaching of temporary natural dams made up of logs. These dams are stable until the logs decay. The decay is partly caused by insects. What are the insects and how do they operate? Likewise a study of organisms which break down leaf litter is necessary to understand the bottlenecks limiting the release of nutrients to sustain regeneration of seedlings. Tracking tree growth and mortality in permanent plots (now nine years old) has involved identifying to species every tree and sapling in the plots. Studies of the role of birds and mammals in fruit dispersal has led to an inventory of fruits, seeds and associated trees and vines. Current work on

S. Sutton/Royal Society

The Royal Society Field Centre at Danum Valley, Sabah.

differences in populations of moths and butterflies in primary and selectively logged forest has to be done at the species level (even if the species cannot in many cases be formally named) and is leading to the discovery of new species and the extinction of the known range of many. Comprehensive studies on the ecology of dung beetles have explained diversity and distribution in terms of niche requirements and highlighted a class of aboreal dung scavengers.

In all these studies knowledge of biodiversity is an essential tool to research, but not an end in itself, given the terms of reference of the SE Asia Rainforest Research Programme.

However, DVFC is by no means solely the preserve of the Royal Society, and a number of other vigorous programmes are under way there. One of these aims to make an inventory of insect biodiversity, with particular reference to ants. This is run by the University Kebangsaan Malaysia (Sabah campus). With access to the upper canopy at a number of points within the reach of the Field Centre, plans are afoot for a long term study of the faunal differences between primary and logged-over sites at the 40–70m level.

The Forestry Research Centre, Sabah Forest Department, maintains extensive collections of plants and insects from the Danum area, which are being actively curated and identified.

(Source: Stephen Sutton, Danum Research Programme Coordinator, The Royal Society.)

Box 5.21 The Institute of Zoology: the Research Division of the Zoological Society of London

The Institute of Zoology is a research institute with over 80 staff, including about 30 post-doctoral staff. Research is focused on topics that benefit the conservation of animal species. These include studies of behaviour, population ecology, conservation genetics, reproductive biology and veterinary science. Emphasis is placed upon the development of scientific principles that can be applied practically to both wild and captive populations in natural and restored habitats.

Projects in ecology include work on the factors responsible for declining populations, especially disease and interspecific competition. Studies of behaviour focus on social signalling and its potential role in overcoming problems of behavioural incompatibility in captive breeding programmes, mating systems and the evolution of sociality. Work in conservation genetics includes the development of molecular techniques to identify taxanomic units for conservation in species ranging from marine mammals to primates and reptiles (see also Chapter One, Box 1.7), and the identification of hybridisation where this threatens native species.

Veterinary projects include clinical investigation of diseases in rare species held in captive breeding programmes and research on the role of disease and pollutants in marine mammal mortality. Studies of reproduction concentrate on the development of assisted reproduction techniques for enhancing breeding and genetic management of threatened species, and the use of non-inevasive tests for reproductive status of larger mammals. Staff are also involved in the development of systems for assessing conservation priorities. At the species level, and in collaboration with IUCN and others, this work has led to the development of internationally threatened species categories. Other studies explore approaches to area conservation priorities based on the analysis of patterns of endemic congruence and tests of various measures of biodiversity.

(*Source*: L.M. *Gosling, Institute of Zoology*.)

Teaching and training by Institute staff is carried out through collaborations with undergraduate and graduate courses at the University of London and other universities. There is also an extensive programme of postgraduate training with over 20 PhD students on the staff, as well as 12 attending an MSc course on Wild Animal Health run in collaboration with the Royal Veterinary College.

The Conservation and Consultancy Division concentrates on the development and implementation of practical conservation in the field. The approach is multidisciplinary incorporating skills from all Divisions of the Society. Particular areas of specialisation are East and Southern Africa, Arabia and the Middle East.

5.3.2 National specialist botanical Societies

The following organisations comprise a total membership of over some 10,000 people, both amateurs and professionals. Their primary aims are to stimulate and promote interest in all branches of their group although activities are predominantly orientated around the flora of the British Isles. Field meetings may include Europe from time to time. In most cases their publications cover their specialist subject on a worldwide basis. In the case of the British Bryological Society there is a positive programme to investigate floras abroad and there have been expeditions to Africa. There is an educational, if not specific training, element in both activities and publications. Their range of interest is as follows: ***Botanical Society of the British Isles*** (flowering plants, vascular cryptogams and charophytes); ***Botanical Society of Scotland*** (all plants and fungi); ***British Bryological Society*** (bryophyta); ***British Lichen Society*** (lichenised fungi); ***British Mycological Society*** (fungi); ***British Pteridological Society*** (pteridophyta); ***British Phycological Society*** (freshwater and marine algae). A further, non-specialist society, ***Plantlife,*** is a recently formed UK plant conservation organisation with a remit to promote the conservation of biodiversity in plants in the UK and Europe by protecting their habitats and by active management.

5.3.3 Botanical and zoological garden organisations

5.3.3.1 Botanic Gardens Conservation International

Set up in 1987 by the World Conservation Union (IUCN) as their Botanic Gardens Conservation Secretariat but now an independent UK-registered charity with over 400 member institutions in 91 countries. BGCI provides technical guidance, data and support for botanic gardens worldwide. It has organized training courses in China, Colombia and Mexico (see **Box 5.27**). A computer database on the rare plants in over 300 institutions is maintained to bring worldwide coordination to the individual efforts of each garden. (See Chapter Six for the exotic material grown in UK gardens). BGCI, through its various activities, has a specific role within the aims of the Biodiversity Convention.

5.3.3.2 Henry Doubleday Research Association

The HDRA conserves crop biodiversity by making varieties of vegetables available to those gardeners who wish to grow their own through its Heritage Seed Programme (which involves nearly 5000 members), currently containing some 700 varieties that are not registered on the UK *National List* (see below). These are maintained at the Association's gardens and through a network of volunteers (members of HDRA) who bulk up selected varieties. Members are encouraged to try new varieties and save their own seed. The Association thus seeks to promote conservation through utilisation; in 1995 it distributed some 20,000 packets of seed.

In the UK, the modern varieties that are registered for sale on the National List have been developed with the needs of intensive commercial operations in mind. Older and unregistered varieties have qualities that appeal to low input systems and home gardeners, but existing legislation forbids their sale. Stocks are thus maintained for future plant breeders (see Chapter Six).

Henry Doubleday Research Association

Attractive beans from the collection of rare vegetable varieties held by the Henry Doubleday Research Association.

Research and development in less developed countries is another fast-growing aspect of their work. Initially concerned with trees that would grow in desert-type conditions (that could be utilised by peasant farmers to provide leaves for fodder for their animals) the Association carried out field trials on Cape Verde Islands. Many useful tree species have been collected in South America, India and the Middle East and seed of over 200 tropical trees and shrubs are in store, including an important collection of seeds of the *Prosopis* tree.

5.3.3.3 National Council for the Conservation of Plants and Gardens

The role of this national body is discussed in Chapter Four. Encouraging the cultivation of rare species and cultivars in a series of National Collections helps to maintain a significant gene bank of material of ornamental (and some useful) plants (see Chapter Six). It consists of regional chapters and mobilises the vast potential resource of gardening enthusiasts.

5.3.3.4 The Royal Horticultural Society

A society that plays a significant role in acquiring, introducing to horticulture and maintaining a large variety of ornamental plants and vegetable varieties. Many growers of the larger, frequently-grown garden ornamental genera (e.g. *Iris*, heather (*Erica*), rose, tulip) form national societies, each to further the study of a particular group. These indirectly contribute to biodiversity studies through maintaining a collection ('gene bank').

5.3.3.5 The Federation of Zoological Gardens of Great Britain and Ireland

This Federation was established in 1966 and represents the interests of 60 zoological and wildlife collections. One of its prime tasks is to help organise, coordinate and publicise the conservation, education and scientific work of its members. Its involvement in co-ordinating the maintenance of animal gene banks in UK zoos is discussed in Chapter Six.

5.3.3.6 Zoological Societies

Two Zoological Societies (Edinburgh and London; see Sections 5.3.1.6 and 5.3.1.8) maintain major zoological gardens for public enjoyment and research. The research work of ZSL is illustrated in **Box 5.21** and its role as an animal gene bank discussed in Chapter Six, Section 6.4. A number of other local or regional Zoological Societies (see **Box 5.9**) exist solely to maintain zoos with the support of a variable membership base.

5.3.4 Ornithological Societies

5.3.4.1 Wildfowl and Wetlands Trust (including Wetlands International Africa, Europe & Middle East R.O.)

The WWT was founded in 1946 by the late Sir Peter Scott. Its collection of wildfowl from around the world, and its reserves are important for biodiversity and conservation studies. In particular, through a captive breeding programme, it saved from possible extinction the Nene or Hawaiian Goose, reduced to a wild population of 30 individuals by 1948. Studies of Nene at WWT Centres continue to support conservation efforts in Hawaii, particularly with respect to habitat enhancement, control of introduced predators plus improvements in captive breeding techniques.

M. Hughes/Wildlife and Wetlands Trust

The Wildfowl and Wetlands Trust 'duckery' on the meadows at Slimbridge, UK.

WWT coordinates the Wetlands International/IUCN Threatened Waterfowl Research Group which works to identify the world's threatened wildfowl species, draws up Action Plans and produces newsheets which promote information exchange between workers in the field. Through its **Wetland Link International** programme it publishes *Wetland Link International News* on initiatives in this subject worldwide. WWT has been involved in field work in Thailand and Sumatra (White-winged Wood Duck), Turkey (White-headed Duck) and Spain (Marbled Teal).

In 1947, WWT initiated national counts of waterfowl, in order to monitor populations, particularly any declines. This scheme was amalgamated with the Birds of Estuaries Enquiry in 1993 to form the Wetland Bird Survey or WeBS which is funded by WWT, BTO, RSPB and JNCC. It aims to provide a scientific basis for the conservation of wildfowl populations and is of enormous value in decision making, both with regard to site designations and planning applications. It is a technique that can be applied to similar areas abroad.

5.3.4.2 Royal Society for the Protection of Birds

The Society has one of the largest amateur memberships of any similiar society (885,000 members) and was founded in 1889 to encourage the better conservation and protection of world birds and the environment. The RSPB is involved in cooperative projects in the N.E. Atlantic area, Europe and West Africa. Its major activities are designed to maintain the UK's rich heritage of wild birds and to increase it where desirable. It considers that conserving habitats is the best way to protect wild birds, by creating and managing nature reserves, and by influencing what happens to the rest of the countryside. It aims to create a favourable climate of opinion to carry out its work in wildlife conservation through an effective educational programme.

5.3.4.3 BirdLife International

Formerly the International Council for the Protection of Birds, with its headquarters in Cambridge, UK, BirdLife International represents a large world-wide constituency of active non-government organisations and motivated individuals. It is the leading authority on the status of the world's birds and their habitats. Its priority targets include monitoring the conservation status of all bird species and important bird habitats world-wide and to assist nations to fulfil their obligations to international treaties and conventions.

BirdLife International has identified 221 areas in the world that have the highest concentration of unique bird species: Endemic Bird Areas (EBAs) (Bibby *et al.*,1992). Comparisons between birds and other life-form groups show that the patterns for birds reflect those for other less well studied animals and plant groups. In 1989 a directory of Important Bird Areas (IBAs) requiring protection in Europe, was produced. A total of 2,444 sites of importance for congregating, threatened or restricted-range bird species were identified. It provides a framework for a coordinated programme which currently covers about 30 countries.

BirdLife International, with partners in the developed world including the Royal Society for the Protection of Birds, has promoted capacity building, encouraged conservation education and helped purchase breeding sites of endangered birds (e.g. the Madeiran Freira). With Fauna and Flora International it promotes annually two major expeditions with an ornithological conservation and biodiversity programme under a competitive scheme sponsored presently by British Petroleum.

5.3.4.4 Other bird societies

Three other UK bird organisations, the **British Ornithologists Union,** the **British Trust for Ornithology**, and the **Scottish Ornithological Club** have substantial membership but are mainly concerned with UK issues including the National Bird Ringing Scheme, Nest Records Scheme, Common Birds Census and Estuaries Enquiry.

5.3 5 Other Societies and Trusts specifically studying and/or conserving animals

As with the botanical societies the remit of these organisations is again predominantly UK orientated but networking and partnerships with counterparts abroad are strong in some cases. Publications are of high standard and usually cover the topics worldwide.

5.3.5.1 Bat Conservation Trust

The BCT, whilst predominently UK-oriented, is promoting biodiversity studies in Europe, the Maldive Islands and other areas especially in relation to studies on Old World fruit bats. It has an active training policy and arranges study tours for overseas personnel to learn more about bat diversity and conservation.

5.3.5.2 Herpetological Conservation Trust

The HCT is similar in organisation and aims to the BCT, contributing, through UK and overseas expertise to biodiversity problems in the broad sense with particular action on European diversity.

5.3.5.3 Rare Breeds Survival Trust

A society promoting the maintenance of living collections of unusual and rare breeds of domestic animals. Its role in maintaining both living and cryopreserved gene banks is discussed in Chapter Six.

5.3.5.4 British Herpetological Society

Whilst much of the activity of BHS members relate to those few British species of amphibia and reptile, the Society is concerned with captive breeding of world species and advises both nationally and internationally on breeding and farming on a sustainable basis. Accordingly, it has information to share.

5.3.5.5 Other specialist animal Societies

These include the **British Arachnological Society, British Dragonfly Society**, and the **Mammal Society**. All three mobilise considerable expertise, frequently giving advice, and, the latter, occasionally funding biodiversity work abroad. In all cases their major work is on animals of their group within the UK, although that is increasingly seen in the context of a wider Europe.

5.3.6 Societies set up to study particular ecosystems

5.3.6.1 British Ecological Society

The BES covers world biomes in its symposia and journals. It also encourages biodiversity research by students and qualified scientists through grants and travel burseries.

5.3.6.2 Freshwater Biological Association

As with its marine counterpart (below) its premises now house a research unit on freshwater biology (see Chapter Four). Its limited research is mainly UK-oriented looking at the spread and ecology of invasive alien macrophytes, and species diversity in algae and aquatic invertebrates.

5.3.6.3 Marine Conservation Society

In carrying out an active programme promoting marine conservation, especially of the UK shores, MCS mobilises its corporate scientific knowledge and technical skills (e.g. in SCUBA diving) in biodiversity surveys abroad, especially in S.E. Asia and other tropical coral reefs.

5.3.6.4 Marine Biological Association of the United Kingdom

Founded in 1884 and four years later it opened the Plymouth Marine Laboratory at Citadel Hill (see JMBA (1987) for fuller account of history). The work of the Association is closely allied to that of the NERC Unit at the PML, but is managed by the MBA Council. There are currently 12 Fellows working at the Laboratory funded by MBA. Its research is predominantly concerned with biodiversity in UK waters (English Channel) and the N.E. Atlantic. The Association has been instrumental in establishing 'The Sir Alister Hardy Foundation for Ocean Science' to continue the sixty-year sequence of continuous plankton recorder surveys in the North Atlantic and North Sea. A strong tradition of work is the physiology and functional biology of invertebrates (presently concentrating on cephalopods and tunicates) which increasingly relates to biodiversity studies. See Annual Report for review of research and other Association matters.

5.3.6.5 Scottish Association for Marine Science

SAMS is the Scottish equivalent of the MBAUK whose research is mainly concerned with organisms in UK waters and the N.E. Atlantic. Like the latter it is closely associated with a NERC Unit that shares the same premises (see Chapter Four).

5.3.7 Organisations with an international brief

A number of non-government organisations have a conservation interest and are involved in the study, conservation and sustainable use of biodiversity on an international scale. They include the following:

5.3.7.1 Coral Cay Conservation

CCC is a non-profit organisation that utilises a large workforce of volunteer divers to assist with coral reef survey programmes and management initiatives worldwide (see **Box 5.22**). On-site training opportunities are provided by allocating places for nationals on each expedition (for example, a joint Fisheries Department/CCC Scholarship Programme in Belize provides scuba training courses and training in marine life identification and survey techniques for up to 50 Belizeans annually. As a result of the success of CCC's work in Belize, a number of other developing nations, including Indonesia and the Philippines have asked CCC to assist them assess their coastal resources.

5.3.7.2 Earthwatch Europe

An organisation with the experience and infrastructure to put mature volunteers into the field under the direction of competent scientists. Several biodiversity programmes are carried out each year, often on a long-term basis.

The survey of S'Albufera is an example (see **Box 5.23)**.

5.3.7.3 Flora and Fauna International

Founded in 1903, this is the world's oldest international wildlife conservation society working to save endangered species from extinction. The Society is especially concerned with the prevention of illegal trade and publishes information and news about wildlife conservation throughout the world. Its main thrust in plant biodiversity conservation is Asiatic bulbs used in the horticultural trade, and studies in tree species used to make wind instruments (rose wood and ebony). It supports, with donated funds, several expeditions (often of university students) to study biodiversity abroad and in conjunction with BirdLife, promotes a major expedition each year (presently funded by British Petroleum).

Box 5.22 Coral Cay Conservation – an integrated programme using a voluntary work-force

The UK-based organisation, Coral Cay Conservation, is currently undertaking projects on Turneffe Atoll and Bacalar Chico in Belize, an archipelago of forested islands, or 'cays', bordered by an extensive fringing coral reef system, and on Danjugan Island in the Philippines. These projects put on record the major habitats found and the outstanding biodiversity of these areas.

Turneffe atoll is home to resident pods of dolphin and herds of the endangered manatee, and is perhaps the last significant offshore nesting site in the Caribbean for the American crocodile (*Crocodylus acutus*). The fringing reefs which surround Turneffe are amongst the richest and most biologically diverse to be found in the Caribbean, with sea-grass lawns and coral fringes. The Bacalar Chico area in Belize is one of the few places in the world where a barrier reef meets mainland. Danjugan Island in the Philippines is home to a number of threatened species including bats, sea eagles, coconut crabs and a number of corals limited in their distribution.

Each year, over 250 volunteer divers (the majority of whom are from the UK) join CCC expeditions to Belize and to the Phiippines to assist with this survey work. A common theme amongst them being that they are willing to fund themselves in order to actively participate in marine conservation projects.

The survey techniques employed by CCC have been specifically designed for the acquisition of semi-quantitative data by non-scientists. The training and survey programme have recently been reviewed and approved by a consultant for the UNDP/GEF (United Nations Development Fund/Global Environment Facility). The positions of each survey site are located using satellite navigation equipment (Global Positioning System). Data is collected on specially designed data recording forms and eventually entered into a computer database for incorporation into a powerful Geographical Information System (GIS) database. Once entered into the GIS, this data can be overlaid onto accurate base maps (produced from aerial and satellite imagery) and combined with other data sets (e.g. socio-economic) to produce composite maps. These can be used by coastal zone planners to devise management plans for the protection and sustainable use of the natural resources. (See Mumby, P.J., *et al.* (1995) Geographical Information Systems: A tool for integrated coastal zone management in Belize. *Coastal Management* 23(2): 111–121).

Coral Cays Conservation

Well-organised team work with volunteers contribute a major resource to this reef survey in Belize.

(*Source*: P. *Raines, Coral Cay Conservation.*)

5.3.7.4 Jersey Wildlife Preservation Trust

The JWPT plays a significant role in biodiversity studies and conservation action. Its research activities at Jersey Zoo are discussed in Chapter Six. It also has a major educational role with training courses in zoo biology, breeding biology, interpretation through zoos and similar topics.(See Section 5.4).

5.3.7.5 Raleigh International

RI organises expeditions undertaking a wide variety of environmental and community projects on ten expeditions around the world each year in which multinational youth development is set within a framework of multidisciplinary programmes responding to host country research priorities. Two large scientific expeditions on the inaccessible island

Box 5.23 Project S'Albufera: a biodiversity model in Mallorca

Project S'Albufera is an example of biodiversity studies in action. The Project is an intensive study of a major coastal wetland in Mallorca, Spain. It aims, through long-term monitoring and data collection by an increasingly international team of scientists, to evaluate the effects of a controlled management regime on the environment, and assess the relative contributions to change made by management and outside factors, natural and man-induced. Priorities for the project are:

1. Multi-disciplinary study of processes going on, affecting or dependent on: *i*) the *Phragmites-Cladium* dominated wetland ecosystem, *ii*) the hydrological system, *iii*) the dune systems, and *iv*) the whole catchment of the Park and adjoining coastal waters.
2. The impact of management and related studies.
3. Environmental and socio-economic studies.
4. Furtherance of data-processing and the data-base potential and methodology, including the achievement of compatibility with, and integration into, international networks and schemes.
5. Long-term monitoring aimed at assessing environmental change.
6. Extension of baseline information.

The development of a Biodiversity Model has been identified as an essential element of this work. The need to ensure that this model is applicable beyond the boundaries of the site, i.e. is internationally replicable and compatible, has also been recognised.

The following have been identified as key elements in designing and implementing an internationally compatible biodiversity model:

1. Collect and use baseline data to identify biodiversity indicators for detailed long-term monitoring.
2. Identify abiotic features the impact of which is likely to alter or change the biodiversity.
3. Implement monitoring regimes incorporating the above factors.
4. Establish databases to hold inventory and qualitative data on biodiversity, management information and physiological factors (such as climate etc.).
5. Develop mechanisms to relate and evaluate relationships between the above, to underpin management and site protection measures.

Although a project designed by British scientists and administered through **Earthwatch Europe** at Oxford, full collaboration was established with the *Parc Natural de S'Albufera* authorities from the start. The involvement of Mallorcan and Spanish nationals in S'Albufera studies was also recognised and implemented. This has proved of vital importance to the Project on several levels, including improved perception of the value and aims of the Project locally, nationally and internationally, ease of access to the site at all seasons, and the introduction of local experience and expertise. Work done by Project S'Albufera is increasingly being used as resource material, for educational as well as for interpretation purposes, within Mallorca and at a European level.

G. Perello/Parc Natural de S'Albufera

View of the main bay at S'Albufera and the wetlands, 1991.

A natural progression of these developments was the transfer of responsibility, at the end of 1993, from Earthwatch Europe to the newly established *Associació Balear d'Estudis Internacionals en Medi Ambient* (*The Balearic International Environmental Research Association*, ABEIMA). Earthwatch Europe continues to provide volunteer and financial support but policy decisions and administration are now in the hands of ABEIMA, a Palma de Mallorca-based organisation headed by a group of eminent Mallorcan conservationists and ecologists.

(*Source*: N. *Riddiford*, *Earthwatch Europe*, *Oxford*.)

of Seram, Indonesia, in 1987 have led to the publication of the *Natural History of Seram* (Edwards 1993), based on the scientific survey work carried out in the island's Manusela National Park. A similar pattern has been followed in southern Chile where, in partnership with Corporacion Nacional Forestal, RI has supported a range of scientific research in the Laguna San Rafael National Park. Expeditions to Guyana, Malaysia, Uganda, Zimbabwe, and Belize are planned for the next few years.

5.3.7.6 Scientific Exploration Society

Originally launched to mobilise logistic expertise present in national defence forces for combined field programmes and expeditions with scientific personnel.

5.3.7.7 Society for Environmental Exploration

The SEE is an organisation that has set up a successful research and training programme under the name of 'Frontier', presently active in East Africa in the coastal forests of Tanzania. Professional biologists are supported by graduate students and volunteers on field programmes lasting up to six weeks, during which local graduates, park staff and other nationals of the host country receive training in research techniques.

5.3.7.8 Tropical Biology Association

The TBA is an organisation for scientists from Europe and tropical countries working together to train graduate biologists. It has its operating headquarters in the UK. The initial development of this organisation is supported by a Darwin Initiative grant (see **Box 5.27**).

5.4 Training in museums, botanical gardens and zoos

The biodiversity expertise residing in many provincial museums in the UK (see also Chapter Six) disprove the idea, often regarded, that museums are solely collections of dead organisms kept only for archival purposes. Curators are often world specialists in particular groups of animals or plants, often one well-represented in their own museum collection. The same curators are capable of training junior curators from both the UK and abroad, and frequently do so.

Similarly, botanic gardens have both scientific and technical expertise and several train students to look after living collections, resulting in a Diploma of Horticulture. Those awarded by both the RBG Kew and the RBG Edinburgh are awards of high distinction. A number of other botanic gardens give similar training and diplomas (e.g. Chelsea Physic Garden, Cambridge University Botanic Garden). More specialised courses given by RBG Kew and RBG Edinburgh are given in Section 5.5.

The same kind of expertise and willingness to share it exists in the zoological garden fraternity. The Federation of Zoological Gardens of Great Britain and Ireland is currently mobilising such training to help developing countries set up more effective zoos. Separate from this, the Jersey Zoological Park, stands out independently as having a major input to biodiversity and conservation training. Located on the larger of the independent Channel Isles, its **International Training Centre for Breeding and Conservation of Endangered Species (ITC)** was opened in 1978, as part of the Jersey Wildlife Preservation Trust founded in 1963 by the late Gerald Durrell. The ITC offers facilities for theoretical and practical training in zoo biology and conservation of endangered species. Two types of courses are given: (i) A twelve-week Training Programme (three courses per year, involving 4/5 weeks lectures/workshops follwed by 7/8 weeks practical work in Jersey Zoo). After completion of this programme, trainees may register for the Diploma in Endangered Species Management in conjunction with the University of Kent at Canterbury. (ii) An annual three-week Summer School, which is an intensive course on the above topic, omitting the practical work. Since 1977, nearly 700 students have been trained at ITC.

Museums (e.g. Manchester, Liverpool and Edinburgh) also make a substantial contribution to training both nationals and visiting students from abroad, using the facilities of their collections (see Chapter Six). While they may not have the student facilities offered by major universities they compensate for this by the specialist libraries and facilities available, the proximity to major biodiversity collections and possibly, the more personal attention (sharing techniques, ideas and knowledge), offered by these smaller, highly focused, institutes. The insect course run at the University College Cardiff jointly with the National Museum of Wales, and the parasitic wasps course, given jointly by The Natural History Museum and London University (see **Box 5.2**) are cases in point. Another university with a major biodiversity collection at hand is Manchester.

The importance of public education should not be ignored, for while it is not providing in-depth biodiversity training at graduate level, nonetheless it is informing the public and stimulating the interests of tomorrow's biologists. Birmingham Botanical Gardens, for instance, runs a high profile education department catering for up to four school groups per day and offering a comprehensive range of subject matter linking-in directly with the needs of the national curriculum and often on a multi-cultural basis. In fact, public education has become increasingly important and consequently well organised in the last twenty years. Most museums, botanic gardens and zoos now offer substantial programmes to cater for all levels and abilities; these include activities and programmes aimed at primary, secondary, tertiary levels and also at non-specialist adults. Artifacts or timber specimens neatly arranged in display cases are now rarely seen; these have been largely replaced by audio-tours, interpretation, role play, multi-cultural classes, video presentations, worksheets and other printed materials. However, despite the increase in communication techniques, the spoken and written words are still amongst the most popular and guided tours by experts are still well attended.

5.5 UK Government promotion of biodiversity training

5.5.1 Direct Government grant-aid

The main thrust of government funding for research and training in the UK into biodiversity goes directly into the three 'major biodiversity centres' (The Natural History Museum, and the Royal Botanic Gardens at Kew and Edinburgh), and through the Research Councils (Natural Environment Research Council and Biotechnology and Biological Sciences Research Council), some of which may be closely associated with university departments and are an integral part of their teaching and postgraduate student training. Other, smaller amounts (as far as biodiversity research is concerned) are channelled through the Ministry of Agriculture, Fisheries and Food, the Scottish Office and the Overseas Development Administration, the latter mainly into the Natural Resources Institute. The interrelationships and research briefs of these institutes are discussed in Chapter Four but their role in training of many graduate students studying for a doctorate degree, either in-post or with outside (or appropriate Research Council) funding, is a major input into science training. As the biodiversity (systematics, taxonomy and ecology) content of their research increases, so does their role as trainers of graduate students and promoters of biodiversity studies.

The Natural History Museum and the Royal Botanic Gardens at Kew and Edinburgh – are the greatest resource for biodiversity study and assessment, and have an increasingly important role to play in training future taxonomists and others studying biodiversity. All three institutes promote the sharing of knowledge as effectively and widely as possible, and link with leading universities to develop coordinated graduate education and postdoctoral fellowship programmes. At RBG, Kew, for example, over 50 PhD projects are co-supervised by Kew staff in collaboration with some 25 universities and Kew funds six PhD students and a Masters student at any one time. Similar numbers of students will be found at the NHM and a proportionately smaller number at RBG, Edinburgh. RBGE runs an MSc/Postgraduate Diploma course jointly with the University of Edinburgh on the Biodiversity and Taxonomy of Plants (see **Box 5.24**). The Natural History Museum offers an MSc course on Biodiversity run jointly with Imperial College (London University) and contributes to a Birmingham University MPhil, course on Botanical Diversity. It also organises short specialist insect taxonomic courses with Imperial College, (see **Box 5.2**). At RBG, Kew, several international diploma courses are run in

Box 5.24 The Biodiversity and Taxonomy of Plants: Royal Botanic Garden, Edinburgh and University of Edinburgh

This MSc course was established in 1992 by the Royal Botanic Garden, Edinburgh and the University of Edinburgh in response to growing demand worldwide for trained plant taxonomists. Forming a bridge between traditional and modern approaches, it equips biologists, conservationists and ecologists with a wide knowledge of plant biodiversity, combined with instruction in the methods of pure and applied taxonomy.

The course consists of formal instruction, practical work, essays, research projects and tutorials covering the following major areas:

- functions and philosophy of taxonomy;
- evolution and biodiversity of the major plant groups;
- elements and origins of biodiversity;
- plant geography and biodiversity;
- surveys and monitoring;
- production and use of floras and monographs;
- use of computers in handling and processing data;
- nomenclature of wild and cultivated plants;
- artificial selection, breeding strategies, genetic manipulation;
- chemical estimation of variation;
- expeditions and plant collecting;
- curation of living collections, herbaria and libraries;
- managing and funding taxonomic institutes; and
- public services and public relations.

There are also workshops and training sessions on the practical aspects of the following: identification, cytology, pollination ecology, computing, numerical taxonomy, cladistic analysis, molecular methods, herbarium and horticultural techniques, and bibliographic methodology.

Fieldwork and visits to other institutes are an integral part of the course. There is a twelve-day field course in Portugal during the Easter break.

Course graduates are employable in a wide range of fields, including:

- survey work in threatened ecosystems;
- assessment of plant resources and genetic diversity;
- basic research;
- management of institutes and curation of collection; and
- university and college teaching.

The course lasts twelve months ending with examinations. On the basis of these exams and other course-work, students then either embark on a five-month research project to qualify for the MSc, or proceed to a third term of taxonomic study and essay-writing to be awarded the Diploma.

conjunction with universities, institutes and agencies; over 125 foreign students are trained at Kew each year. The seventh course on Herbarium Techniques (see **Box 5.25**), for which selected overseas students are supported by British Council, Darwin Initiative and other funds, was completed in 1995. The School of Horticulture runs the prestigious three-year 'Kew Diploma', and short-term courses covering subjects such as botany, plant photography, garden design and botanical illustration, are also available.

Many staff of these institutes transfer their skills by teaching or by providing training or supervision at all levels from schools to post-doctoral, through formal and informal courses, seminars, workshops and conferences. At the RBG Edinburgh, the existing DHE course, which has been running for many years, was replaced in October 1995 by a specialist HND course entitled Plantsmanship and Plant Collection Management, run in conjunction with the Scottish Agricultural College.

Box 5.25 International Diploma Course in Herbarium Techniques, Royal Botanic Gardens, Kew

This eight week course held at Kew aims:

- to give each participant the knowledge and skills to become a proficient herbarium technician;
- to give participants the insight to select methods most appropriate to the needs of their own institutes;
- to enable participants to understand the principles of herbarium management;
- to enable participants to appreciate the value of information available in their herbarium and make it accessible to a wider audience; and
- to enable partcipants to gain a wider understanding of conservation and development in other countries.

Although there are no restrictions on age or qualifications, the participants must have a reasonable command of English and be employed in the herbarium of a recognised institution. A comprehensive coverage of herbarium techniques and managment is given. This is backed up by guided and individual practical sessions giving ample opportunity to learn and practise a variety of skills. In addition, a selection of ancillary subjects relating to herbaria and taxonomy is included together with topics of wider interest.

The course also contains an optional aspect. This is designed to give each participant practical experience in a subject relevant to the herbarium. Participants are asked to indicate (with the agreement of their home institution) their choice from the list of 'Options':

- simple library management;
- illustration (plant drawing, maps and diagrams);
- check-lists;
- plant identification (practical experience);
- supplementary herbarium techniques (e.g. rearranging the collection according to a new publication. or the extraction of loans).

Courses have been held in 1987, 1988, 1990, 1991, 1992, 1994 and 1995, each accommodating 9(–12) participants. So far a total of 73 participants representing 47 countries, have attended. Participants have obtained funding from a total of 36 sponsors. The British Council and the Commonwealth Science Council took part in the initial planning of the course, and since then have continued their sponsorship, promotion and goodwill. In both 1994 and 1995 the *Darwin Initiative for the Survival of Species* granted funds to allow 5 Darwin Scholars to attend the course.

The Herbarium Handbook, a text book especially designed to accompany the course was first published in 1989 and a fully revised and much expanded edition was published in late 1992. A Russion edition, sponsored by the *Darwin Initiative*, was published in 1995.

The course newsletter, *'TechniQues'*, functions to help keep a network of practitioners sharing similar problems and experiences. It is distributed to all past participants and candidates on the waiting list. So far there have been four issues (1, January 1991; 2, October 1992; 3, March 1994; 4, May 1995).

(*Source*: D. *Bridson, Royal Botanic Gardens, Kew.*)

CAB International is an intergovernmental organisation (but with origins and a headquarters in the UK; see also Chapter Four, Section 4.10) with a strong remit to disseminate its considerable knowledge of biodiversity, especially in entomology and mycology. Selected courses taken from its training programme for 1994-1995 (CABI, 1994), as an example to show the extent and range of what is available is given in **Box 5.26**. It must be stressed that these courses may not be held every year and potential candidates should contact the Training officer at CABI.

The university programmes listed above (see **Box 5.3**) obtain their main funding from exchequer funds through the Universities

Box 5.26 Education and training at CAB International institutes, 1994–1995

INTERNATIONAL INSTITUTE OF ENTOMOLOGY (IIE)

IIE and the Natural History Museum have collaborated to provide a large resource personnel with a great depth of taxonomic knowledge in all insect Orders. Two courses recently held are:

Identification of insects and mites for agriculture and forestry (3 weeks, regional Malaysia, Sri Lanka, East Africa).

Applied taxonomy of insects and mites of agricultural importance (7 weeks).

INTERNATIONAL MYCOLOGICAL INSTITUTE (IMI)

IMI offers a wide range of courses and tuition opportunities for individuals and, since 1985, has trained 783 people from 90 countries. Students are able to work at IMI through our collaboration on several MSc courses run by UK universities. Most notable are the MSc in Biotechnology specializing in 'Fungal Technology' with the University of Kent and the MSc in 'Pure and Applied Plant and Fungal Taxonomy' with the University of Reading.

A selection of courses from IMI's current programme

Techniques Courses
Modern methods for the identification of bacteria and filamentous fungi (2 weeks).
Mycological and culture collection techniques (3 days).

Specialist Courses
Identification of entomopathogenic fungi (3 days).
International course on the identification of fungi of agricultural importance (6 weeks).
Advanced Courses (usually 3-day courses)
Identification of *Aspergillus* and *Penicillium* species
Identification of *Colletottichum* species
Identification of *Fusarium* species
Identification of *Pythium* and *Phytophthora* species

INTERNATIONAL INSTITUTE OF PARASITOLOGY (IIP)

Identification of plant nematodes of economic importance (6 weeks).

(*Source*: CAB *International, Wallingford.*)

Funding Council, and through grants-in-aid from the Research Councils and the Royal Society, although targeted research is receiving increasing amounts of income from contracts and sponsorship.

5.5.2 The Department of the Environment and the Darwin Initiative for the Survival of Species

As creators of national legislation, such as the *Wildlife and Countryside Act* 1981, and signatories of international directives such as the Biodiversity Convention, Habitats Directive and Berne Convention, the UK Government must have the organisational capacity and ability for their implementation. This is done through Government Departments such as the Department of the Environment (DoE) and country agencies or statutory bodies such as, in the UK, English Nature, Scottish Natural Heritage, Countryside Council for Wales and DoE (Northern Ireland). The biodiversity programme and strategy for the UK is substantial but is not within the scope of this book (but see DoE, 1994).

The concept and creation of the Darwin Initiative for the Survival of Species was announced by the Prime Minister at the United Nations Conference on Environment and Development (UNCED) in Rio de Janeiro in 1992. Essentially the idea of the Initiative was to draw upon British strengths in the field of biodiversity to assist with the conservation and sustainable use of the world's biological resources. To help refine the idea into a series of recommendations and criteria for selection an advisory committee, chaired by Sir Crispin Tickell, was established. The report compiled by the committee (DoE, 1993) listed about ten recommendations for types of project that would be suitable for funding and also gave some possible examples.

One of the key recommendations was that the Initiative should be directed primarily towards the establishment of collaborative projects in countries poor in resources and rich in biodiversity, including training in taxonomic and other techniques. These projects should ideally be based on established links between British institutions and institutions in those countries, or on links which are being developed, and could include the improvement of access to data held in British institutions.

To date, four 'rounds' of grants worth in total £12 million have been awarded (see **Box 5.27**), 31 projects were funded in 1993, totalling in excess of £3.5 million, and 22, 33 and 30 respectively in subsequent years. The Initiative encourages projects which involve additional amounts of matching funding or resources from charitable organisations, countries concerned, international economic organisations or industry and commerce.

The 116 projects funded to date are almost as diverse as the diversity they seek to record and conserve! In many cases collaboration has not only been sought with institutions overseas but has included collaboration between British institutions, often between a university and a research institution. For instance, Aberdeen University has teamed-up with RBG Kew in a study to identify the differences between protected and unprotected headwaters in terms of rheophyte biodiversity, and the University of Durham has joined forces with the RGS in a project to evaluate the biodiversity of an arid land area in Jordan and the development of an environmental management plan.

Much of the funding has been devoted to training. In the first round of the Initiative, for instance, the British Ecological Society was awarded a substantial grant to set up the Tropical Biology Association. The purpose of the Association was to (i) run a series of field courses for a combination of European and African students at two established field stations in Uganda and (ii) to help form a federation of European and African universities and institutes which will provide the existing expertise and will, in future, fund the courses by their annual subscriptions.

All aspects of biodiversity are eligible for funding and already grants have been given for studies of fungi in Ukraine, fish in the Ruwenzori, insects in Belize, onions in the Sahel, mangroves in Malaysia, forests in Brazil, sharks, rays and swordfish in East Malaysia, *Melaleuca* wetlands in Vietnam and Thailand, mink and otter in Russia and endangered tree species in Chile – to mention just a selection. **Box 5.27** overleaf summarises the total projects up to 31 March 1995.

5.6 Funding for training

The Darwin Initiative for the Survival of Species mentioned above gives considerable grant-in-aid for funding training *per se*. Overseas students, often mature scientists, funded in this way spend a period of their training in the UK and may continue in their own country under the supervision of their UK sponsor. In **Box 5.27** those projects with substantial training are emphasised.

5.6.1 General information

The information given below is on sources of financial assistance for overseas students coming to the UK for training. They have been taken from the British Council information sheet on the subject (British Council, 1994). Full reference to other useful books may be found in Section 5.7 below. British Council support for biodiversity studies and programmes generally are given in **Box 5.28**.

The following is a list of the major sources of financial assistance available to overseas students, mainly at postgraduate level. For the majority of the schemes students will have to obtain further information in their own country. In addition, British Council offices overseas (see Appendix 3 for addresses) normally publish a specific information sheet detailing awards available in that country.

Students should only come to the UK or embark on a course of studies having first made sure that they have the funds to pay academic fees and living expenses to cover the whole course for which they have registered. It is difficult, if not impossible, to arrange financial support once they have left their home country.

Box 5.27 Projects funded by the UK Darwin Initiative for the Survival of Species, 1993–1995
Projects are mostly funded over a period of three years.

Contact's name and address	Project synopsis	Training potential	Starting date & funding	Location
1. University of Aberdeen: Royal Botanic Gardens, Kew				
Dr M.D. Swaine Department of Plant and Soil Science University of Aberdeen Aberdeen AB9 2UD	To identify the differences between protected and unprotected headwaters in terms of rheophyte biodiversity by contrasting river systems (eg. logged and unlogged forest reserves in forested and deforested areas and between high and low rainfall forest types).	An opportunity for Ghanaian staff to work in an international herbarium	1993 £49,800	Ghana
2.Botanic Gardens Conservation International				
Dr Peter Wyse Jackson Descanso House 199 Kew Road Richmond Surrey TW9 3BW	To provide technical assistance, logistical support and advice to botanic gardens and arboreta in developing areas by means of a network system and with a resource centre based in UK.	Providing guidelines & information packs.	1993 £146,350	Africa, SE Asia, Latin America and the Caribbean
3. University of Bradford				
Prof. John MacArthur Development and Project Planning Centre University of Bradford Bradford BD7 1DP	The project will fund 3 Darwin Fellows each year on a six week international course on Planning of Projects for Biodiversity Conservation. The course will cover project identification, design and costing, finance, economic analysis and justification, report preparation, presentation skills, and basic management concepts.	Predominantly a training project	1993 £43,800	UK
4. British Ecological Society				
School of Biological Sciences University of Bristol Woodland Road Bristol BS8 1UG	To set up the Tropical Biology Association (i) to run a series of field courses for a combination of European and African students at two established field stations in Uganda and (ii) to help form a federation of European and African universities and institutes which will provide the existing expertise and will, in the future, fund the courses by their annual subscriptions.	Predominantly a training project	1993 £388,200	Africa
5. Scott Polar Research Institute, University of Cambridge: Moscow State University: World Conservation Monitoring Centre				
Dr Gareth Rees Scott Polar Institute Lansfield Road Cambridge CB2 1ER	To develop methods, using satellite remote sensing data, of measuring the physical, chemical and biological properties of tundra vegetafion on the Kola Peninsula and to assess the impact of industrial pollution.		1993 £89,200	Arctic Russia
6. University of Cambridge, Faculty of Economics: World Conservation Monitoring Centre				
Timothy Swanson Faculty of Economics and Politics University of Cambridge Austin Robinson Building Sidgwick Avenue Cambridge CB3 9DD	To survey the range and qualities of diverse genetic resources produced commercially, analyse the data and devise management regimes for sustainable commercial activities in resources.		1993 £84,000	Worldwide
7. Coral Cay Conservation: University College of Belize, Government of Belize				
Peter Raines The Ivy Works 154 Clapham Park Road London SW4 7DE	To survey the marine and terrestrial habitats of the Turneffe atoll in order to construct GIS maps of the area for future management initiatives. The project also aims to establish a Marine Research Centre on the atoll to strengthen the UCB marine studies programme.		1993 £186,500	Belize

Box 5.27 *continued*

Contact's name and address	Project synopsis	Training potential	Starting date & funding	Location
8. The Durell Institute of Conservation and Ecology, University of Kent				
Prof. Ian Swingland The Durrell Institute of Conservation and Ecology The University of Kent Canterbury CT2 7PD	To develop a concensus for biodiversity management in an area, and assess requirements for a regional database. To train 6 Fellows to MSc level in Conservation Biology (one year course) and 12 short-term Fellows in Conservation Biology and Biodiversity Management.	Basically a training project.	1993 £135,000	UK, Indonesia, Zimbabwe and either India or Brazil
9. University of East Anglia: Nepali Government's Department of National Parks and Wildlife Conservation				
Dr Diana Bell School of Biological Sciences University of East Anglia Norwich NR4 7TJ	To examine the reasons for loss of tropical grasslands in the Indian sub-continent, to explore the effects of different management regimes on habitat biodiversity and formulate longterm management plans for remnant tall grasslands which allow appropriate levels of its sustainable utilisation by local people.		1993 £104,100	Nepal
10. The Federation of Zoological Gardens of Great Britain and Ireland; various zoos both in UK and overseas				
Peter J.S.Olney Director Federation of Zoos of G.B. & I Regents Park London NW1 4RY	To carry out a feasibility study of the contribution that British zoos could make to species conservation in poorer countries. It will also aim to improve overseas zoo's, captive management and breeding programmes and help them to raise local awareness in conservation matters.	To train people from poorer countries using the facilities and expertise within the Federation.	1993 £	Worldwide
11. Institute of Terrestrial Ecology: National Biodiversity Data Bank, Kampala: Kings College London				
Robin M.Fuller Institute of Terrestrial Ecology Monks Wood Huntingdon Cambridge PE17 2LS	To build a database for Sango Bay, Uganda, on the shore of Lake Victoria, using satellite images to map the land cover. Databases on wildlife will be incorporated into a geographical information system.	To transfer technology to construct and integrate biodiversity databases.	1993 £139,400	Uganda, Lake Victoria
12.International Mycological Institute				
Dr D W Minter International Mycological Institute Bakeham Lane Egham Surrey TW20 9TY	To assess the N.G. Kholodny Botanical Institute's collections of fungi (estimated at 50,000 specimens) and to encourage further development of Ukrainian expertise in biodiversity measurement and monitoring, and to produce a preliminary checklist of fungi of the Ukraine.	IMI will train staff in computer databasing.	1993 £82,100	Ukraine
13. International Mycological Institute: International Institute of Entomology: International Institute of Parisitology				
Prof. David L.Hawksworth International Mycological Institute Bakeham Lane Egham Surrey TW20 9TY	Systematic topics relating to the groups of organisms covered by the Institutes, i.e. plant bacteria, fungi (including lichens), insects and other arthropods, plant nematodes and animal helminthoids, will be studied in the project.	To train 22 Darwin Fellows to develop regional networks of BioNET on their return.	1993 £305,300	IMI, UK personnel from predominantly tropical countries
14. Kings College, London: Makerere University; Ugandan Freshwater Fisheries Research Organisation				
Dr Roland Bailey Kings College London Division of Life Sciences Campden Hill Road London W8	To investigate the diversity and distribution of fish faunas and aquatic ecosystems in W. Uganda. To input into the Ugandan National Biodiversity. To examine fish diets in relation to the food available and obtain data on native fish in perennially accessible habitats.	Training of local staff in fieldwork and databasing	1993 £42,400	Uganda, Ruwenzori mnts.
15. Natural History Museum				
Dr Christopher H.C.Lyal Department of Entomology Natural History Museum Cromwell Road London SW7 5BD	To examine the effect of forestry intervention practices on plant-associated insect biodiversity, especially those on mahogany and chicle and their relatives. The pest status of insects on selected tree species will be assessed and the natural enemies of these insects studied.	Some training will be provided.	1993 £81,500	Belize

Box 5.27 *continued*

Contact's name and address	Project synopsis	Training potential	Starting date & funding	Location
16. Natural History Museum: Instituto Nacional de Biodiversidad				
Dr Ian Gauld Biodiversity Co-ordinator Natural History Museum Cromwell Road London SW7 5BD	To produce (as part of the training programme) keys to the Banchine ichneumonoids (parasitic wasp) and construct a database relating the abundance and distribution of the fauna to land use changes.	To train a Costa Rican in insect identification and in the production of field guides.	1993 £46,500	Costa Rica
17. Natural Resources Institute: ICRISAT, India: ICARDA, Syria				
Dr P.C.Stevenson Chemical Ecology Resource Centre Natural Resources Institute Chatham Maritime Kent ME4 4TB	To investigate wild species of chickpea against different pathotypes (races) of *Fusarium oxysporum*. Plant exudates and extracts will be analysed to identify compounds which are responsible for resistance of plants to the disease.		1993 £73,200	Syria and India
18. Natural Resources Institute: Horticulture Research International				
Dr Nigel Poulter Natural Resources Institute Chatham Maritime Kent ME4 4TB	To encourage the production of seed of tradition onion landraces selected by farmers in the Sahel. Seed of local varieties will be multiplied, so that the varieties can be evaluated in trials within the region and internationally.	West African staff will be trained in collection and conservation of seed stocks.	1993 £143,000	West Africa: Burkina Faso & Cote d'Ivoire
19. Oxford Forestry Institute, University of Oxford: Natural History Museum				
Dr John Burley Department of Plant Sciences University of Oxford South Parks Road Oxford OX1 3PS	To give the necessary skills to deal with political and public pressures on conservation issues, to develop stategies for sustainable use and work practically in biodiversity exploration, evaluation and conservation.	A training programme over six weeks for 10 forest scientist and managers each year	1993 £85,000	UK students from Malaysia & India
20. Plymouth Marine Laboratory: Malaysian Science University, Penang				
Dr R.M.Warwick Plymouth Marine Laboratory Prospect Place Plymouth PL1 3DH	To provide information and training techniques on mangrove areas which will lead to developing management guidelines on the conservation of the Merbok mangrove ecosystem.	Some training will be given.	1993 £194,000	Malaysia
21. Royal Botanic Gardens, Kew				
Diane M.Bridson Herbarium Royal Botanic Gardens Kew Richmond Surrey TW9 3AE	To fund five students already employed in herbaria in their own countries to attend Kew's International Diploma Course (eight weeks). An additional four week period allows the scholars to continue with a project of mutal benefit to themselves and their home institution.	A training programme in herbarium techniques.	1993 £40,000	UK
22. Royal Botanic Gardens, Kew: Federal University of Pernambuco, Recife				
Dr Simon Mayo Herbarium Royal Botanic Gardens Kew Richmond Surrey TW9 3AE	To survey and evaluate plant biodiversity in the highly threatened Brejo forests of North-east Brazil. Brejo forests are important in protecting watersheds. They represent the major centres of regional biodiversity and make a vital contribution to the economy of the semi-arid region, the poorest part of Brazil.	To train local people and contribute to local teaching programmes.	1993 £140,300	N.E. Brazil
23. Scientific Exploration Society				
John Hunt Scientific Exploration Society Expedition Base Motcombe Shaftesbury Dorset SP7 9PP	To carry out a botanical and zoological survey of Namdapha National Park, in a site of special scientific interest and a tiger reserve.	Training will be given in the collection and preservation of specimens.	1993 £124,200	Arunchal Pradesh, N.E. India

Box 5.27 *continued*

Contact's name and address	Project synopsis	Training potential	Starting date & funding	Location
24. Scottish Agricultural College: M S Swaminathan Research Foundation				
Dr Robert Finch Plant Science Department SAC, Auchincruive Ayr KA6 5HW	To collect and survey, using molecular methods, the genetic diversity of *Porteresia coarctata*, a salt tolerant, wild relative of rice of mangrove ecosystems and to establish a reliable propagation system.		1993 £50,900	India
25. UK Dependent Territories Conservation Forum				
Sara Cross Co-ordinator c/o 14 Goose Acre Cheddington Nr. Leighton Buzzard Beds. LU7 0SR	To prepare and implementan action plan, which will identify organisational needs and environmental project priorities in the UK. Dependencies, whilst at the same time strengthening existing UK technical and scientific support in order to meet those needs.		1993 £25,000	UK Dependent Territories
26. School of Agriculture and Forest Sciences, University of Wales: University of West Indies, Mona				
Dr John Healey School of Agricultural and Forest Sciences University of Wales Bangor Gwynedd LL57 2UW	To quantify the effects of the invasive alien tree species, Pittosporum undulatum on biodiversity in primary forests in the Blue Mountains National Park, to assist the Park in responding to the invasion by determining the severity and urgency of the threat it poses and the most cost-effective methods for its control.	National park staff and local communities will be involved.	1993 £34,000	Jamaica
27. School of Agricultural and Forest Sciences, University of Wales, Bangor: Royal Botanic Gardens, Kew: Limbe Botanic Gardens				
Dr John Healey School of Agricultural and Forest Sciences University of Wales Bangor Gwynedd LL57 2UW	To investigate the natural vegetation dynamics and regeneration mechanisms of individual tree species and compared with the natural regeneration that occurs following disturbance created by local use of the forest, utilising both scientific and social science approaches.	Contains a training element for local people.	1993 £92,500	Cameroon, Mt Cameroon Forest Reserve
28. World Conservation Monitoring Centre: InBIO, Costa Rica				
Harriet Gillett World Conservation Monitoring Centre 219 Huntingdon Road Cambridge CB3 0DL	To gather baseline data on the status, distribution and utilisation of plant genetic resources of Central America as a basis for their conservation and sustainable use. Particular focus will be on the wild progenitors and landraces of agricultural crop plants.	A broad based consortium of agencies will be established to build capacity in institutions within the region.	1993 £90,300	Central America
29. World Wide Fund for Nature				
Hamish Aitchison WWF UK Panda House Catteshall Lane Godalming Surrey GU7 IXR	To support 8 Darwin Scholars/Fellows from developing countries on the certificate course in Environmental Education at Jordanhill College, Glasgow. The course is designed to strengthen capacity building, to develop programmes of environmental education, to train experienced teachers, curriculum developers and conservation educators.	Predominantly a training project	1993 £155,000	UK
30. WorldWide Fund for Nature				
Dr Alan Hamilton WWF Panda House Catteshall Lane Godalming Surrey GU7 IXR	To train ethnobotanists and other specialists from developing countries to work on practical conservation and development issues with local.people. Two key field workers will carry out fieldwork and advisory visits. Also to record local knowledge in the coastal forests of East Africa and in the Ruwenzori National Park, Uganda.	Predominantly a training project	1993 £368,500	East Africa including Kenya and Uganda.

Box 5.27 *continued*

Contact's name and address	Project synopsis	Training potential	Starting date & funding	Location
31. Tropical Marine Research Unit, University of York				
Dr Rupert Ormond Director, Tropical Marine Research Unit University of York York YO1 5DD	To assist National and Marine Park departments in the Indian Ocean and Red Sea to protect their coral reef areas and to understand the factors responsible for the diversity of fish which the reefs support and cooperate with park staff in investigating selected management problems.	Professional reef biologists supported by UK volunteers will provide training for park staff.	1993 £112,200	Kenya and Egypt (Red Sea)
32. Asian Wetland Bureau: the Government of the Kingdom of Cambodia				
Dr Nather Khan Asian Wetland Bureau Institute of Advanced Studies University of Malaya Lembah Pantai 59100 Kuala Lumpur Malaysia	To assist in the conservation and management of wetlands and their biodiversity in Cambodia through inventories of the wetlands and developing of a national wetland action plan.	Training and awareness building will be a high priority.	1994 £75,000	Cambodia
33. University of Birmingham : Southern African Development Community: the International Plant Genetic Resources Institute				
Dr Nigel Maxted School of Biological Sciences University of Birmingham Edgbaston Birmingham B15 2TT	To provide through Fellowships the skills and experience necessary to order, collect, conserve and utilise their native botanical diversity, including seed conservation, genebank management, and conservation data management.	Three courses, one per year, 60 Darwin Fellows	1994 £115,000	Southern African students in UK
34. World Wide Fund for Nature/International Institute for Environment and Development				
Ms Margaret I.Evans Avensac 32120 Maurezin France	To develop policies to promote the in situ conservation of wild and traditional indigenous food species, both plant and animal, and to increase access to them in a sustainable manner with particular emphasis on Ethiopia and Mexico.		1994 £60,000	Ethiopia & Mexico
35. The Foundation for Ethnobiology				
Dr Conrad Gorinsky The Foundation for Ethnobiology North Parade Chambers 75 Banbury Road Oxford OX2 6PE	To study human usage and perception of the greenheart tree (*Ocotea rodiaei*), and associated ethnobiological issues and to include the examination of issues relating to sustainable development and intellectual property rights.	A training element will be included.	1994 £148,000	S.America
36. Foundation for International Environmental Law and Development:Secretariat of the Convention of Biological Diversity; UNEP				
Mr Phillippe Sands Department of Law School of Oriental and African Studies, London University 46/47 Russell Square London WClB 4JP	To appoint four Darwin Fellows to work on the preparation of background documents on international legal aspects of the implementation of the Biodiversity Convention. involving practical legal assistance.	Basically a training programme.	1994 £148,000	UK
37. Institute of Hydrology				
Mr R C Johnson Institute of Hydrology Unit 2, Alpha Centre Innovation Park Stirling FK9 4NF	To investigate the consequences of river catchment management on the aquatic biodiversity of rivers and riparian zones in the Himalaya, in three regions with contrasting climate, geology and agriculture over a range of altitudes. The spatial patterns of aquatic biodiversity, stream water chemistry and fluvial sediments will be studied.	Exchange of ideas and results with local organisations will result.	1994 £171,800	Himalaya

Box 5.27 *continued*

Contact's name and address	Project synopsis	Training potential	Starting date & funding	Location
38. Institute of Terrestrial Ecology(N) : Cameroon Agronomic Research Institute: Fountain Renewable Resources (UK)				
Dr Julia Wilson Institute of Terrestrial Ecology Bush Estate Penicuick EH26 OQB	To identify, through field visits and socioeconomic surveys, native fruit trees used by local communities. Germplasm will be collected and raised at the nursery in S. Bakundu Forest Reserve, for testing in experimental agroforestry plots.	Technicians and villagers will be trained in low-technology propagation techniques	1994 £131,700	Cameroon
39. Institute of Zoology, Zoological Society of London				
Dr Andrew Balmford Institute of Zoology Regent's Park London NWI 4RY	To employ a Ugandan postdoctoral scientist to identify which taxa are reliable indicators of patterns of diversity in 55 Ugandan forests, by analysing an existing data base covering the distribution of birds, mammals, trees and butterflies.	The project will provide training for local scientists.	1994 £48,300	Uganda
40. International Centre of Landscape Ecology, University of Loughborough				
Dr Max Wade Department of Geography Loughborough University Loughborough Leicestershire LE1 1 3TU	To identify and evaluate the biodiversity of the lowland wet grassland resource in order to produce guidelines for its conservation and sustainable management.	To teach trainers to disseminate the information among practitioners.	1994 £77,000	Czech Republic and Estonia
41. International Mycological Institute: Research and Development Centre for Biology, Bogor, Indonesia				
Dr David Smith International Mycological Institute Bakeham Lane Egham Surrey TW20 9TY	Courses will include taxonomy and identification, collection, isolation, preservation techniques and culture collection management. A collection will be established in Indonesia for the ex-situ conservation of local microbial biodiversity.	Basically a training programme for three Darwin Scholars.	1994 £59,400	Bogor, Indonesia and UK
42. International Waterfowl and Wetlands Research Bureau				
R Crawford Prentice IWRB Slimbridge Gloucester GL2 7BX	To strengthen capacity of the existing nature conservation agencies for wetland conservation in Central and Eastern Europe, through training courses and the establishment of a training centre in Ukraine.	Predominantly a training project.	1994 £135,600	Eastern Europe & Ukraine
43. The Natural History Museum				
Dr Geoff Boxshall Department of Zoology The Natural History Museum Cromwell Road London SW7 5BD	To facilitate the identification of larval and adult crustaceans from the plankton of the brackish-water coastal lagoons of Southern Brazil particularly of commercially important decapod crustaceans and the developmental stages of parasitic copepods from local fish.	Involves training Darwin Fellows in taxonomic techniques and in culture methods, and establishing a database.	1994 £59,800	S.Brazil; S.Atlantic
44. The Natural History Museum : Instituto National de Biodiversidad, San José				
Dr Malcolm J Scoble Department of Entomology The Natural History Museum Cromwell Road London SW7 5BD	To provide training in identification and inventory assembling for a Costa Rican scientist, through study of a large group of moths belonging to the family Geometridae.	A training project for a Darwin Fellow.	1994 £42,500	UK & Costa Rica
45. The Natural History Museum : Instituto Nacional de Pesuisas da Amazonas				
Dr N E Stork Department of Entomology The Natural History Museum Cromwell Road London SW7 5BD	To study the species richness of the insect fauna of the Reserve Ducke area near Manaus. One student will focus on beetles of the forest canopy and the other on beetles on the forest floor	Two Brazilian PhD students will be trained in beetle taxonomy	1994 £83,700	Brazil

Box 5.27 *continued*

Contact's name and address	Project synopsis	Training potential	Starting date & funding	Location
46. The Natural History Museum				
R.I. Vane Wright Department of Entomology The Natural History Museum Cromwell Road London SW7 5BD	To record and assess biodiversity of the Western Ghats using the WORLDMAP computer program. Databases for butterflies, hawkmoths, bats and palms will be made during the project by Indian and British specialists.	The project will involve intensive training of three Indian collaborators.	1994 £44,900	India: Western Ghats
47. University of Nottingham : Earthwatch Europe				
Dr Robin C. Brace Department of Life Sciences University of Nottingham University Park Nottingham NG7 2RD	To survey the ecology and conservation of the birds, insects and small mammals, of natural forest islands in the savanna of the Estaci6n Biologica del Beni, a lowland MaB Reserve, and an assessment of the effects of cattle grazing.	At least 5 Bolivian biologists and 6 Bolivian students will be working alongside each of the three, 9-person teams of fieldwork volunteers.	1994 £95,600	Bolivia
48. University of Oxford, Department of Zoology				
Dr Amanda Vincent Department of Zoology University of Oxford South Parks Road Oxford OXI 3PS	To study of the biology of Indo-Pacific seahorses, assess the socioeconomic importance of the seahorse fishery (particularly for local fisherfolk) and initiate management plans for seahorse conservation.		1994 £135,100	Phiippines & Vietnam
49. The Royal Botanic Garden, Edinburgh				
Dr D.F. Chamberlain The Royal Botanic Garden Inverleith Row Edinburgh EH3 5LR	This project will provide specialist training to students from China in horticultural techniques, especially classification of *Rhododendron*. Assistance will also be given on the construction of appropriate databases and training in their use.	Basically a training project.	1994 £53,400	UK
50. The Royal Botanic Garden, Edinburgh				
Dr S. Lindsay Royal Botanic Gardens Inverleith Row Edinburgh EH3 5LR	To study the extent of *Adiantum asarifolium* in the wild and consider, if needed, re-introduction from cultivated material.	Local involvement.	1994 £39,800	Mauritius & UK
51. Royal Botanic Gardens, Kew				
Diana Bridson The Herbarium Royal Botanic Gardens, Kew Richmond Surrey TW9 3BW	To develop and run a course with Russian scientists in Moscow or St Petersburgh for herbarium managers and curators. This will help the participants to identify and prioritise the needs of their institution where collections are physically in danger and much of the data is inaccessible.	Predominantly a training project.	1994 £63,600	Russia
52. Royal Geographical Society				
Nigel Winser Deputy Director Royal Geographical Society Kensington Gore London SW7 2AR	To study.the terrestrial invertebrates of the Mkomazi Game Reserve, an *Acacia-Commiphora* savanna ecosystem and determine the effects of buming and grazing pressure on diversity.	The project will involve training Tanzanian staff.	1994 £118,00	N.E.Tanz-ania

Box 5.27 *continued*

Contact's name and address	Project synopsis	Training potential	Starting date & funding	Location
53. Tropical Biology Association: University of Bristol				
Dr Leon Bennun School of Biological Sciences University of Bristol Woodlands Road Bristol BS8 IUG	To establish a 30-month research fellowship at a research and teaching institution in Kenya or Tanzania in which the Darwin Fellow will undertake research on an applied topic in biodiversity conservation.	The Darwin Fellow will spend three months attached to an appropriate institution in the UK	1994 £34,700	Kenya or Tanzania
54. Botanic Gardens Conservation International				
Julia Willison Botanic Gardens Conservation International Descanso House 199 Kew Road Richmond Surrey TW9 3BW	BGCI and the Nanjing Botanical Garden, China, will run a two week Darwin Initiative education training course for staff working in Chinese botanic gardens. The course will develop education programmes and act as a catalyst for the development of environmental education programmes in botanic gardens throughout China.	Predominantly a training project	1995 £13,000	China
55. University of Durham and Royal Geographical Society: Royal Society for the Conservation of Nature, Jordan				
Dr Roderic Dutton Centre for Overseas Research and Development University of Durham Science Laboratories South Road Durham City DH1 3LE	To assess the biodiversity of north-east Badia for the protection and/or re-introduction of indigenous species, including establishing the reintroduction of the Arabian oryx. Throughout its work it will pay particular attention to the views of the local communities.	Predominantly a training project	1995 £163,000	Jordan
56. Earthwatch Europe				
Dr Robert Barrington Earthwatch Bolsyre Court 57 Woodstock Road Oxford OX2 6HU	To provide twenty Fellowships each year for three years for southern African protected area managers enabling them to join British-led biodiversity field projects. Each placement will last for two weeks. Fellows will also learn the practical details of organising a field research projects and how volunteers can be integrated into such programmes.	Predominantly a training project	1995 £137,000	South–Africa
57. International Centre for Conservation Education				
Mr Mark Boulton ICCE Greenfield House Guiting Power Cheltenham GL54 5TZ	To produce, print and distribute a Basic Ecology Guide for use by trainers and educators in the field of biodiversity/environmental education in developing countries. The guide will be offered to non-specialist trainers, educators and group leaders.	Training of NGO personnel.	1995 £30,000	Africa
58. International Centre for Conservation Education				
Mr Mark Boulton ICCE Greenfield House Guiting Power Cheltenham GL54 5TZ	To facilitate a seminar, bringing together senior Wildlife Club personnel from seven African countries to share previous experience and build new networks and capacity for the future. To prepare a wildlife clubs' handbook to guide the development of Club organisation.	Preparing training material.	1995 £10,000	Worldwide
59. IUCN Shark Specialist Group: Sabah Fisheries Department; WWF Malaysia				
Sarah Fowler The Nature Conservation Bureau Ltd 36 Kingfisher Court Hambridge Road Newbury Berkshire RG14 5SJ	To investigate the biodiversity, distribution and conservation needs of elasmobranchs (sharks, rays, sawfish) in rivers, estuaries and inshore waters of Sabah. The study will address their socio-economic importance, the need for fisheries management, protected areas and education of local people, and will inform decision-makers.	Education of local people	1995 £76,000	East Malaysia

Box 5.27 *continued*

Contact's name and address	Project synopsis	Training potential	Starting date & funding	Location
60. University of Leicester				
Prof Andrew Millington Department of Geography University of Leicester Leicester LE1 7RH	To develop a methodology for mapping forest and related habitats in three key national parks in Bolivia using a methodology which combines the analysis of remotely sensed data and field survey; and to monitor changes in the spatial distribution of these habitats.	Training of Bolivian personnel	1995 £114,000	Bolivia
61. London Ecology Unit: Chilean Ministry of Housing and Urban Development; University of Chile				
Prof David A. Goode London Ecology Unit Bedford House 125 Camden High Street London NW1 7JR	To raise awareness of biodiversity and produce a Biodiversity Action Plan for the city of Santiago. The project will address the use of native species in open space management in the city and techniques for the management of wildlife habitats for nature conservation in urban areas.	Training in ecological survey methods	1995 £77,000	Chile
62. Natural History Museum: Fundacion Moises Bertoni				
Dr Sandra Knapp The Natural History Museum Cromwell Road London SW7 5BD	Direct a programme of sustainable use in the buffer zone around Mbaracayú Forest Nature Reserve in eastern Paraguay. This project will link this work by selecting groups from plants and medically important insects for an inventory of the reserve.Training includes collection and preliminary identification techniques.	Train two professional taxonomists and up to six parataxonomists.	1995 £206,000	Paraguay
63. Overseas Development Institute: Agricultural Research Department, Zimbabwe				
Elizabeth Cromwell Overseas Development Institute Regent's College Inner Circle Regent's Park London NW1 4NS	To develop, use and publicise a replicable methodology for identifying farming communities, and individual households within them, who are interested and able to conserve crop genetic resources in situ.		1995 £69,000	Zimbabwe
64. Royal Holloway Institute for Environmental Research				
Prof Edward Maltby Royal Holloway Institute for Environmental Research Royal Holloway College University of London Huntersdale Callow Hill Virginia Water GU25 4LN	To investigate the biodiversity of *Melaleuca* forests in the Mekong Delta, Vietnam and the environmental linkages responsible for its maintenance, in order to formulate guidelines for optimal sustainable management.	Training of local scientists, local workshops	1995 £170,000	Vietnam and Thailand
65. University of St Andrews: University of the West Indies				
Dr Anne E. Magurran School of Biological and Medical Sciences University of St Andrews Bute Building Fife KY16 9TS	To quantify current levels of biodiversity in fish communities in Trinidad and Tobago, especially to evaluate the decline in biodiversity over recent decades and to assess the impact of current threats such as deforestation, pollution and siltation. Attention will be paid to the distribution and abundance of indicator species.		1995 £87,000	Trinidad
66. World Conservation Monitoring Centre				
Harriet Gillett World Conservation Monitoring Centre 219 Huntington Road Cambridge CB3 0DL	To gather data on the status, distribution and utilisation of plant genetic resources of Central America as a basis for their conservation and sustainable use.		1995 £67,000	Mexico & Costa Rica

Box 5.27 *continued*

Contact's name and address	Project synopsis	Training potential	Starting date & funding	Location
67. Lancaster University: Universities of Tirana, Albania, Riga, Latvia; Botanical Institutes of Ufa, Russia; Prühonice, Czech Republic and Bratislava, Slovakia				
Dr John Rodwell Unit of Vegetation Science Lancaster University Lancaster LA1 4YQ	To raise awareness of the importance of biodiversity among the natural and semi-natural plant communities of eastern Europe and to lay foundations for their sustainable management and establish database for a Red Data Book.		1995 £125,000	Latvia, Russia, Czech and Slovak Republics and Albania
68. University of Oxford: Institute of Terrestrial Ecology				
Dr David W. Macdonald Department of Zoology University of Oxford South Parks Road Oxford OX1 3PS	A project with Tallinn Zoo, Estonia and the Central Forest Biosphere Reserve, Russia, to study the mechanics underlying the diversity of semi-aquatic carnivores, mink and otter.		1995 £125,000	Russia, Estonia and Belarus
69. University of Bradford				
Prof John D. MacAuthur Development and Project Planning Centre University of Bradford Bradford BD7 1DP	This project will continue the funding from the first round project to enable six students to attend the six week international course on Planning of Projects for Biodiversity Conservation. The course will cover project identification, design and costing, finance, economic analysis and justification.	A training project.	1995 £45,000	Worldwide
70. Durrell Institute of Conservation and Ecology				
Prof Ian Swingland The Durrell Institute of Conservation and Ecology Kent Research and Development Centre The University Canterbury Kent CT2 7PD	Three trainees each from Tanzania, Eritrea, Zaire and Madagascar will study biodiversity management (Msc Conservation Biology course) at DICE. At the end of the year's course they will return to their country and undertake field programmes under the co-supervision of their home institution and DICE. The implementation phase will be particularly focused on taxanomic groups which have received relatively little conservation attention in the countries concerned.	A training project.	1995 £124,000	Tanzania, Zaire and Madagascar
71. University of Edinburgh: Royal Botanic Garden, Edinburgh				
Dr Adrian Newton Institute of Ecology and Resourse Management University of Edinburgh Darwin Building Mayfield Road Edinburgh EH9 3JU	A project to help develop the appropriate conservation methodologies for the Maya region of South East Mexico. It will include more detailed training for three members from the Colegio de la Frontera Sur in the UK. Research will focus on the conservation and use of indigenous plant species by local people, the impacts of forest fragmentation on biodiversity and the conservation of endangered tree species.	A training project.	1995 £50,000	Mexico
72. Field Studies Council: Natural History Museum				
Dr Stephen Tilling Field Studies Council Preston Montford Montford Bridge Shewsbury SY4 1HW	To provide a short course for five selected students from South East Asia to attend a seven week course in the UK on preparing AIDGAP material for identifying difficult groups of plants and animals	Predominantly a training and education project.	1995 £103,000	South East Asia
73. Natural History Museum				
Dr Paul Eggleton Department of Entomology The Natural History Museum Cromwell Road London SW7 5BD	To study termite diversity in unlogged, logged and regenerated forest in the Danum Valley Reserve, and quantify the production of CH_4 and CO_2 from defined assemblages by direct measurement of gas fluxes from representative species and associated mound/soil systems.	Malaysian students will be trained in the UK.	1995 £15,000	Malaysia

Box 5.27 *continued*

Contact's name and address	Project synopsis	Training potential	Starting date & funding	Location
74. Royal Botanic Garden Edinburgh: University of Edinburgh				
M.F. Gardner Royal Botanic Garden Edinburgh Inverleith Row Edinburgh EH3 5LR	To develop a collaborative research programme based at the field station on Chiloe Island aimed at the in situ and ex situ conservation of endangered tree species in a nearby forest reserve.	Provide training & for scientists to visit the UK.	1995 £87,000	Chile
75. Royal Botanic Gardens, Kew: Forest Research Institute, Kepong-Kuala Lumpur; Forest Research Centre,Sandakan				
Mr Barrie Blewett Herbarium Royal Botanic Gardens Kew Richmond Surrey TW9 3AE	To carry out a 3-week training course for herbarium techniques in Malaysia. It will cover a range of herbarium techniques and fieldwork skills.	Predominantly a training project.	1995 £15,000	Malaysia
76. Royal Botanic Gardens, Kew				
Dr Colin Clubbe School of Horticulture Royal Botanic Gardens Kew Richmond Surrey TW9 3AE	A training project directed at technicians from the UK Dependent Territories and other countries for which a two month course will provide training in the propagation and maintenance of rare and endangered species.	Predominantly a training project.	1995 £117,000	UK Dependent Territories and others
77. British Bryological Society: Makerere University				
Nicholas G. Hodgetts Joint Nature Conservation Committee Monkstone House City Road Peterborough PE1 1JY	A research project with Uganda to make an inventory of the bryophytes in the Afromontane region (particularly the Bwindi-Impenetrable Forest, Mgahinga Forest and the Ruwenzori Mountains).		1995 £45,000	East Africa
78. International Institute of Entomology (an Institute of CAB International)				
Dr Valerie K. Brown International Institute of Entomology 56 Queen's Gate London SW7 5JR	To quantify the effects of selective logging on the diversity and abundance of insect herbivores in Guyana. The effects of logging on the composition of the tree seedling community, the chemical defences of the seedlings and the insects associated with them will be identified.	Train students in field sampling methods, analysis and interpretation.	1995 £140,000	Guyana
79. Marine Conservation Society: National Aquatic Resources Agengy, Sri Lanka; Coastal Resources Management Project				
Dr Elizabeth Wood Marine Conservation Society Hollybush Chequers Lane Eversley Hook Hants RG27 0NY	To carry out research on coral reef fish and invertebrate populations and reef condition. Field Officer and Research Assistant will be funded through the Initiative. Management stratergies for the ornamental fishery to devise in collaboration with conservation agencies, local communities, exporters and collectors.	To give training courses in reef survey and monitoring.	1995 £128,000	Sri Lanka
80. Natural History Museum				
Dr Nigel Merrett Department of Zoology The Natural History Museum Cromwell Road London SW7 5BD	To foster research on sustainable development of hitherto neglected deep demersal fish resources of the Maldive Islands in which a Maldivean biologist will undertake a 6-week tailored course in deep demersal fish taxonomy in London. On his return to the Maldives he will begin to record and assess material collected recently in order to publish a check list of the deep demersal fishes of the Maldives.		1995 £15,000	Maldive Islands

Box 5.27 *continued*

Contact's name and address	Project synopsis	Training potential	Starting date & funding	Location
81. Plymouth Marine Laboratory: Mahidol University, Thailand				
Dr A.J.S. Hawkins Plymouth Marine Laboratory West Hoe Plymouth Devon PL1 3DH	To assess gene flow and genetic diversity within each main species of native oyster sampled throughout Thailand. The genetic consequences of pollution in population surviving within contaminated areas will also be examined.		1995 £72,000	Thailand
82. Society for Environmental Exploration: University of York: National Environment Commission, Mozambique				
Ms Eibleis Fanning The Society for Environmental Exploration 77 Leonard Street London EC2A 4QS	To survey the marine habits and resource use patterns of the inshore coastal zone between Pemba Bay and Ibo Island in Cabo Delagado province in northern Mozambique. The study will make recommendations for the future sustainable management of the area and for identifying key sites of biodiversity for protection.	Training for sixteen Mozambique students and scientist.	1995 £152,000	Mozambique
83. Botanic Gardens Conservation International				
Dr Peter Wyse Jackson Botanic Gardens Conservation International Descanso House 199 Kew Road Richmond Surrey TW9 3BW	A project to create a biodiversity database system for the Indonesia Network for Plant Conservation (InetPC) and the Indonesian Botanic Gardens (Kebun Raya Indonesia) for their living plant collections.		1995 £130,000	Indonesia
84. Natural History Museum				
Mr J.R. Press Department of Botany The Natural History Museum Cromwell Road London SW7 5BD	To produce two field guides, one ethnobotanical and the other on plants in agriculture, documenting the practices of the Tawahka peoples of Honduras. The guides will be illustrated and bilingual (Spanish/Misquito).		1995 £41,000	Honduras
85. Royal Society for the Protection of Birds				
Jim Stevenson The Royal Society for the Protection of Birds The Lodge Sandy Bedfordshire SG19 2DL	To compile a directory of the sites of ornithological importance in Tanzania and to feed the results into the Tanzanian national Biodiversity Action Plan under the requirements of the Biodiversity Convention.		1995 £84,000	Tanzania
86. University of York				
Prof Charles Perrings Department of Environmental Economics and Environmental Management University of York Heslington York YO1 5DD	To strengthen capacity in the analysis of, and appropriate change in East, Central and Southern Africa. The project will also identify the scope for complementary regional conservation strategies under the Biodiversity and Desertification Conventions, and will make recommendations on both.		1995 £91,000	Kenya, Zambia and Botswana

Students should be aware that unauthorised employment or claiming certain state benefits will break the conditions of the student visa and, if they do this, they may be asked to leave Britain before the end of a course.

Information on fellowships, grants, etc., for university staff in a Commonwealth country who wish to carry out research, make study visits or teach at a university in another Commonwealth country, are given in *Awards for University Teachers and Research Workers*. (ACU, biennial); and *Education Grants Directory* (DSC, biennial). The latter lists over 1,100 educational charities. Most will be only for small grants for very specific courses or projects. *The Grants Register* (Macmillan, biennial) includes information on a variety of scholarships, fellowships and research grants, exchange and vacation study opportunities, professional and vocational awards. *Study Abroad* (UNESCO, biennial) lists scholarships to study in over 100 countries world-wide including Britain, arranged by country, and including international awards.

5.6.2 Applying for funding

Competition for grants to study in Britain is extremely fierce. Most grant-making organisations require applicants to meet specific requirements such as: age; sex; nationality; and subject of study. Many of the UK government awards also require them to be nominated by their own government before final selection for the award. If they do not meet all the eligibility requirements of a grant an applicant will not be considered.

The following should also be considered when applying for funding:

- Applications for full awards must normally be made at least one year before the course starts. The deadline for requests for short-term help varies from one grant-making body to another.
- The application should be clear and concise. If the grant making body has an application form this must be used.
- The majority of grant-making charities, listed in the publications referred to, only give small amounts of money which would not cover the full cost of a course.

5.6.3 Awards for postgraduate study and research

The EU Programme entitled 'Co-operation with third world countries and international organisations' provides funding in the field of biodiversity for research groups in developing countries. A further EU Programme, 'Training and Mobility of Researchers' provides funding for the training within the EU, of individual scientists from developing countries.

5.6.3.1 British Council Fellowships

These awards are for postgraduate study or research. British Council Directors overseas may decide in what subject areas they wish to offer their awards, and they are responsible for the selection of candidates.

Awards range from short attachments to PhD research. The benefits payable to Fellows may vary from small grants up to fully-funded awards which cover fees and living expenses.

The Fellowships are awarded by the British Council Director overseas; students must apply to the British Council office in their own country. No information is available in the United Kingdom. Those considered for an award will be expected to attend an interview in their own country.

5.6.3.2 Commonwealth Scholarships

These scholarships are offered to citizens of Commonwealth countries or British dependent territories who are normally under thirty-five and permanently resident within those areas. They are for postgraduate study or research so applicants should normally hold a university degree or equivalent qualification. If however, there are no undergraduate courses in a particular subject in the applicant's own country or regional university it may be possible for him/her to apply to do a first degree course under this scheme.

Awards are for one to three years and usually cover the cost of travel, tuition fees and living expenses. There may, in some cases, be additional allowances for books or clothes. An allowance may be paid to help with the cost of maintaining a scholar's spouse.

These scholarships are advertised and students should apply to the Commonwealth Scholarship Agency in their own country. Agency addresses are listed in Appendix II of the Commonwealth Universities Yearbook. For those already already in Britain, the Association of Commonwealth Universities can help with general information concerning the Scheme, but it cannot issue application forms to overseas students.

5.6.3.3 The British Chevening Scholarships (formerly the FCOSAS)

The principal aim of The British Chevening Scholarships is to provide study awards in the

Box 5.28 The British Council

The British Council's purpose is to promote a wider knowledge of the UK and to encourage scientific, technological and educational co-operation with other countries. The Council works in 228 cities in 109 countries and plays a significant role in aid delivery on behalf of the British government. Through its global network of offices the Council forges long-term relationships with key institutions and individuals, and supports sustainable and environmentally sound economic and social development. The Council maintains science qualified staff in many of its overseas offices, supported by a science team based in the UK.

Programmes of professional short courses and training schemes in the UK

The Council manages training programmes for approximately 12,000 overseas professionals coming to study in the UK, many of these in the environment field. Clients include ODA, FCO, UN, EC and overseas governments. An increasing proportion come to study biodiversity conservation and related subjects.

It further organises around 90 professional short courses every year, showcasing British expertise in areas of science and technology. Many of these courses are in environmentally-related fields (**see Box 5.3**). Examples include:

- Maintaining Biodiversity: planning for sustainable development (Natural History Museum and Oxford University);
- Biodiversity and change in vegetation: the European experience. (Lancaster University); and
- Protected landscapes: conservation through development (International Centre for Protected Landscapes).

Visitorships

The British Council supports and arranges many visitorships and study tours to the UK each year. Approximately 50 visitors will come specifically to learn about UK strengths in biodiversity studies. Similarly, many UK specialists travel overseas each year under British Council auspices on consultancy and planning missions in the same subject areas.

Workshops and seminars abroad

The British Council organises many overseas workshops and courses on environmental subjects: Biodiversity is becoming an increasingly popular subject area. Examples include: Biodiversity for sustainable development in South East Asia (Indonesia) and the Indo-British workshop on biodiversity conservation (India). The Council also supports Darwin Initiative Projects. For example: Supporting participants in the Tropical Biology Association Training Project. (See Box 5.27).

Higher education links

The British Council supports approximately 80 formalised links in the environment field between research and higher education institutions in the UK and overseas, including eleven links in Biodiversity studies. These links promote joint research, exchange of persons, new training initiatives. Examples include:

- Kew Gardens – Mexico Ecological Institute (setting up a floristic database);
- Aberdeen University – San Marcos University, Peru (avifauna and flora conservation);
- Newcastle University – Aligargh University, India (endangered species conservation in rural and urban environments).

The British Council's Environment Projects fund

A central environmental fund was established in 1991 to support national or regional activities that contribute to environmental and developmental objectives in Developing countries and east and central Europe. More than 60 projects have been funded, including biodiversity conservation and research projects, for example:

- integrated coastal zone management (Ghana);
- developing environmental education centres (Sierra Leone);
- Arctic environmental database (Russia);
- Himalayan biodiversity symposium (India);
- sea turtle conservation project (Egypt);
- establishment of indigenous tree nurseries (Ethiopia);
- *Acacia albidia* conservation (Senegal);
- International Climate Change conference (Zimbabwe).

United Kingdom for present and future leaders, decision-makers and formers of opinion, and also particularly able students from countries with whom the United Kingdom's economic relations are expected to develop. The awards are normally made for formal courses at postgraduate level in most subject fields and preference is given to candidates already established in a career. Awards may be given to cover all or part of the costs of the period of study.

Some British Chevening Scholarships are jointly funded with leading industrial and commercial firms and grant-giving foundations. These are usually advertised locally in those countries where the awards are to be offered.

Applications and enquiries should be made to the British Embassy, High Commission or British Council office in the applicant's own country.

5.6.3.4 ODA Shared Scholarship Scheme (ODASS)

This scheme is jointly funded by the Overseas Development Administration (ODA) and certain British higher education institutions. It is to help students of high academic calibre in developing Commonwealth countries who would benefit from higher education. To be eligible, an applicant must be from a developing Commonwealth country and not be employed by their government or parastate organisations. They must have good English and should not be living or studying in a developed country. Awards are generally for taught postgraduate courses although undergraduate courses not available locally may be considered. Preference will be given to subjects related to economic and social development but could include postgraduate courses in biodiversity studies and sustainable utilization.

Further information and application forms are available from the British institutions participating in the scheme who will select those applicants whom they think are of the highest calibre. A list of participating institutions can be obtained from the British High Commission in the country concerned.

5.6.3.5 Overseas Research Student Awards Scheme (ORS)

These are awarded on a competitive basis to postgraduate students who undertake full-time study for a higher degree as registered research students at British universities. Academic merit and research potential are the criteria governing selection. Each application is marked on the basis of the selectors' assessment both of the candidates' achievements and also of their promise and research potential as attested by a referee's report.

All postgraduate research students attending higher education institutions funded by the Higher Education Funding Council for England (HEFCE), the Scottish Higher Education Funding Council (SHEFC), the Higher Education Funding Council for Wales (HEFCW) and the Department of Education for Northern Ireland, are eligible to apply.

Initially the awards are for one year, but they are renewable for a second and third year subject to satisfactory progress. These awards make up the difference between 'home' and 'overseas' student fees.

Application forms are available through the British higher education institution which has accepted, or provisionally accepted the student for the course.

5.6.3.6 International agencies and organisations

Award schemes for postgraduate study and research are also funded in some countries by international agencies and organisations such as UNESCO or the European Union. For information about these students should apply to the offices of these bodies in their own country or through their Government Education Department. Information is also available in '*Study abroad*'.

5.6.4 Awards for other training – technical cooperation training

5.6.4.1 Technical cooperation training (TCT)

Training is available in a wide range of subjects to develop the knowledge and skills needed for economic, social and technical development. In many countries awards are offered exclusively for training which is directly related to development projects which are supported by the British government. There are TCT programmes in over 100 countries and over

4,000 new awards were being offered worldwide in 1994/95. Applicants must be nominated by their own governments, and should normally be under forty-five years old. Awards are normally for post-experience training and priority is given to those who will themselves become trainers. In certain countries undergraduate training may be considered where suitable courses are not available locally but not for the study of humanities, fine arts or cultural subjects.

Awards may be offered for up to three years. Normally the full costs, including air fares, are covered but there is some variation between different countries.

Applicants must be nominated by their own government to the local British Council office, Embassy or High Commission.

5.7 References and bibliography

ABRC, (1977). *Taxonomy in Britain*. The Advisory Board for the Research Councils. HMSO, London.

ACU, (1994, biennial). *Awards for university teachers and research workers*. Association of Commonwealth Universities, London.

ACU, (1995a, annual) *Commonwealth Universities' Yearbook*. Association of Commonwealth Universities, London.

ACU, (1995b, annual) *Awards for postgraduate study at Commonwealth Universities*. Association of Commonwealth Universities, London.

Baines, J & James, B. (eds) (1994). *Directory of environmental courses* 1994–1996. The Environment Council, London.

BBSRC, (1995). *Biotechnology and Biological Sciences Research Council*. 1994/1995 *Handbook*. BBSRC, Swindon.

Bibby, C.J., Collar, N.J., Crosby, M.J., Heath, M.F., Imboden, Ch.,Johnson, T.H.,Long, A.J., Stattersfield, A.J. and Thirgood, S.J. (1992) *Putting biodiversity on the map: Priority areas for global conservation*. International Council for Bird Preservation, Cambridge.

Blackmore, S. (1995) Welcome to the network. *Systematic Biology Network Newsletter* 1: 1.

Bridson, D.B. and Forman, L, (eds). (1992). *The herbarium handbook*. (2 edn). Royal Botanic Gardens, Kew.

British Council, (1994). *Sources of financial assistance for overseas students*. (BC Information sheet). British Council, London.

CABI, (1994) *Training courses* 1994–1995. CAB International, Wallingford.

Claridge, M. F. and Ingrouille, M. (1992). Sysytematic biology and higher education in the UK. In *Taxonomy in the 1990's*: 39–48.

DoE, (1993). *Darwin Initiative for the Survival of Species: Report and recommendations of the Advisory Committee*. Department of the Environment, London.

DoE, (1994). *Biodiversity: The UK Action Plan*. Department of the Environment/HMSO, London.

DSC, (1994, biennial). *Education grants directory*. Directory of Social Change, London.

Edwards, I.D., (1993). *Natural History of Seram*. Intercept, Andover.

Europa, (1994, annual). *The world of learning*. Europa Publications, London.

HoL, (1992); *Systematic biology research*. House of Lords Select Committee on Science and Technology (HL Paper 22–I). HMSO, London.

Horne, J. and Zeitlyn, M. (1995, biennial). *The directory of university expertise and facilities: Biosciences*. Oakland Consultancy and Publishing Services, Cambridge.

Knight, W.J. and Holloway, J.D. (eds). (1990). *Insects and rain forests of South-east Asia (Wallacea)*. Royal Entomological Society, London.

Krebs, M. (1992). *Evolution & Biodiversity – the New Taxonomy*. Natural Environment Research Council, Swindon.

Lee, A.J. (1992). *The directorate of fisheries research: Its origins and development*. Ministry of Agriculture, Fisheries and Food, Lowestoft.

MBA, (1987). The history of the Marine Biological Association of the United Kingdom. *J. Marine Biology* 67: 463–506.

Macmillan (1994, biennial). *The Grants Register*. Macmillan, London.

UCAS, (1995). *The UCAS Handbook: 1995 Entry*. Universities and Colleges Admissions Services, Cheltenham.

UNESCO, (1994, biennial). *Study abroad*. UNESCO, Paris.

Zuckerman, Lord. (1976). *The Zoological Society of London 1826-1976 and beyond*. (Proc. Sympos., 25-26 Mar. 1976). Academic Press, London.

SPECIMEN DATA BANKS FOR BIODIVERSITY STUDIES

6.1 The role of specimen banks in biodiversity studies

6.2 Museums and herbaria in the UK

6.2.1 The Natural History Museum

6.2.2 The Royal Botanic Gardens, Kew

6.2.3 The Royal Botanic Gardens, Edinburgh

6.2.4 The International Mycological Institute

6.2.5 The National Museums of Scotland, Edinburgh

6.2.6 The National Museum of Wales, Cardiff

6.2.7 Ulster Museum, Belfast

6.2.8 National Museums of Merseyside: Liverpool Museum

6.2.9 Institute of Oceanographic Sciences

6.2.10 Municipal and other museums

6.3 Botanic gardens and plant gene-banks

6.3.1 Royal Botanic Gardens, Kew

6.3.2 Royal Botanic Garden, Edinburgh

6.3.3 Crop plants and UK plant gene banks

6.3.4 Plants of ornamental value

6.3.5 Plants of international conservation value – *Ex situ* collections

6.3.6 Seed and tissue conservation

6.4 Zoological gardens: importance of living animal collections

6.5 Collections of microorganisms

6.6 References and bibliography

Chapter authorship

This chapter was researched and edited by Clive Jermy *and* Martin Sands *with contributions and comments from* L. Alderson, M. Ambrose, M. Angel, R. Barnett, H. Buckley, J. Cherfas, H. Chesney, Q. Cronk, S. Davis, G. Douglas, J Edmondson, S. Edwards, S. FitzGerald, P. Hackney, F. Howie, T. Irwin, E. Leadley, G. Legg, D. Long, P. Morgan, P. Olney, C. Pettitt, D. Rae, R. Rowe, M. Shaw, M. Stephenson, N. Stork, M. Turner, K. Walter, G. Whalley, M. Wilkinson, T. Wilkinson, S. Winser *and* P. Wyse Jackson.

6.1 The role of specimen banks in biodiversity studies: general comments

This chapter discusses the different forms of data that have been amassed in the UK, in some cases over a considerable period of time, and which now can be used to help those in countries abroad who are considering making inventories of their own biodiversity. *The Guidelines for Country Studies on Biological Diversity* (UNEP, 1993) provides detailed recommendations on what information needs to be collected to support these studies in a number of categories and suggests ways in which this data might be collected, analysed and managed (see also this Volume, Chapter Two). Amongst the suggestions is the recommendation that data gathering can be significantly advanced if the main out-of country sources are accessed early in the planning stage.

From earliest times the inquisitive nature of man has lead to the collection and cataloguing of nature's curiosities. Initially, their relationships and classification was a challenge and their origins mattered little but, as the potential for travel increased, their relationship to geography became more important. Private collections assembled by those with the education and resources to do so, were left to posterity through state or municipal ownership. Invariably the specimens collected by these early naturalists came from areas beyond their national boundaries. Their study and cataloguing initiated the science of taxonomy and developed man's approach to the biological diversity that surrounded him. The total number of natural history specimens in the world's museums and herbaria has been estimated to be well over 2 billion (see **Box 6.1**).

Whilst the academic study of taxonomy and systematics has made the greatest use of these collections, a practical application has been to produce manuals and identification guides to the flora and fauna of a country or region. (See the work of the Natural History Museum and the Royal Botanic Gardens, this volume, Chapter Four and Sections 6.2 and 6.3 below). Thus the museums and herbaria are, in themselves, substantial databases of the world's biological diversity. Very few of them are indexed in detail but there is now an added value attached to this information if it can be made accessible to those who need it. Furthermore, and most important, unidentified material collected in surveys can be compared with named material in the museum or herbarium. Voucher material collected in biodiversity surveys has to be preserved for later determination; Volumes Two and Three of this manual elaborate on these aspects. Once correctly named these biological specimens should be added to the appropriate national, or local collection (e.g. in a park headquarters) to help future identification, and also as a tangible record of an organism's presence at a stated time and place.

Information on major collections and their locations in the world's herbaria can be found in Holmgren *et al.* (1990) and Vitt *et al.* (1985). Major insect and spider collections of the world have been listed in Arnett & Samuelson (1986). Ritchie (1987) reviewed the current status and future prospects of African collections. Preserved entomological reference collections, to be found within 85 tropical countries, are listed by Hawksworth & Ritchie (1993). Similarly the last-mentioned work also lists collections of plant nematodes and animal helminths in the tropical countries. More detailed information on non-living biological collections held in the UK is given in Section 6.2.

It is not only collections of preserved specimens that are important to biodiversity studies. Living plants, whether as crops or ornamentals, or scientific and educational collections in botanic gardens, as well as their animal counterparts on farms and in zoos (including aquaria), form part of a viable biodiversity database and a 'gene bank', which can be used by plant and animal breeders to improve crops and domesticated animals. Animals and plants 'in captivity' (generally called *ex-situ* collections) are also valuable in conserving endangered species. Both, with the right understanding of ecology and breeding behaviour, can be re-instated into their former habitats once the threats affecting them are removed. Similarly, and often more easily, microorganisms such as strains of algae, fungi, bacteria, protozoa, phages and plasmids are also kept as live cultures, for pathological, medical and other reasons (e.g. genetic engineering). They too add to our picture of biodiversity (see Section 6.5).

Box 6.1 Natural science collections: Estimated worldwide holdings
Interim figures are based on actual holdings in 100 museums sampled 1991/92. Data researched by F. Howie, Natural History Museum, London

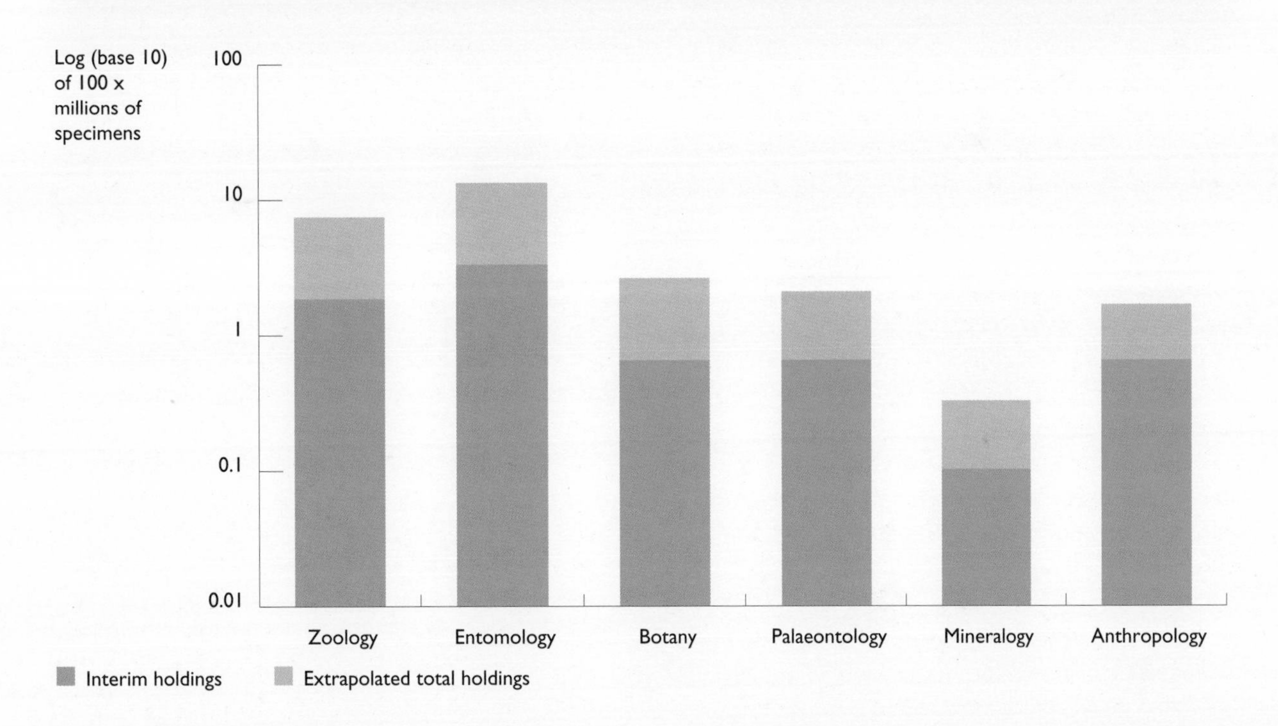

There are about 1,600 botanic gardens in the world (Wyse Jackson, 1995). The increasing rate of loss of plant diversity and wild habitats worldwide has encouraged many botanic gardens to become important conservation centres. Many new botanic gardens are being opened or developed to act as centres for such plant study and education, particularly of plants native to their own regions. In the context of this chapter, their collections are a database of information about plants from a given area.

Two international organisations collect and disseminate information about botanical gardens: the International Association of Botanic Gardens (IABG), and the Botanic Gardens Conservation International (BGCI). The latter, an initiative of the World Conservation Union (IUCN) and formerly its Botanical Gardens Conservation Secretariat but now independent, has its headquarters in the UK (see Chapter Five). It plays an active role in promoting conservation issues in botanic gardens, and maintains a database of living plant collections of its members. The two major botanic gardens in the UK, the Royal Botanic Gardens, Kew and the Royal Botanic Garden, Edinburgh, maintain their own extensive databases but work closely with BGCI, as does the World Conservation Monitoring Centre at Cambridge.

While the majority of botanic gardens can be defined as public gardens which maintain large, wide-ranging collections of live plants – for scientific research, conservation, education or simply enjoyment, other living collections may be maintained solely for their genetic potential. On the international level the International Plant Genetic Resources Institute (IPGRI) coordinates genetic research in a number of crop plant research centres and international gene banks, and is itself part of a larger consortium, the Consultative Group on International Agricultural Research (CGIAR) (see Chapter Four). UK plant genetic resources consist of both gene banks and technical knowledge generated by their own research, and contribute not only to the bodies mentioned above, but also to many UK initiatives sponsored by the Ministry of Agriculture, Fisheries and Food (MAFF) and appropriate Research Councils.

There are some 900 major zoological gardens in the world (Olney & Ellis, 1993). As with botanic gardens, there is an international fraternity that promotes an interchange of ideas and information in the form of the International Union of Directors of Zoological Gardens (IUDZG), the World Zoo Organisation (WZO) with its eight regional organisations and the twenty national associations encompassing over a thousand zoos. The Species Survival Commission of IUCN has set up a Captive Breeding Specialist Group which promotes conservation activities in zoos and into which UK zoologists make a significant input. Unlike plants, animals have greater health problems when maintained in captivity and considerable resources have to be found to cover this aspect. Accordingly fees for visiting zoos are therefore normally higher than for visiting botanic gardens. Collections of living animals may also be kept in specialised conditions such as aquaria, insectaria, butterfly houses, or more simply as breeding populations of specific breeds of domesticated animals. The role of zoos is greatest in maintaining live healthy animals of rare or endangered species for breeding and subsequent translocation into the appropriate wild site.

The UK has substantial holdings of both plant and animal 'gene banks' in botanic and zoological gardens and other specialist collections. These are discussed in Sections 6.3 and 6.4 below.

6.2 Museums and herbaria

A number of UK museums and herbaria hold and maintain historically important national collections of international importance. Those of The Natural History Museum (formerly the British Museum (Natural History)), the two national herbaria at the Royal Botanical Gardens at Kew and Edinburgh, and leading museums elsewhere in the UK are described below. The Linnean Society of London (see Chapter Five) is the guardian of the internationally important plant and animal collections of Linnaeus (see **Box 6.2**), purchased from Linnaeus' widow in 1783. While not directly widening our knowledge of a country's biodiversity, the greatest value of the Linnean collections resides in the many base-line ('type') specimens, originally described by Linnaeus himself and consequently of particular importance to taxonomic study.

There are over 200 public or private museums in the UK with natural history collections, and over 50 of these hold significant collections from overseas. The majority of collections will be bequests and donations, often from competent naturalists who worked abroad. Many were experts in their respective plant or animal group, and published widely. For that reason, their collections frequently included 'type' specimens and much cited material. Other naturalists of the period would send material to their local museum on their return from various expeditions, and these valuable voucher specimens may now be found in the many provincial museums around the UK.

With the exception of a few of the larger museums, restrictions on museum staff resources usually prevent country holdings of even selected taxa being extracted and sent for biodiversity studies abroad. Currently, only in isolated instances are data in electronic form and therefore easy to transfer, although this is a matter in the forefront of many institutes' development plans. Those scientists preparing country studies, inventory listing, etc. are welcomed and are advised to visit the collections where a curator has confirmed substantial relevant holdings. Where material is loaned for short periods to *bona fide* institutes abroad, the museum will usually cover the expense of dispatch, but borrowers should be aware of the later cost that they will have to bear when it is returned.

An assessment of natural science collections held by UK museums, their contents value, and staff and other resources needed to maintain them, was undertaken under the auspices of the Biological Curators Group (Hancock & Morgan, 1980). Specialist collections will reflect the research carried out in an institute and their acquisition policy when collecting material from the wild. In such institutions the guidelines described in Chapter Two will be followed in principle.

Those plant collections at Kew and The Natural History Museum in many instances will complement each other, reflecting as they do the interests and energies of past directors and leading staff (see below). Similarly, several universities will have built up collections based on research projects of staff and students. (e.g. Mediterranean plant collections at the University of Reading, and the Greenland plants in the University of Lancaster). In other universities collections are maintained for teaching purposes only. During the past 25 years university research and teaching in taxonomy has reduced drastically in favour of anatomical and molecular sciences while maintenance of

Box 6.2 Linnaeus and the Collections

As young men training to be physicians Linnaeus and his friend Petrus Artedi set themselves the task of classifying and naming all living things. After Artedi's untimely death in 1735 Linnaeus worked on alone and eventually provided a concise usable survey of all the world's plants and animals as then known. He distinguished and named over 8,000 species of plants and 4,400 species of animals comprising 828 molluscs, 2,100 insects, 477 fishes, and numerous birds and mammals including *Homo sapiens* whom he classified with the primates.

His publications, by reason of their encyclopaedic scope, helped to establish and standardise the consistent binomial nomenclature for species which is used today. His *Species Plantarum* (1753) and his *Systema Naturae*, 10th edition, volume 1 (1758) have accordingly been accepted, by international agreement, as the official starting points for botanical and zoological nomenclature. This confers a unique scientific importance on the specimens used by Linnaeus, many of which are in his personal collections now treasured by the Linnean Society. These include: 14,000 plants, 3,198 insects, 1,564 shells, 158 dried fishes, a library of 1,600 volumes, and manuscripts and papers relating to almost every aspect of his scientific work, together with over 3,000 letters. The herbarium, especially important for its richness of 'type' material, is used by botanists from all over the world.

Linnean Society of London

Some of the Linnean collection in Burlington House, London.

The **Linnean Society of London** is proud to have the responsibility of maintaining the Linnaean collections and library for the benefit of science and upholding the tradition of international scientific interest so well manifested by Linnaeus himself. However, in the interests of their conservation, they are housed in an atmospherically controlled strong room where, by prior arrangement, they can be, and regularly are, consulted for research.

(*Source*: G. Douglas, *Linnean Society*.)

teaching (and other) specimen collections has been given a low priority. Fortunately with the acceptance of the importance of biodiversity this trend is being reversed (see Chapter Five).

6.2.1 The Natural History Museum

The British Museum was founded in 1756 through the bequest to the nation of the vast collection of books, antiques, coins, minerals, fossils, and botanical and zoological specimens amassed by Sir Hans Sloane. The natural history and geological collections, now housed at South Kensington became the British Museum (Natural History), and since 1989, the Natural History Museum. It includes the zoological collections of Walter Rothschild and the museum which houses them at Tring, Hertfordshire, just north of London.

The NHM, in total, holds some 67 million specimens of which approximately 6 million are plants and fungi. Life science research and the relevant collections are divided between three departments – Botany, Zoology, and Entomology (which also houses the research staff of the International Institute of Entomology, a member of CAB International; see Chapter Four). Details of the Museum and its collections can be found in Anon (1904), Günther (1912) and Stearn (1981). They are world-wide and most groups of plants and animals are well represented. There is a summary listing in the database issued by the South Eastern Collections Research Unit (Bateman *et al.*, 1993. See 6.2.10).

In the Botany Department, the lichen herbarium contains 400,000 specimens of which 10,000 are types. The bryophytes and algae, formerly housed at RBG Kew, are now

The Natural History Museum

A few of the 334 volumes of pressed plants in the historic Sloane Herbarium, NHM.

on permanent loan at the NHM. The total bryophyte collection is the most comprehensive in the world (over 730,000 specimens) and contains some 20,000 'types' (see Vitt *et al.*, 1985). There is also a collection of 18,000 specimens of Myxomycota, but other fungi are now on permanent loan to RBG Kew. The algae collections have worldwide representation of Charophyta (14,000); the marine collections (200,000 herbarium sheets; 11,200 specimens in liquid preservative; 17,000 slides) specialise in Antarctica, Indian Ocean and N.E. Atlantic (see Tittley & Sutton, 1984). The diatom collections include 175,000 microslides, fossil material and 25,000 bottles, the result of a merging of international collections (see Williams, 1988). The pteridophyte herbarium is rich in material from tropical Africa, S.E. Asia, Malaysia, and New Guinea including many cytological vouchers. Flowering plant collections (which include 33,000 'types') are comprehensive with much material from Boreal and Arctic areas, Himalaya, tropical Africa and Central America. There is also a pollen slide collection with specialities in Acanthaceae, Asteraceae and Solanaceae.

The zoological collections at the NHM are as extensive as other groups. The mammal collection (350,000 specimens) is a major resource of importance to the international community. Examples of most described taxa are represented. The collection is also of historic importance containing the early collections of Albert Seba, upon whose specimens many of Linnaeus' names were based (see Gray, 1888; Lyddekar, 1912-1916; Napier & Jenkins, 1970-1990; Oldfield-Thomas 1906). Of the other vertebrates there are 110,000 reptiles, 90,000 amphibia and over one million fish. (Boulenger 1885-89; 1889; 1893-96).

The insect and spider collections are curated by the Entomology Department and the insect orders are summarised in **Box 6.3**. Of the other invertebrate phyla in the Zoology Department, the molluscan and crustacean collections are the most substantial. There are about 8 million specimens of mollusca, including 15,000 cephalopods, approximately 10% of which are 'types'; coverage is worldwide (see Gray, 1849). The largest and most important single collection is that of Hugh Cuming, comprising marine, land and freshwater shells mainly from Polynesia, the Philippines and S. America. Additionally there are major collections of land snails from Africa, India and Australia, and of marine molluscs from various voyages such as the 'Discovery' and 'Challenger'. The collections of sponges and bryozoa are also extensive (>100,000), and in both cases between one-third and one-half of the specimens are types.

The crustacea collections number 6.5 million and are significant for research worldwide (see White, 1847; Bell, 1855; Bate, 1862; Thurston & Allen, 1969). Aschelminthes and parasitic nematodes number some 223,000 specimens and are one of the most comprehensive and largest collections in the world. Similarly, the Platyhelminthes collection is a major repository for type specimens. Other groups of significance include Echinodermata (80,000 specimens) (Clark, 1925; Gisten, 1928), and Annelida and Progonophora (c. 300,000) (see Johnston, 1865).

Close links with universities both in the UK and abroad are maintained and graduate students are encouraged to study for higher degrees using the collections available. Researchers abroad can also apply to borrow collections for taxonomic studies in their own institutions, but such workers wanting to study a considerable amount of material are encouraged to visit the Museum.

The Natural History Museum

An entomologist working on the collection at The Natural History Museum, London.

Box 6.3 Insect collections at the Natural History Museum
Data from AMCRU/M&GC survey 1989 (SECRU, 1993) and P. Hammond (NHM, 1995)

ORDER OR GROUP	APPROX NO OF SPECIMENS	% SPECIES DESCRIBED	NUMBER OR (%) OF TYPES	GEOG AREAS (SPECIAL COLLECTIONS)	USEFUL REFERENCES AND NOTES
Apterygota	50,000	?	200	Worldwide	
Coleoptera (beetles)	>8 mill.	?	75,000	Worldwide incl. Broun, (New Zealand) & Wollaston (Atlantic Islands), Himalayas, India. Central America.	
Diptera (flies)	2 mill.	40	(25)	Worldwide	Theodor (1967) Townsend (1990). Species and genera represented on a computer database
Homoptera – Sternorrhyncha (aphids), Thysanoptera and Psocoptera	2.2 mill.	50	(40) 9,000		Eastop & Lambers (1976) Mound & Halsey (1978)
Hemiptera (bugs)	1 mill.		15,000	Old World	
Hymenoptera (ants, wasps and bees)	3 mill.		21,000	Worldwide Old World	
Isoptera (termites)	300,000	60	(20) 400	Worldwide Old World	
Lepidoptera (macrolep)	3.5 mill.	70	39,000	Old World	Walker (1854–1866) Hampson (1898–1920)
Lepidoptera (microlep.)	1.5 mill.	c. 60	(50) 36,000	Worldwide	Walker (1854–1866) Clarke (1955–1970)
Lepidoptera (butterflies)	2.6 mill.	95	(75)	Worldwide	
Neuropteroid orders (include. dragonflies)	460,000	60	4,000	Worldwide (Commonwealth, Europe)	Davies & Tobin (1985)
Orthopteroid orders (grasshoppers etc)	350,000	50	(18) 5,000	Worldwide	Kirby (1904–1910)
Phthiraptera (lice)	200,000	70–80	(high) 1,800	Worldwide	Hopkins & Clay (1952)
Siphonaptera (fleas)	258,000	57	36	Worldwide	Hopkins & Rothschild (1953–1971); Mardon (1981); Smit (1987)

Such visits from overseas are welcomed but projects should be discussed in advance with the appropriate Head of Department.

Details of the research programme at the Natural History Museum is given in Chapter Four and information on its library resources in Chapter Seven.

6.2.2 The Royal Botanic Gardens, Kew

Both the Herbarium and Library at Kew were founded in 1852 with the presentation of the collections of the British botanist, W.A.Bromfield, to which the extensive herbaria and libraries of George Bentham and Sir William Hooker were added in 1854 and 1867 respectively. In 1879 the entire mycological herbarium (20,000 specimens) was received as a gift from the Rev. M.J. Berkeley, and M.C. Cooke was appointed to curate the Fungi and Algae.

Today the Kew Herbarium, probably the most comprehensive and best-curated in the world, consists of nearly seven million pressed and dried specimens of vascular plants and fungi (c.700,000), including over 275,000 'types', with ancillary collections of material preserved in liquid and of dried fruits and seeds. The herbaria of bryophytes and algae are on permanent loan to the Natural History Museum (see above). The collections, reflecting an active programme of collecting over recent years, and enriched by gifts and specimen exchanges, show an extremely wide coverage, both systematically and geographically. New specimens arrive at the rate of some 45,000 a year and at least 30,000 specimens per year are dispatched to, or received from, some 150 herbaria worldwide as part of a reciprocal exchange and loan programme.

A Palynology Unit, which has conducted extensive pollen surveys in several families, including the Coniferae, Burseraceae,

Sapotaceae, Leguminosae, Acanthaceae and Palmae, has built up a reference pollen collection. The mycological collections are unique both in their world-wide range and in their wealth of authentic and type material. Herbarium collections are increasingly supported by a living culture collection, giving emphasis to the wood-decaying Basidiomycota (both temperate and tropical).

Also at Kew is the Centre for Economic Botany which curates a collection of more than 73,000 items of useful plant products and artefacts, a proportion of which formed the world's first Museum of Economic Botany founded by Sir William Hooker in 1846. Making reference to the collections, the Centre researches various aspects of non-crop plants of economic and potential economic value. The Centre runs the Economic Botany Bibliographic Database with more than 150,000 references which is continually augmented. It is also responsible for the Survey of Economic Plants for Arid and Semi-Arid Lands (SEPASAL) (see **Box 6.4**), which collates and disseminates data on 6,000 under-utilised species from the dry tropics. An important element of the Centre's work is in dealing with the large number of public and commercial enquiries it receives, especially concerning poisonous plants and fungi. A joint project with the Poisons Unit of Guy's & St. Thomas' Hospital Trust has been established to produce an interactive database for the identification of poisonous plants and fungi. A CD-ROM for plants in Britain and Ireland was published in 1995.

For living collections see Section 6.3.1; for information on research at Kew see Chapter Four and Chapter Seven for library resources.

Royal Botanic Gardens, Kew

A sample of the seed bank collections at RBG, Kew.

For further information on the history of the Gardens and Herbarium see Desmond (1995).

6.2.3 The Royal Botanic Garden, Edinburgh

The collections in RBGE of around two million preserved, mostly herbarium, specimens, has its origins in the middle of the 19th Century as an amalgamation of the herbaria of the Botanical Society of Edinburgh and of the Botany Department of the University of Edinburgh; approximately 25,000 are 'types'. This collection has a world-wide representation but it is particularly well endowed with material from the Himalaya, China, S.W. and S.E. Asia and Southern Africa, reflecting past and present research interests. The Scottish collections of all the major plant groups, and of the fungi, represent an important resource for the study of the biodiversity of Northern Britain. There is an active acquisitions policy across all plant groups.

For the living collections of the Royal Botanic Garden, Edinburgh, see Section 6.3.2, and for its research programme see Chapter Four; for library resources see Chapter Seven. See Hedge & Lamond (1970) for index to collectors in Edinburgh Herbarium, and Fletcher & Brown (1970) for a history of the Gardens and Herbarium.

D. Long/RBG, Edinburgh

Party of local botanists, guide and sherpas with the RBG, Edinburgh, team collecting for the Flora of Bhutan Project.

Box 6.4 Survey of Economic Plants for Arid and Semi-Arid Lands (SEPASAL)

F. Cook/ECOS/RBG, Kew

The roots of this *Acacia nebrownii* contain potable water, Botswana.

Arid and semi-arid environments occupy approximately one-third of the world's land surface, including about 50% of the surface area of developing countries. Of the one in six (850 million) people who live in arid and semi-arid lands, more than 80% live in rural areas and are dependent upon agriculture and/or animal husbandry. Many also rely on local plant resources to supply a range of basic commodities, such as food, fodder, fuel and medicines. Sustainable utilisation of plant resources is a critical factor in preventing damage to the environment and improving the quality of life of millions of people.

However, information on useful plants of drylands is often patchy and scattered, and traditional knowledge on the value or management of plant resources varies from place to place. The value of a plant species for a particular purpose may, for example, only be realised in a small part of its geographic range. To address these problems a database (SEPASAL) has been developed and is being maintained at the Royal Botanic Gardens, Kew, with the support of the Clothworkers Foundation. SEPASAL brings together diverse traditional and academic knowledge on useful plants of drylands to enable evaluation and assessment of plant species, and to help transfer knowledge between areas and disciplines, and ultimately to facilitate development and improvement of lives.

At present, the database contains information on approximately 6,000 useful dryland species, excluding major crop species. Data held include: scientific and vernacular names, trade names, geographical distribution (to country or state level), life form and life cycle, habit, uses according to the Economic Botany Data Collection Standard (Cook, 1995), site and climate tolerances (e.g. soil moisture, pH, salinity, moisture, rainfall, frost tolerance, altitudinal range).

Database fields allow information to be recorded on species evaluations, processing, management, cultivation and seed sources. Each data item can be linked to a data source. The database can store photographic images of plants and plant products.

The records can be searched on many of the fields listed above, and can provide assessments of the economic value and potential of individual plant species.

Alongside the SEPASAL database, Kew also maintains an Economic Botany Bibliographic Database which currently contains citations of more than 150,000 references dealing with plants of economic value (including those of arid and semi-arid lands).

The value of the SEPASAL database depends on its use as a research tool for identifying plants which can contribute to sustainable development. SEPASAL welcomes information on useful plants for incorporation into the database. If you are engaged in producing a catalogue of plants used in a certain area, or have expertise in a particular family of plants and could update SEPASAL records on taxonomy, distribution or uses made of particular taxa, or if you would like data from the database for a particular project in your country, contact: SEPASAL, Royal Botanic Gardens, Kew, Richmond, Surrey TW9 3AB. email: SEPASAL@rbgkew.org.uk

Box 6.4 Continued

Some examples of recent enquiries of the SEPASAL database

Users include aid agencies and development organisations, including the World Bank, FAO, the International Centre for Research in Agroforestry (ICRAF), OXFAM and the Catholic Fund for Overseas Development (CAFOD), as well as university departments and many individual research workers, foresters and farmers.

Some recent enquiries have included the following:

- Ethiopian plants with edible fruits;
- salt tolerant plants from Pakistan useful in land reclamation;
- fuelwood plants occurring in regions with rainfall below 300 mm;
- nitrogen-fixing trees and shrubs in the Sahel which can also be used as stock-proof hedges;

F. Cook/ECOS/RBG, Kew

Overgrazed sand dunes in South Botswana.

- Brazilian plants used in water purification;
- evergreen, thorny shrubs for areas above 1,500m;
- essential oil plants used in perfumes;
- plants introduced to Arizona which are forage legumes; and
- medicinal plants from Morocco.

(*Source*: S. Davis, *Royal Botanic Gardens, Kew*.)

6.2.4 The International Mycological Institute

Founded in 1920 as the Commonwealth Mycological Institute, the IMI was formerly housed on Kew Green, Richmond and moved to a purpose-built and refurbished building at Egham, Surrey, in 1992. The herbarium collections at Egham contain over 362,000 dried reference specimens representing over 32,000 different species from 151 countries. It has extensive holdings of living cultures of fungi, yeasts and plant bacteria (see 6.5 below).

IMI is one of four institutes of CAB International and its extensive activity in research and training is described in Chapters Four and Five.

6.2.5 The National Museums of Scotland, Edinburgh

The non-British collections at NMS, Edinburgh, are particularly rich in birds (42,000 skins, 15,000 egg-clutches) from W. Palearctic, W. Africa, Brazil, Mascarenes and Antarctic. In addition, there are 2,200 skins/skeletons of mammals from Madagascar and Africa (primates), N.E. Atlantic (cetaceans), Antarctic (pinnipeds), Seram (rodents), and 45,000 fish, mainly from the N.E. Atlantic mesopelagic waters. Invertebrates include 80,000 molluscs, 350,000 insects (especially Middle East/Arabia/Africa), 24,000 Crustacea and 10,000 Echinodermata with particular strengths in Arctic and Antarctic material.

6.2.6 The National Museum of Wales, Cardiff

The zoological collections, established in 1907, are international in scope, numbering three million specimens, of which two million are marine and palaearctic Mollusca. They are especially strong historically with nomenclatural types, and figured and cited material. In recent years research interests have increased the marine invertebrate collections with material from the Indian Ocean and China Sea. In the plant collections, bryophytes, especially from Europe, are substantial.

Box 6.5 A unique facility for the study of oceanic biodiversity: the 'Discovery Collections' and the 'Sir John Murray Collections'

The World's oceans comprise the largest biome on Earth, and yet remain poorly known. Their living resources are, almost without exception, heavily over-exploited, yet the impact of this exploitation on ocean diversity is generally considered to be of little consequence. Even so, ocean faunas have far greater disparity with representatives of twice as many metazoan phyla as are found in terrestrial ecosystems. In spite of the volume of the life zone within the waters (the ocean being some 180-fold greater than that of the terrestrial biome), the richness of the species inhabiting the water is over 100-fold lower. One possible reason is because there are only about 4,500 species of microscopic planktonic plant, compared with 250,000 terrestrial green plants many of which are large and complex. Even so, at any one place as many different species can be caught in the ocean as in any terrestrial habitat, but the stirring of the ocean by currents eddies ensures that the pelagic species have far more extensive geographical ranges than their terrestrial counterparts.

In contrast, on the floor of the deep-sea recent data imply that the species richness may, rather unexpectedly, be as high as that of the richest of terrestrial habitats. The taxonomic challenge of understanding how this richness has evolved and continues to be maintained in the oceans provides an ideal testing ground for hypotheses concerning the relationships between ecological function and diversity.

The UK holds one of the most comprehensive collections of sea-bed samples and cores divided between the NHM and the IOS, and this provides the basis for research suggested above. It also has the potential for studies of ocean warming and the extent and nature of marine pollution over the last 100 years.

The collection at the **Institute of Oceanographic Sciences**, Southampton, dates from 1926 with material collected as part of the 'Discovery Investigations' from the Southern Ocean, thus now representing nearly seventy years of data from almost 10,000 stations, including the North Atlantic, Southern and Indian Oceans. Samples have been collected with a range of trawls, corers and traps from depths ranging from the sea surface to 5,540m. Stations locations and sampling details have either been published in the form of station lists, or (since 1965) are held in a computer data-base. There is also a biological data-base of pelagic species at IOS, which includes over 5 million identifications of 1,700 taxa involving over 75,000 records.

H. Buckley/Natural History Museum

The Discovery Collections: material from deep sea cores now held in the Natural History Museum.

Collections are held at the **Natural History Museum** and consist of samples from some 30,000 localities. All oceans are represented, the approximate proportions being Atlantic 40%, Pacific 35% and Indian 25%. The most important component is the Sir John Murray Collection which includes the 'HMS Challenger' 1872–76 sea-bed samples, material from the John Murray Expedition 1933–34 to the Indian Ocean and the more recent Denis Curry Collection of sediments and cores from the English Channel and Western Approaches. Material is spread over the Mineralogical, Zoological and Palaeontological Departments. The collection is entered on a computer database, ocean localities based on one degree squares (Marsden Squares), with date, depth, ship name, station number, method of collection, nature of material and reference to published work and slides. A printed version and a microfiche for Library use has been also produced.

(*Source*: M. *Angel*, IOS, *Southampton*; H. *Buckley*, NHM, *London*.)

Box 6.6 Significant smaller herbaria in the UK with international holdings
(Source: Index Herborium (Holmgren et al. 1990))

MUSEUM	FOUND DATE	TOTAL SPECIMENS	ANGIO-SPERMS	FERNS	MOSSES	ALGAE	LICHENS & FUNGI	PARTICULAR COLLECTION OR AREAS
Bradford Municipal Museum	1879	70,000	●	●	●			Europe & N. America
Birmingham University, School of Biological Science	1882	150,000	●	●	●			Worldwide Solanaceae; and crop plants in general a specialist collection
Bristol City Museum	1823	30,000	●	●				Historical Broughton Jamaica includes types
Cambridge University, School of Botany	1761	550,000	●	●	●	●	●	Worldwide; much Eastern European and Russian; Darwin material
Cambridge: British Antarctic Survey	1969	40,000			●	●	●	Antarctic esp. South Georgia; South America
Huddersfield,Tolston Museum	1925	22,000	●	●	●		●	Worldwide
Lancaster University, Plant Science Dept.	1964	44,000	●	●	●			Mainly N. Hemisphere and Arctic
Leicester City Museum Service	1849	90,000	●	●	●		●	Europe; worldwide lichens
Leicester University, Plant Biol.Dept.	1945		●	●	●			Mainly European
Liverpool, Merseyside Museums	1802	305,000	●	●	●	●	●	Sri Lanka, Himalaya & SE Asian; historic colls of J.E.Smith, T. Velley & J.F. Royle
London, Linnean Society	1730	33,800	●	●	●	●	●	Many type of Linneus and his contemporaries; J.E. Smith Herbarium
London, South London Botanical Institute	1910	100,000	●	●	●	●	●	Europe; A.O. Hume herb.
Manchester University (formerly Manchester Museum)	1860	1,000,000	●	●	●	●	●	Spruce Amazonia, Andes; worldwide especially Europe
Newcastle upon Tyne University Plant Sciences Dept.	1900	12,000	●	●		●		General flowering plants;tropical ferns (a specialist research collection) North Sea algae at marine station
Newcastle upon Tyne, Hancock Museum (now part of the University)	1829	62,000	●	●	●			Contains several important foreign collections sold off by Linnean Society
Nottingham City Museum	1857	50,000	●	●	●	●	●	North America, Asia, Europe
Oxford University, Forest Institute	1924	175,000	●					Predominantly woody tropical plants, and Africa especially
Oxford University Plant Sciences Dept. Fielding/Druce Herbarium	1621	375,000	●	●	●	●	●	Worldwide especially arctic and South America Historical, pre 1800
Southampton University, Biology Dept.	1960	15,000	●					Worldwide Leguminosae (Viciaea); a specialist collection

6.2.7 Ulster Museum, Belfast

The Ulster Museum holds collections of around 350,000 zoological and nearly 250,000 botanical specimens. The non-British collections contain some 60,000 insects and considerable Lepidoptera from Europe and Malaysia; also Diptera from the Falkland Island. Mollusca collections of some 8,000 specimens include materials from N. America and Pacific. The mammals include c.50 species of marsupials and the Tasmanian wolf. The herbarium contains significant collections of European vascular plants and bryophytes, and William Harvey's marine algae from the Southern Hemisphere. North and West African flowering plant material is also well represented.

6.2.8 National Museums of Merseyside: Liverpool Museum

The natural history collections at the Liverpool Museum are both national and international in extent and importance. The plant collections include 12,000 marine algae, mainly from N.E. Atlantic and the Baltic Sea; 95,000 European and 31,000 extra-European phanerogams, including collections made in the Himalaya and S. India by J.F. Royle; 11,000 ferns with particular strength in Sri Lanka and S.E. Asia; and 23,000 bryophytes which include material from the West Indies.

The zoological collections include 52,000 bird and mammal skins which are second only to the NHM in importance; 2,500 amphibia and reptiles, mostly African, fish material from the 'Challenger' Expedition; and 700,000 insects, mostly of British and Palaearctic provenance. Marine invertebrates and shells are also of international importance.

6.2.9 Institute of Oceanographic Sciences

The IOS (see Chapter Four) maintains the 'Discovery' Collection which is the largest ocean plankton collection in Europe representing nearly seventy years of data from 10,000 stations in North Atlantic, Southern and Indian Oceans. Part of the collection is housed at the NHM, London (see **Box 6.5**).

6.2.10 Municipal and other museums

Municipal (i.e. public) museums in the UK are mainly institutions for education rather than research but many have trained biologists employed as curators competent in preserving and maintaining local or national collections. These staff are also broad-based naturalists who often play a significant part in local biodiversity studies, and many, being specialists in their selected disciplines, can also contribute to biodiversity studies at a world level. Professional bodies and coordinating organisations are the Museums' Association and the Biological Curators Group. Funding for such museums and other collections is for the most part the responsibility of either the local municipal government, or in the case of those attached to universities (e.g. Manchester Museum) through the government grant-in-aid to that university. Additionally the Area Museum Councils (funded by the Museums and Galleries Commission) can support special projects.

The natural science collections held at provincial museums in the UK are vast when taken as a whole. A definitive catalogue of those extensive herbaria of British plants that have been donated to municipal museums throughout the UK has been published by Kent and Allen (1984). The significant herbaria with international collections are listed in Holmgren *et al.* (1990) (see **Box 6.6**). Similar listings for animals have not been made but would probably have a similar spread.

It would be invidious, from the information now available, to attempt any listing of museum holdings that would be of use to those abroad reviewing their country's biota. As described in the next paragraph, considerable information is being assembled on a national database. For anyone making enquiries, the names of those that have collected in a country, rather than detailed geographic locations or even systematic names, could lead one to a significant holding in a UK museum.

Several museums have outstanding collections (see below for published lists of holdings). To give examples of a few, even small museums like the Booth Museum at Brighton, has worldwide collections of insects containing several hundreds of type specimens; land snails from Pacific islands; and Burmese bird eggs. Cambridge University Zoological Museum has substantial collections linked with former academic members: Darwin's Beagle voyage, including the fish collections, 30,000 birds from California and Australia, including 600 'types'; and Indian Ocean (especially Maldives) marine invertebrates (Crustacea, corals). The zoological collections in the Oxford University Museum have been documented in detail by Davies and Hull (1976). Manchester Museum has historic collections of Richard Spruce from South America, and algal collections and notebooks of Dawson Turner. The Glasgow Museum at Kelvingrove has good material of N. India and Himalaya and New Zealand including skins of three extinct bird species. The insect collections again contain many types, as do the excellent collections of mollusc shells. Those museums privately endowed, like the Hancock

Art Gallery and Museums, Glasgow

The New Zealand Huia (*Heteralocha acutirostris*) was last seen in 1907, and now considered extinct, possibly due to the European insatiable desire for curios. Specimens like this one, on display at the Art Gallery and Museum, Glasgow, are important for scientific study and record.

Box 6.7 Geographical representation of natural history collections in the museums of the East Midlands Museums Service Area *(Source: G. Whalley, Nottingham Museum, 1995)*

ORGANISM:		PLANTS	INSECTS	OTHER INVERTEBRATES	VERTEBRATES	GENERAL ZOOLOGY
Total*		427	419	187	119	5
African & Indian Ocean Region	Africa –unspecified	6	10	4	9	
	Algeria		2	2		
	Angola		2			
	Burundi		1			
	Cameroon		1			
	Central Africa		4	1		
	East Africa		5	2	1	
	Egypt		3		3	
	Ethiopia				1	
	Ghana		3			
	Guinea				1	
	Indian Ocean			1		
	Kenya		3		1	
	Madagascar	1	6	2	1	
	Malawi		6			
	Mascarene Islands	1			1	
	Mauritius	1		3	1	
	Morocco	2	2		1	
	Nigeria		2			
	Reunion Islands		1			
	Senegal			2		
	Sierra Leone		4		1	
	Somalia		1			
	South Africa	6	12	3		
	Tanzania	1	1		1	
	Uganda		1			
	West Africa		5	2		
	Zaire	2				
	Zimbabwe		2		1	
American Region (Middle)	West Indies – general	4	1	5	1	1
	Antigua					
	Antilles			2		
	Barbados			1		
	Costa Rica		4			
	Cuba		2			
	Dominican Republic		1			
	Grenada		1			
	Guatemala		2	2		
	Honduras		2			
	Jamaica	5	3	2		
	Martinique	1		2		
	Mexico	5	6	2		
	Nicaragua		1			
	Panama		4	3		
	Trinidad	1	3		1	
American Region (North)	North America – unspecified	5	12	1	4	
	Alaska (USA)		1		1	
	Bahama Islands			1		
	Bermuda Islands			2		
	Canada	3	3	3	4	
	Greenland	5			2	
	United States	10	6	2	8	
American Region (South)	South America – unspecified	2	14	4	3	2
	Argentina	1				
	Bolivia		3	2		
	Brazil	5	9	2		
	Chile	1	2	3		
	Colombia	1	5	2		
	Ecuador		3	2		
	Falkland Islands	1				
	Galapagos			2		
	Guyana		1	1		
	Paraguay		4			
	Peru	3	7	2		
	Surinam		3			
	Uruguay	1				
	Venezuela	1	3			

*Each unit represents either a whole or part collection/consignment and may be one or many specimens/species.

Box 6.7 *continued*

ORGANISM:		PLANTS	INSECTS	OTHER INVERTEBRATES	VERTEBRATES	GENERAL ZOOLOGY
Antarctic Region	Antarctica				1	
	Sandwich Islands			3		
	South Georgia	1				
Asian Region (Main)	Asia – unspecified					
	Andaman Islands		1			
	Borneo	1	4	2		
	Burma (Myanamar)		7		1	
	Central Asian States	3	2		3	
	China	2	8	3	3	
	Cocos Islands		1			
	Hong Kong	1				
	India	6	25	2	11	
	Indian Ocean			1		
	Indonesia – general		2	1		
	Java	4	4	2		
	Sulawesi		4			
	Sumatra	1	1			
	Timor		1			
	Moluccas Islands			2		
	Japan	4	4	2		
	Korea		1			
	Malaysia	1	9			
	Nepal	1				
	North China Sea			1		
	Pakistan		1			
	Philippines		3	5		
	Sabah	1				
	Sikkim	1	3			
	Singapore	2	2	2		
	Sri Lanka	1	9	5		
	Taiwan		1			
	Thailand	1	2			
	Tibet		1			
	Vietnam		1			
Asian Region (Middle East)	Middle East – unspecified	4	1	2		
	Cyprus		1			
	Iran	1				
	Israel	1				
	Jordan		1			
	Lebanon	2	1			
	Palestine		1			
	Red Sea			3		
	Saudi Arabia	1				
	South Yemen	1	2			
	Syria	1	1			
	Turkey (also see European Turkey)	4	1			
Australasian Region	Australia – unspecified	9	11	8	9	1
	Tasmania	3	1		1	
	New Zealand	20	5	5	4	1
Pacific Region	Pacific Ocean – general	1		6	1	
	Fiji	3		2	1	
	Hawaii	1				
	New Britain	2	1			
	New Caledonia			2		
	New Guinea	3	8		1	
	New Hebrides	5	1			
	New Ireland	1				
	Polynesia	1				
	Samoa	2	1			
	Society Islands	1		2		
	Solomon Islands	1		2		
	Tahiti	1		2		
	Tuamotu	1		2		
	Vanuatu	5	1			

Box 6.7 *continued*						
ORGANISM:		PLANTS	INSECTS	OTHER INVERTEBRATES	VERTEBRATES	GENERAL ZOOLOGY
European Region	Europe –unspecified	21	11	8	11	
	Andorra	1	1			
	Austria	15	4	2	1	
	Azores	1				
	Balearic Islands	3	1			
	Belgium with Luxembourg	2				
	Bulgaria	4	1			
	Canary Islands	2	1	2		
	Cape Verde Islands					
	Corsica	5	2	3		
	Crete	5	2	1	1	
	Czech Republic and Slovakia	1	1			
	Denmark	1		2		
	Faroe Islands					
	Finland (also see Lapland)	1			1	
	France	26	15	3	1	
	Franz Joseph Zemlya Is.	3				
	Germany	9	10	3	3	
	Greece	12	2	2	1	
	Iceland	2	1	3	3	
	Italy with Sicily	18	8	3	1	
	Yugoslavia (former)	11	2	2		
	Lapland		2	1	3	
	Malta	4	1	2	2	
	Netherlands	6	2			
	Norway	13	3	4	2	
	Poland	5			1	
	Portugal	10	3			
	Romania	4		2		
	Russia (European)	5	2			
	Sardinia	3				
	Sicily (see Italy)					
	Spain including Gibraltar	21	9	3	3	
	Spitzbergen	4				
	Sweden (also see Lapland)	6	2		2	
	Switzerland	18	14	3		
	Turkey (European)	4	1			

Museum in Newcastle upon Tyne, had money to purchase larger collections that came to the auction rooms in the nineteeth century and during this time bought several plant collections left to, and ultimately sold by, the Linnean Society of London.

A major project to enter UK museum holdings of natural history material into a central database is that being organised by the Museum Documentation Association (MDA), in a programme called FENSCORE. The origins and development of FENSCORE (standing for Federation for Natural Science Collections Research) and the operational strategy adopted are covered in Pettitt (1986, 1991, 1992). This national database contains entries for most of the natural science collections held in museums in the UK, although information for Wales and Northern Ireland is currently being assembled. Each entry in the database gives the name by which the collection is known, the nature of the material present, the geographical areas of the world from which it was collected, the size of the collection, and the period over which it was assembled. The holding institution is noted, together with additional information such as whether taxonomic types were present. All entries are indexed for major taxonomic group(s) contained, geographical area(s) covered, and period collected. Various regional sections of the database are being updated on a co-operative basis. The National Database is funded at present by the UK Museums and Galleries Commission and collection of data is organised on a regional basis through the Collection Research Units of the Area Museums Councils (see **Box 6.8**).

The Database is held by the MDA which can conduct searches for researchers on behalf of FENSCORE. A charge may be made for enquiries from commercial organisations or from outside the UK (relevant addresses are given in Appendix Two). It is planned to make the principal fields of the database available on the internet in 1996. At present the data on

Box 6.8 Regional Collection Research Units of FENSCORE

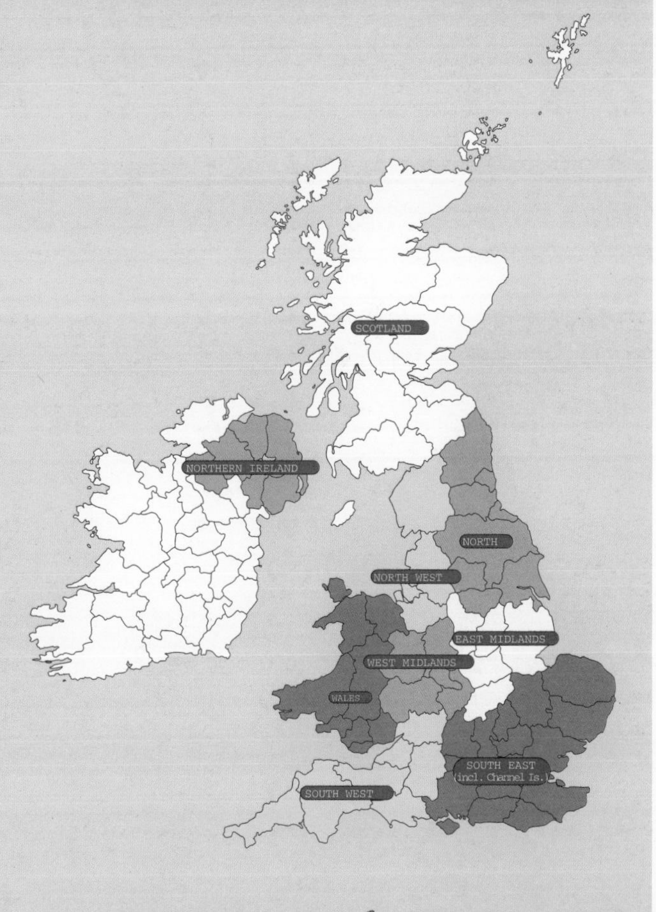

Area Museum Council for the South West
Hestercombe House
Ceddon Fitzpaine
Taunton TA2 8LQ

Council of Museums in Wales
32 Park Place
Cardiff CF1 3BA

Northern Ireland Museum Council
189 Stranmillis Road
Belfast BT9 5DU

East Midlands Museums Service
Courtyard Buildings
Wollaton Park
Nottingham NG8 2AE

Museum Documentation Association
Lincoln House
347 Cherry Hinton Road
Cambridge CB1 4DH

North of England Museums Service
House of Recovery
Bath Lane
Newcastle upon Tyne NE4 5SQ

North West Museums Service
Griffin Lodge
Cavendish Place
Blackburn BB2 2PN

Scottish Museums Council
County House
20–22 Torpichen Street
Edinburgh EH3 8JB

South East Museums Services
Ferroners House
Barbican
London EC2Y 8AA

West Midlands Area Museums Service
Hanbury Road
Stoke Prior
Bromsgrove
Worcestershire B60 4AD

Yorkshire & Humberside Museums Council
Farnley Hall
Hall Lane
Leeds LS12 5HA

collections on six areas of the UK has also been published as printed Registers (see Davis & Brewer, 1986; Hancock & Pettitt, 1981; Hartley *et al.*, 1987; Stace *et al.*, 1988; Bateman *et al.*, 1993; Walley, 1993). The holdings of the East Midlands of England database (Walley, 1993) is one of the most sophisticated and the table in **Box 6.7** has been prepared specially to illustrate the geographical areas covered in this sample of collections. These detailed geographic codes have now been added to all the earlier records in the database. The one prepared by the South Eastern Collections Research Unit (SECRU) contains over 19,000 records of natural science collections held within museums, universities and colleges, and in private hands, within the counties of South Eastern England and the Channel Island States (including a summary of the collections in The Natural History Museum, London). It has been published both as a bound hardcopy, on microfiche, and on 3.5in. (90mm) and 5.25in. (133mm) computer discs.

6.3 Botanic gardens and plant gene-banks

Many live plants, representing tens of thousands of species, are grown in the UK in botanical gardens and arboreta, in special research collections and for enjoyment and personal interest in private gardens. Whilst some of these plants will be of unknown provenance and of limited interest to those studying the diversity of a given area, others, fully documented, can be of immense value, especially if they are an endangered species.

There are some 60 botanic gardens in the UK details and major collections of which are summarised in *Botanic Gardens and Arboreta of Britain and Northern Ireland* (BGCS, 1990). General coordination of holdings, technology and development of UK botanic gardens is now provided by PlantNet, a recently formed association of UK botanic gardens (see Appendix 2 for addresses of this and the larger gardens) and a group working closely with the Botanical Gardens Conservation International (see Chapter Five).

National arboreta holding tree collections also play a specific role in *ex situ* conservation. Westonbirt Arboretum and Bedgebury Pinetum (Forestry Authority) both in England, and Castlewellan Arboretum in Northern Ireland, include collections of 3,869, 251, and 1,200 taxa, respectively. Collectively the UK botanic gardens hold several thousand accessions of threatened trees, including some that are now extinct in the wild.

Most UK universities have small botanic gardens, many of which have developed as botanical supply units for teaching. They contain specialised collections of wild and cultivated plants of exotic species which are, or have been, used for biodiversity research. Other universities, especially the older established ones, have gardens of historic importance (e.g. Oxford Botanic Garden which is the oldest botanic garden in the UK, being founded in 1621; that at Cambridge in 1762). These are collections of plants brought together mainly for teaching plant diversity (see Chapter Five), although Cambridge has made a speciality of growing endangered plants of Eastern England, the UK and Europe generally. Other university gardens have a role in *ex situ* conservation (see **Box 6.13**).

6.3.1 Royal Botanic Gardens, Kew

The Gardens at Kew have been part of British history for more than 300 years and were established as the Royal Botanic Gardens in 1759. Wakehurst Place dates back to late Norman times, although its exotic collections were commenced in the mid 19th century. The Living Collections Department now manages one of the world's largest documented *ex situ* collections of well-curated living plants (80,000 accessions; 36,000 taxa). This is used to support international and national programmes of science and conservation as well as to provide a valuable teaching resource and a major visitor attraction. The Gardens have been enjoyed by the public since their earliest days; today RBG Kew strives to maximise the pleasure and education of visitors to Kew and Wakehurst Place. It also aims to optimise long-term revenue and support, by programmes of promotion and improvement of horticultural displays and landscapes as well as visitor services. Kew's glasshouses cover two hectares, and include four listed buildings of great historical importance – the Palm House (with a basement for marine collections and displays), the Temperate House, the Waterlily House and the Architectural Conservatory (House No. 1). Not far from the Palm House, the other main tropical plant display area is the Princess of Wales Conservatory divided into sections recreating ten climatic zones. RBG Kew also maintains quarantine facilities and has micro-propagation facilities in which many endangered species are propagated by tissue culture (see **Box 6.14**).

A. Jackson/Royal Botanic Gardens, Kew

The Royal Botanic Gardens, Kew, satellite garden at Wakehurst Place where exotics thrive in moss-rich native woodland.

The Gardens at Kew include many thousands of trees and extensive shrub collections. The combined arboreta at Kew and Wakehurst Place contain over 26,000 accessions; the Kew collection being arranged mainly in a systematic manner, and those at Wakehurst Place following a phytogeographic system. Decorative and herbaceous plantings include many major features such as the Alpine and Bulb Display House, the systematic Order Beds, the Rock Garden, the replica 17th century Queen's Garden and, at Wakehurst Place, species-rich walled gardens and water gardens. Important extensive reference collections of live plants are held for national and international research which also involves the almost daily shipment of plants or plant parts to researchers and other users around the world. See also Chapter Four for further information on research at Kew.

6.3.2 Royal Botanic Garden, Edinburgh

The RBG at Edinburgh now comprises four gardens throughout Scotland: Inverleith in Edinburgh, Younger in Argyll, Logan in Wigtownshire, and Dawyck in Peeblesshire. The Living Collections Department is responsible for all of these, maintaining plants used in research and for *in situ* and *ex situ* conservation of rare and endangered species. Full documentation and data recording makes efficient use of information technology to disseminate horticultural knowledge through publication and electronic information exchange (Walter *et al.*, 1995). Living collections records are maintained on BGBASE™, a database rapidly becoming a standard throughout botanic gardens and herbaria and museums, libraries and zoological collections (see **Box 6.9**).

Links with other institutions in other countries broaden and extend the representation of the plants available for study at RBG Edinburgh. On average, 1000 plants or plant propagules are sent to other institutes each year. The living collection is particularly well represented in the families Ericaceas (esp. *Rhododendron*) , Orchidaceae and Rosaceae. The collections of Pinaceae and Liliaceae are also particularly extensive, holding 97% and 81% of the world's species respectively. China, S.E. Asia (including Taiwan, Malaysia and Papua New Guinea), the U.S.A., Himalayan region

The Royal Botanic Garden at Edinburgh.

(including Nepal and Bhutan) and the Mediterranean region (including Spain and Turkey) are particularly well represented in the living collections. At present these collections represent 29 families, 993 genera, and 16,693 species. Total accessions are 39,055, a number growing by over 3,000 accessions per year; making the RBGE collections the fifth largest in the world.

The Garden's acquisitions policy requires that the living collections are augmented by expeditions, and by collaboration, cooperation and seed exchange with other institutions in areas which match RBGE's research, education and conservation interests, and whose climatic and habitat conditions make their plants suited to cultivation in one or other of the gardens. The differences in soils, climates and other conditions between the four RBGE sites provide a greater diversity of habitats than that available to any other UK botanic garden. On completion of research projects, de-accessioning of plant material takes place; surplus material is then distributed for research and conservation work elsewhere. For a concise history of the Royal Botanic Garden, see Fletcher & Brown (1970). See also Chapter Four for further information on research at RBG Edinburgh.

Box 6.9 BG-Base™: a biological database for collections management

BG-BASE™, version 4.0, is a database application designed for managing biological information in four categories: taxonomy, distribution, conservation, and collections management (living and preserved). It is now widely used around the world in over 50 botanical, horticultural and zoological research institutions, universities, libraries, museums, and in biodiversity monitoring centres such as WCMC (see Box 7.14) where it is used to track the distribution and conservation status.

BG-BASE™ is composed of several separate linked modules: for managing information on living plant collections; gene banks; herbarium and other museum specimens; taxonomy; nomenclature; bibliography; images; gazetteers; distributions; floristic and faunistic surveys; protected areas; conservation status; institutional and individual contacts; and membership, education and development programmes.

BG-BASE™ is built using the relational database management system Advanced Revelation, which allows for variable length fields and records (each can vary in length from one byte to a maximum of 64,000 bytes), thereby eliminating the need to truncate data with blanks to fill a predetermined field length. The system is mouse-driven, has pull-down menus and pop-up selection boxes. Context-sensitive help, which can be translated into languages other than English, is available in virtually all parts of the system. BG-BASE™ manipulates images and maps through its optional graphics module. It can be linked with both GIS and CAD systems, and drives a wide variety of printers, engravers, embossers, and bar code devices. Data may be converted from and into various other formats, including ASCII, DBASE, and Lotus 1-2-3.

(*Source*: *K S. Walter, Royal Botanic Garden, Edinburgh*.)

6.3.3 Crop plants and UK plant gene banks

The conscious and unconscious selection of desirable forms has been a feature of crop improvement since the birth of agriculture. The removal of individuals not possessing desirable attributes means that the selection process also leads to the gradual loss of genetic diversity. Continual cycles of selection over many centuries has left most major food crops with a tiny fraction of the genetic variability found originally in their progenitor species. Whilst agronomically useful characteristics have been improved by this process, genes, not subject to selection, have diminished in number. Without doubt, resistance to pests and diseases comprise the most significant category of genes affected in this way. Their loss has rendered many crops extremely vulnerable to attack following exposure to new pathogens or pathotypes.

One of the best known and most tragic examples of the consequences of such genetic erosion were the famous Irish potato famines of 1845-6. Susceptibility of the potato to the late blight disease, *Phytophthora infestans*, resulted in complete failure of the crop in Ireland and many parts of mainland Europe following a widespread epidemic of the fungus. This had devastating social and economic implications for the people living in affected areas. It seems likely that the possibility of recurrence stimulated the first serious breeding effort on a major crop aimed specifically at introducing resistance to disease (Hawkes, 1990). Work initiated over fifty years ago by Professor J.G. Hawkes, in which careful study, extensive exploration and international cooperation, lead to the setting up of comprehensive gene banks for potato (*Solanum*) in the UK, Holland and Peru. The story is elaborated in Chapter Four, Box 4.7.

Plant gene banks are clones and populations, varieties, cultivars and land-races of particular useful species which are kept to maintain stocks of genetic diversity for future crop breeding. Material for such plant breeding is held in the UK in over 50 carefully maintained units (see **Box 6.10**). Curators of these collections, which include a number of important private organisations maintaining specialist collections of plants, meet periodically under the auspices of MAFF as the

UK *Plant Genetic Resource Group*. Full international liaison is maintained through this group with CGIAR, and IPGRI and its associated international crop gene banks.

An example of a private organisation is the Henry Doubleday Research Association which conserves crop biodiversity by making varieties of vegetables available to those gardeners who wish to grow their own through its Heritage Seed Programme, currently containing some 700 varieties. These are maintained at the Association's gardens and through a network of volunteers (members of the Association; see Chapter Five, Box 5.9) who bulk up selected varieties. The Association thus seeks to promote conservation through utilisation.

Most of the major crop gene bank collections are held by Research Council or MAFF-funded units, e.g. the John Innes Centre, the Institute of Grassland and Environmental Research, and the Scottish Crop Research Institute. Details are laid out in **Box 6.10**. Another independent organisation sponsored by MAFF and BBSRC is Horticultural Research International. The organisation's main aim is meet the particular needs of the UK horticulture industry through effective research, development and technological transfer.

The HRI Genetics Resources Unit (GRU) was established in 1980, designated as an IBPGR base store with a remit for the collection, conservation, characterisation, documentation and research study of a range of vegetable crops and their wild relatives. The HRIGRU is now active in various crop working groups and networks including the European Cooperative Programme *Allium* and *Brassica* groups and an informal *Daucus* group. The collections encompass a wide range of small-seeded, mainly outbreeding vegetables (current/obsolete cultivars and landraces) and associated wild taxa and are listed in **Box 6.10**. The GRU shares the workload and resources for some crops by maintaining complementary collections and mutual safety duplicates with the Centre for Genetic Resources in the Netherlands. Also the GRU maintains passport and some minimal characterisation data as defined by international crop working groups. Information is available, generally in print form, upon request. The European database for *Allium* (onion, leek, garlic and wild species) is maintained by HRIGRU.

HRI also carries out fundamental and strategic research into the assessment and exploitation of biodiversity. Molecular studies are undertaken to assess the level of genetic diversity on species and landrace collections of vegetable and tree fruit genera. Reference populations are maintained for variation in morphology, development, juvenility, flowering time, fruit quality characters and pest and disease resistance.

HRI at Wellesbourne has an extensive collection of edible Basidiomycota stored in liquid nitrogen. This includes an extensive worldwide collection of commercial and wild stocks of *Agaricus bisporus*, the principal cultivated mushroom.

International projects of HRI including the screening of *Phaseolus* beans and their pathogens in Africa; the collecting and characterising of landraces of short-day adapted onion from West Africa in collaboration with a Natural Resources Institute Darwin Initiative project and scientists in Burkino Faso (see Chapter Four); and coordinating a project entitled ' Selection of drought tolerant fruit trees for summer rainfall regions of Southern Africa and India'. This includes the collection and assessment of indigenous *Ziziphus*, *Sclerocarya* and *Strychnos*.

6.3.4 Plants of ornamental value

Stock collections of hardy ornamental plants are maintained by the HRI and include taxa from a number of genera including *Acer*, *Buddleia*, *Berberis*, *Mahonia*, *Rosa*, *Sambucus* and *Syringa*. A collection of *Narcissus* (old varieties and wild species). However, the most important resource is the approximate 60,000 taxa of ornamental plants, mainly cultivars or exotic species, available through commercial nurseries in the UK, and cultivated in numerous public and private gardens. Biodiversity in this NGO sector, is encouraged by the national Royal Horticultural Society (see Chapter Five) and was developed in a positive way under a unique scheme formed in 1978 ultimately to become the National Council for the Conservation of Plants and Gardens (NCCPG).

The NCCPG, nationally and through its Regional Groups, is conserving over 600 collections of garden plants. The main aim of the scheme is,

Box 6.10 Plant gene banks held in the United Kingdom
Source: The UK Plant Genetic Resource Group

FIELD CROPS AND FORAGES

COLLECTION	ACCESSIONS	ADDRESS
Oat Collection (*Avena*)	4,360	Institute of Grassland and Environmental Research Welsh Plant Breeding Station Plas Gogerddan Aberystywth Dyfed SY23 3EB
Rhizobium Collection	536	
Grasses and legumes	9,287	
BBSRC small grain cereals Collection (wheat, barley and oats)	22,500	John Innes Centre, Norwich Research Park Colney Lane Norwich Norfolk NR4 7UH
Pea Collection (*Pisum*)	2,650	
BBSRC wild barley Collection (*Hordeum spontaneum*)	14,822	
Field bean and related species Collection	270	
UK wild Triticeae species Collection	1,200	
UK wheat and barley Precise Genetic Stocks Collection	8,000	
Faba bean and related species (*Vicia*)	754	Scottish Crop Research Institute Invergowrie Dundee DD2 5DA
Barley Collection (*Hordeum* spp.)	1,661	
Wheat (Triticeae) anaploids and barley (*Hordeum* spp.)	138	
UK field pea and pea Registration Collection	1070	Scottish Agricultural Science Agency East Craigs Edinburgh EH12 8NJ
Peas - UK *Pisum* cultivar Collection	3,120	
Cereal crops seed Collection	3,150	
Barley Collection (*Hordeum* spp.)	1,000+	Scottish Agricultural College West Mains Road Edinburgh EH12 8JN
Fibre flax and oil-bearing linseed	350	Scottish Agricultural College Cleeve Gardens Oakbank Road Perth PH1 1HF
Fibre flax and linseed	Approx 200	Northern Ireland Horticultural and Plant Breeding Station Loughgall Armargh BT61 8JB
Hop Collection (*Humulus lupulus*)	Approx 600	Horticultural Research International HRI Department of Hop Research Wye College, Wye Ashford, Kent TN25 5AH

VEGETABLES

COLLECTION	ACCESSIONS	ADDRESS
Commonwealth Potato Collection (*Solanum*)	1,250+	Scottish Crop Research Institute Invergowrie Dundee DD2 5DA
Potato dihaploid Collection	1,500	
Cultivated potato Collection	Approx 700	
UK potato cultivar Collection	1,160	Scottish Agricultural Science Agency East Craigs Edinburgh EH12 8NJ
UK cultivar Collections (comprising calabrese, sprouting broccoli, turnips and swede)	1,109	

Box 6.10 *Continued*

VEGETABLES *(continued)*

COLLECTION	ACCESSIONS	ADDRESS
The Vegetable Gene Bank (includes *Allium*, *Brassica*, *Raphanus*, *Daucus* and *Lactuca*) plus 3 Specialist Collections namely Lettuce Bremia Differentials, *Brassica* European Clubroot Differentials, and the *Brassica* S-allele lines)	11,720	Horticulture Research International Wellesbourne Warwick CV35 9EF
Near Isogenic Lines (NILs) of Tomato	c. 900	Horticulture Research International Worthing Road Littlehampton, West Sussex BN17 6LP

FRUIT

COLLECTION	ACCESSIONS	ADDRESS
Apples, pears and quinces	c. 700	Horticultural Research International West Malling Kent ME19 6BJ
Cherry, plum and related species (*Prunus*)	c. 300	
Raspberries, blackberries and related spp. (*Rubus*)	125	
Strawberries (*Fragaria*)	387	Scottish Crop Research Institute Invergowrie Dundee DD2 5DA
Blackcurrants and related species (*Ribes*)	860	
Cranberries, blueberries and related species and woody perrenials (*Vaccinium*, *Hippophae*, *Aronia*, *Amelanchier*, *Rosa*, *Sambucus*)	90	
National Fruit Collection (includes top fruit, bush fruit and ornamentals)	c. 4,000	National Fruit Collection Wye College Brogdale Road Faversham Kent ME13 8XZ

MISCELLANEOUS

COLLECTION	ACCESSIONS	ADDRESS
The National Willows Collection (*Salix*)	c. 1000	Institute of Arable Crops Research (IACR) Long Ashton Research Station Bristol BS18 9AF
Seed of wide range of tropical tree and shrub species, with extensive collections of *Leucaena*, *Giricidia*, *Calliandra* and several African *Acacia*	c. 4000	Oxford Forestry Institute Department of Plant Sciences University of Oxford South Parks Road Oxford OX1 3RB
Seeds of major components of ecological groupings in Britain	3,000	The NERC Unit of Comparative Plant Ecology Department of Animal and Plant Sciences University of Sheffield Sheffield S10 2UQ
Seeds of wild species from the UK (including most covered by Schedule 8 of the Countryside Act) and dry zone species of potential economic value	9,080	RBG Kew Seed Bank Wakehurst Place, Ardingly Nr Haywards Heath West Sussex RH17 6TN
Vetches and Peas (Vicieae - includes *Lathyrus*)	c. 3,500	Department of Biology University of Southampton Medical and Biological Sciences Building Bassett Crescent East Southampton SO9 3TU

through registered, but usually privately owned, National Collections, to conserve a gene pool of the many thousands of garden plants, bred and introduced by explorers, nurserymen and amateur gardeners. The plant collections are predominantly cultivars, but also contain species and thus supplement the gene pools held in botanic gardens.

The reason for the establishment of this scheme is the demise over the years of the many garden plants grown in UK gardens. Reasons for this include difficulties in propagating some plants which make them more costly to produce. Other factors include fashion (e.g., newly bred varieties may become commercial successes overnight and older varieties may quickly be forgotten), disease (e.g. hollyhocks with hollyhock rust, elms with Dutch elm disease), and genetic instability. Many new cultivars are possibly not as stable as their predecessors and some plants have been vegetatively propagated for many centuries.

The scheme developed modestly at first but by 1982 had one hundred collections. Today, it has about six hundred National Collections (NCs) registered. These contain a total of some 50,000 plants; approximately 38,000 cultivars and 12,000 species, covering 325 genera. The collections range vastly in size. For example, the NC of 160 *Astilbe* includes many old cultivars that have been located in the Czech Republic, Slovakia and Latvia, to name but three sources, together with a large number of the old cultivars raised by the German grower, Arend, that are no longer in his nursery. The *Narcissus* NC comprises 100 species and over 2,400 cultivars. Several collections are now being used as a source for active biodiversity and taxonomic studies. A sample of generic collections are given in **Box 6.11**. For a full list of genera maintained under NCCPG, with their guardians and location see NCCPG (1996).

Box 6.11 Selected examples of NCCPG National Collections

GENUS	NO OF SPECIES	NO OF CULTIVARS
Acacia	50	4
Allium	165	4
Aquilegia (inc Semiaquilegia)	65	55
Begonia	219	110
Betula	75	21
Campanula	176	123
Cistus	66	19
Dendrobium	83	144
Echeveria	50	250
Echinocereus	52	119
Eucalyptus	61	
Fuchsia	79	20
Geranium	178	118
Hypericum	61	15
Iris	80	170
Nerine	30	600 – 800
Paphiopedilum	91	several thousand
Passiflora	88	27
Penstemon	87	118
Pleione	58	173
Primula (Asiatic)	approx 110	
Quercus	154	75
Salix	245	213
Sisirhynchium	68	approx 150
Tillandsia	99	23

Other major collections are held by members of British national societies for ornamental plants such as fuchsias, pelargoniums and geraniums, chrysanthemums, roses and irises. The UK also has numerous specialist societies for bulbs and alpines. The expertise and living collections of these members are a further resource.

6.3.5 Plants of international conservation value – Ex situ collections

Several UK botanic gardens play an important international role in propagating and maintaining plants that are rare or endangered in the wild. Many such plants are distributed widely to other botanic gardens. The ultimate goal of such collections is to re-introduce the material back into the wild once the threat to the species and its habitat have been removed. This is done in collaboration with the country concerned. **Box 6.12** gives an example of such work undertaken on the flora of St Helena.

The existence in botanic gardens of rare or uncommon plants from well-documented sources is in itself a database of biodiversity information. The international botanic garden fraternity, latterly under the leadership of Botanical Gardens Conservation International (BGCI), has encouraged the listing of such endangered species (as defined by IUCN) for each garden under a standard format that can be easily transferred to BGCI. The list of such plants grown in UK gardens is too extensive to list here but a sample is given in **Box 6.13**.

6.3.6 Seed and tissue conservation

For sexually reproducing plants seed 'banks' are the most efficient and effective method of long-term storage provided the source of the seed and its parent plant has been checked. Cold storage (–10°C to –20°C) of seeds for some species increases long-term viability for at least 100 years, but for other species seed

Box 6.12 Genetic screening to conserve *Trochetiopsis* on St Helena

Trochetiopsis (Sterculiaceae) is an endangered woody genus endemic to the South Atlantic island of St Helena. There are two species:
St Helena redwood (*T. erythroxylon* (Forst.) Marais), which became extinct in the wild, and
St Helena ebony (*T. ebenus* Cronk), thought to be extinct until the rediscovery of two plants in 1980.

Trochetiopsis is represented in the collections of at least seven botanic gardens. Confusion of records during transfer of seed and cuttings between institutions has resulted in loss of provenance and collection data. It is known that **ebony** was represented in botanic gardens only by one of two extant plants until 1993.

Cuttings were collected from one of the two plants in 1980 and from the other plant in 1982. Since then, propagation of ebonies by cuttings and seed has led to over 2,000 plants being planted at various sites on the island, representing both re-introductions and introductions. The '1982' plant genotype is probably under-represented in the plantings.

All **redwood** plants now known are descended from a single wild tree that died around 1960. Since then multiplication of the redwood has been through self-pollinated seed. Over 100 trees have been re-introduced to several sites.

The conservation of *Trochetiopsis*, until recently, has been carried out without any genetic information. Current research is examining genetic diversity by means of isozyme and RAPD analysis, in order to answer the following questions:

1. What is the present level of genetic variation within and between the two species?
2. To what degree does introgressive crossing, which may hybridise the pure species out of existence, occur in mixed planted populations?
3. Is representative sampling being carried out for the benefit of future re-introductions?
4. What genotypes are represented in botanic garden collections; and can botanic gardens provide an additional source of genetic variation in the case of the redwood?

Perhaps not surprisingly, little genetic variability has been found within both species, although they do not exhibit unique isozyme and RAPD markers. No additional variation was found in botanic garden material for the redwood. An experiment investigating the fitness of redwood seedlings from self- and cross-pollinations provides direct proof that the redwood is suffering from inbreeding depression which has severe effects on fitness. While fitness can be redeemed to some degree by crossing, the situation maybe so severe that the extinction of the redwood may be inevitable.

Q. Cronk/RBG, Edinburgh

Trochetiopsis ebenus **on St. Helena.**

Introgressive hybridization is occurring in the mixed plantings. what at first may seem disastrous may well turn out to be the salvation for the redwood, if not the genus. The potential of the hybrid can be realised in several ways:

1. To use the F_1 hybrid in a breeding programme to backcross to the redwood in an attempt to instill vigour.
2. The F_1 and later generation hybrids grow quicker, larger and have greater flower and seed production than either the redwood or ebony. As such they are good for planting, having been shown to be successful both on the degraded, dry outer parts of the island and on the wetter central parts.
3. Hybrid plants produce a hard, fine grained wood, intermediate in colour between the rich mahogany of the redwood and the darker wood of the ebony. As such they may find use in the local craft industry.

The ease with which the hybrids now grow should not deter attempts to conserve the pure species. (There is no way of knowing whether the vigour will remain and it is possible that future hybrids may revert more to type). The conservation of the ebony and redwood must maximise the genetic diversity within each of the pure species. Until now this has not been carried out. Future plantings must ensure that the pure species and the hybrids are isolated. Without such action, the redwood is likely to become extinct within one or two generations and the future of a pure ebony is doubtful.

(*Source:* R. *Rowe and* Q. *Cronk, Plant Sciences Department, University of Oxford.*)

Box 6.13 Some rare and endangered species of vascular plants in cultivation in UK botanic gardens from wild provenance (excluding Kew and Edinburgh)

SPECIES	COUNTRY OF ORIGIN	BOTANIC GARDEN
AMARYLLIDACEAE		
Crinum mauritianum	Mauritius	CGG
Galanthus plicatus	Romania, Ukraine	CGG
Leucojum nicaeense	France	CGG
Narcissus longispathus	Spain	CGG
Pancratium canariense	Canary Is.	LIVC
ASCLEPIADACEAE		
Caralluma burchardii	Canary Is.	RNG
Ceropegia ceratophora	Canary Is.	RNG
Ceropegia dichotoma	Canary Is.	RNGR, BCW, CGG
Ceropegia fusca	Canary Is	RNG, CGG
Frerea indica	Maharashtra, India	RNGR
BORAGINACEAE		
Anchusa cespitosa	Greece	STA
Echium giganteum	Canary Is.	CGG
Echium pininana	Canary Is.	CGG
Echium simplex	Canray Is.	LIVC
Echium wildpretti	Canary Is.	LIVC
Lithodora zahnii	Greece	CGG
CAMPANULACEAE		
Brighamia citrina	Hawaii	LIVC, RNGR, RNG, CGG
Campanula incurva	Greece	CGG
Campanula raineri	Italy	CGG
Campanula rupicola	Greece	CGG
Canarina canariensis	Canary Is.	CGG
Edraianthus pumilio	Yugoslavisa	CGG
Musschia aurea	Madeira	LIVC
Musschia wollastonii	Madeira	CGG
Nescodon mauritianus	Mauritius	RNG
Pratia irrigua	Tasmania, Australia	CGG
COMPOSITAE		
Argyroxiphium kauense	Hawaii	LIVC
Beradia subacaulis	France, Italy	CGG
Centaurea clementei	Spain	RNG
Centaurea jankae	Romania	CGG
Cheirolophus junonianus	Canary Is.	CGG
Cheirolophus tagannaensis	Canary Is.	RNG
Commidendrum rugosum	St Helena	CGG
Lactucosonchus webbii	Canary Is.	CGG
Palaeocyanus crassifolius	Malta	CGG
Santolina viscosa	Spain	CGG
Sonchus gummifer	Canary Is.	CGG
Tanacetum ptarmiciflorum	Canary Is.	LIVC
CUPRESSACEAE		
Chamaecyparis formosensis	Taiwan	HILL
Chamaecyparis funebris	Vietnam	HILL
Cupressus abramsiana	California	HILL, WESB
Cupressus chengiana	Gansu, Sichuan, China	HILL, WESB
Cupressus gigantea	Xixiang Zizhou, China	HILL
Cupressus goveniana	California	PERRO
Cupressus stephensonii	California	HILL, WESB
Juniperus cedrus	Canary Is., Madeira	WESB, RNG
Juniperus phoenicea	Egypt	WESB
Juniperus potaninii	Sichuan, China	WESB
Microbiota decussata	former USSR	WESB

Box 6.13 *Continued*		
SPECIES	**COUNTRY OF ORIGIN**	**BOTANIC GARDEN**
CUPRESSACEAE *continued*		
Tetraclinis articulata	Malta, Spain	RNG
Thuja koraiensis	Jilin, China, N. Korea, S. Korea	HILL
EUPHORBIACEAE		
Euphorbia baioensis	Kenya	RNGR
Euphorbia bravoana	Canary Is.	LIVC
Euphorbia schoenlandii	South Africa	RNG
Euphorbia woodii	Natal, South Africa	RNG
IRIDACEAE		
Crocus angustifolius	former USSR	CGG
Crocus imperati	Italy	CGG
Crocus versicolor	France, Italy	CGG
Romulea tempskyana	Greece	CGG
LILIACEAE		
Aloe bellatula	Madagascar	CGG
Aloe buhrii	Cape Prov., South Africa	LIVC
Aloe ciliaris	South Africa	CGG
Aloe suzannae	Madagascar	CGG
Dracaena draco	Canary Is., Madeira, Cape Verde Is.	GBG, LIVC, RNG, CGG
Fritillaria carica	Greece	CGG
Fritillaria conica	Greece	CGG
Fritillaria davisii	Greece	CGG
Fritillaria elwesii	Greece	CGG
Fritillaria involucrata	France, Italy	CGG
Fritillaria obliqua	Greece	CGG
Fritillaria pinardii	Greece	CGG
Fritillaria rhodocanakis	Greece	CGG
Haworthia maughanii	Cape Prov., South Africa	RNGR
Haworthia truncata	Cape Prov., South Africa	RNGR
Muscari gussonei	Italy, Sicily	CGG
Scilla litardierei	'Yugoslavia'	RNG
Semele androgyna	Canary Is., Madeira	CGG
MALVACEAE		
Abutilon menziesii	Hawaii	LIVC
Kokia kauaiensis	Hawaii	CGG
PLUMBAGINACEAE		
Armeria pseudarmeria	Portugal	CGG
Armeria welwitschii	Portugal	CGG
Limonium arborescens	Canary Is.	LIVC
Limonium companyonis	France	CGG
Limonium paradoxum	Ireland, UK	BRIST
Limonium redivivum	Canary Is.	LIVC
Limonium transwallianum	Ireland, UK	STA, CGG
SAXIFRAGACEAE		
Saxifraga cochlearis	France, Italy	RNG, CGG
Saxifraga diapensioides	France, Italy, Switzerland	CGG
Saxifraga florulenta	France, Italy	CGG
Saxifraga latepetiolata	Spain	CGG
Saxifraga moncayensis	Spain	CGG
Saxifraga nervosa	France, Spain	CGG
Saxifraga paradoxa	Austria, 'Yugoslavia'	CGG
Saxifraga pickeringii	Madeira	CGG
Saxifraga portosanctana	Madeira	CGG
Saxifraga reuterana	Spain	CGG
Saxifraga rosacea	Ireland	CGG
Saxifraga tombeanensis	Italy	CGG
Saxifraga vayredana	Spain	CGG

Box 6.13 *Continued*

SPECIES	COUNTRY OF ORIGIN	BOTANIC GARDEN
ZAMIACEAE		
Encephalartos bubalinus	Kenya, Tanzania	CGG
Encephalartos gratus	Malawi, Mozambique	CGG
Encephalartos laurentianus	Angola, Zaire	CGG
Microcycas calocoma	Cuba	BRIST
Macrozamia macdonnellii	N. Territory, Australia	STA

KEY

BRIST	University of Bristol Botanic Garden, Bristol
CGG	Cambridge University Botanic Garden, Cambridge
GBG	Glasgow Botanic Gardens, Glasgow
HILL	The Sir Harold Hillier Garden and Arboretum, Ampfield
LIVC	City of Liverpool Botanic Gardens, Liverpool
PERRO	Thorp Perrow Arboretum, Bedale
RNG	The Harris Garden, Reading University
RNGR	Gordon Rowley Succulent Collection, Reading University
STA	Botanic Gardens, University of St Andrews
WESB	Westonbirt Arboretum, Westonbirt

(*Information compiled from the* Botanic Gardens Conservation International *database by* E. *Leadley* 1994. *Species distribution information courtesy of* WCMC, *Cambridge*.)

stores need to be replenished more regularly. An internationally recognised seed bank has been set up at the Royal Botanic Gardens, Kew, and on a smaller scale at RBG, Edinburgh, where similar facilities for pteridophyte spores are also being developed.

The Seed Conservation Section of RBG, Kew based at Wakehurst Place, now holds the world's most diverse collection of wild-source seeds banked to international standards (nearly 4,000 species, in 178 families), particularly targeting the British flora (in collaboration with English Nature) species useful to man from arid and semi-arid lands. Research on seed physiology in the Section examines the biology and storage potential of individual species (including those with non-bankable seeds) and assesses optimum seed-banking conditions. Kew collectors, in collaboration with host government authorities, gather new seeds in many countries of the world. Kew has two full-time core-funded seed collectors active in the field for eight months of each year. While sufficient seed is always kept to ensure long-term conservation, samples are made available, through a seed-list (Linington, 1994), to other researchers and development projects who are required to observe any restrictions limiting the use of particular seed batches.

The Forestry Authority (UK) at Alice Holt Lodge, Hampshire, maintains a seed collection of 600 samples representing 200 species of trees with timber potential in temperate forests.

Low temperature (cryo-) preservation of parts of plants, such as meristem tips, buds or stem tips, in glass phials, at or near −196°C in liquid nitrogen is becoming increasingly used in botanic gardens. With more than 1,000 taxa being maintained at RBG Kew in this way (see **Box 6.14**), the low temperature slows down the growth rate of such material and enhances long-term conservation.

6.4 Zoological gardens: importance of living animal collections

There are over 70 major zoos and aquaria in the UK with collective experience of the maintenance and management of a wide array of species. These house over 64,000 vertebrates (mammals: 14,200, birds: 25,779, reptiles: 4,014, amphibians: 1507, fishes: 19,036). Historically, most collections have focused on the larger vertebrates but there is an increasing emphasis on smaller ones, such as reptiles and

G. Patterson/NCCPG

A National Collection of *Crocosmia* in a private garden in Devon, England.

fish. A few zoos and most aquaria have living exhibits of invertebrates but these are mainly of common aquatic and terrestrial invertebrates such as insects and molluscs.

The Federation of Zoological Gardens of Great Britain and Ireland (FZGGBI) organises and co-ordinates all species conservation programmes in the zoos and similar collections of its 62 UK members (see Appendix Two for selected addresses). The majority of the animals in zoos are captive bred and individual animals are seen as part of a whole population, a population which is kept genetically and demographically healthy. Priorities for species are based on their status and management in captivity and in the wild, and are the responsibility of the Federation's Taxon Advisory Groups (TAGs). TAGs set the priorities for each species and are responsible for the development and maintenance of breeding programmes and studbooks. FZGGBI's Joint Management of Species Committee organises and co-ordinates all species programmes.

Many of these breeding programmes are part of the larger European activities maintained by the European Association of Zoos and Aquaria (EAZA). The problems of maintaining viable populations of animals are very different from those for plants and microorganisms and are generally much more costly. The capacity of many zoos to maintain diverse collections is therefore limited by financial constraints.

There are over one hundred threatened species which have internationally managed breeding programmes organised under the auspices of the Captive Breeding Specialist Group (CBSG) a unit of the Species Survival Commission (SSC). For practical purposes, however, many more species are managed cooperatively on a regional or national basis. Captive breeding is organised primarily by the analyses of studbook data. **Box 6.15** describes the captive breeding programme for the Diana Monkey and shows the importance of accurate records from studbooks. **Box 6.16** lists some of the more significant collections of other species involved in conservation programmes. There is also close liaison with similar associations throughout the world, such as the SSC Re-introductions Specialist Group which concerns itself with re-introduction programmes of both animals and plants.

The London Zoo (owned by the Zoological Society of London) and the Jersey Wildlife Preservation Trust's zoo both play leading roles in this international work of captive breeding and re-introductions. The scientific programme of the former is discussed in Chapter Five, Section 5.3.1.8 and in Chapter One, Box 1.6. The JWPT reports annually in its journal *Dodo* of its extensive programme in training students (in conjunction with the Durrell Institute of Conservation and Ecology, University of Kent; see also Chapter Five, Section 5.4) field collecting, veterinary work and captive breeding. Their new programme will include consultations for captive propagation and health programmes in habitat countries, health assessments for wild populations, and the development of collection and handling protocols for all types of biomaterials for use by scientists worldwide (Pearl, 1995). This will include a series of care workshops for the Asiatic elephant in India, Indonesia and Myanamar.

Elsewhere in the UK are more specialist collections which have positive conservation roles. One example is that of the Wildfowl and Wetlands Trust (see Chapter Five, Section 5.3.4.1). Another is the collection of snakes and other reptiles brought together by dedicated naturalists for the West Midlands Safari Park. Such collections can make a significant contribution to biodiversity studies.

The FZGGBI is reviewing in detail the contribution that UK zoos could make to species conservation, and thereby biodiversity

Box 6.14 Cryopreservation research at the Royal Botanic Gardens, Kew

Royal Botanic Gardens, Kew

Young plantlets of *Cosmos atrosanguineus* recovered from frozen encapsulated tissue on agar.

The preservation of genetically stable tissue for future propagation is of fundamental importance for breeders and conservationists alike. Cryopreservation, or the storage of biological material at or near –196°C in liquid nitrogen (LN), provides the capability and is a valuable tool for *ex situ* conservation.

For cryopreservation, tissues with small, highly cytoplasmic cells are normally chosen as these are relatively stable and more likely to be capable of direct regeneration of shoots than more differentiated tissues. The formation of callus and adventitious shoot regeneration may increase the risk of aberrant genetic variation and are best avoided. The tissues most often used include meristems and zygotic or somatic embryos. Cell survival through freezing is improved by the prevention of ice crystal formation within the cell, which can prove lethal. Non-lethal desiccation of the cells prior to freezing is often used. The tissues require protection from the effects of severe desiccation, which is achieved by encapsulating the meristems in alginate on thin strips of filter paper, adapted from Fabre and Dereuddre (1990). The alginate strips are cultured in a liquid pretreatment medium containing a high molarity of sucrose prior to freezing. The use of more toxic protectants such as dimethyl sulphoxide is thus avoided. Further dehydration is required before freezing to reduce available water in the cell, two methods are described here: The first requires the use of controlled liquid nitrogen freezing for slow, two step freezing. The sucrose-protected, sterile alginate strips are put into 2 ml cryovials and placed in the freezing chamber. The temperature is dropped from 0.3°C/min to –35°C, and ice nucleation, first occurring in the intercellular fluid, draws water from the cytoplasm, thus dehydrating the cell. Sufficient dehydration has occurred by –35°C to prevent large ice crystals forming when the strips are plunged directly into LN.

Air drying the alginate strips is the second method. Once precultured on high sucrose media, the strips are placed in sterile air flow for 6-8 hours before being placed in cryovials and plunged into LN. By decreasing the temperature rapidly (c. 200°C/min), the cells freeze too quickly for large ice crystals to grow. Small non-lethal crystals form and the integrity of the cell is maintained. Thawing is rapid to prevent ice recrystalisation. The cryovials are plunged into a water bath at 40°C for 2 min. Thawed strips are placed on recovery media in the dark for seven days to reduce damage from oxidative stress (Benson *et al.*, 1989). During this period the sucrose concentration is reduced to 3%. Auxins and cytokinins are used to induce direct growth of the meristems.

Royal Botanic Gardens, Kew

The micropropagation growth room at Kew.

These techniques are being used at Kew for endangered plant material. One example is *Cosmos atrosanguineus*, the Mexican chocolate flower, which is a herbaceous perennial, endemic to mountains of central Mexico. It is believed to be extinct in the wild, but is common in cultivation. Individuals appear to come from a narrow genetic base, possibly one original source plant. This species is self incompatible and therefore at least two genotypes are required for the production of viable seed. Safe storage of clonal cultures is essential while tests for variability are carried out. Cryopreservation is being used to provide a safe base storage system.

By cryopreserving *Cosmos*, individual clones may be maintained in a genetically stable state, reducing the dangers of contamination or the genetic selection possible in standard tissue culture systems.

Individuals are being genetically fingerprinted using RAPD (random amplified polymeric DNA) technology in an attempt to find genetically distinct clones for cross compatibility tests. The ultimate aim is seed production for reintroduction of a self sustaining population in the wild.

(*Source*: *T. Wilkinson, Royal Botanic Gardens, Kew.*)

Box 6.15 The Diana monkey studbook

The Diana monkey (*Cercopithecus diana diana*) is categorised as **vulnerable** in the Red Data Book (IUCN, 1990) its captive situation is not yet secure and it is on CITES Appendix I (animals whose international trade is banned). Although the studbook deals primarily with the management of the captive population, information is also given on distribution and status in the wild and on the Tiwai Island Project in Sierra Leone, which has involved studies of the wild Diana monkey group.

M. Stephenson/Edinburgh Zoo

A Diana monkey at Edinburgh Zoo.

'World Herd'

This studbook dealt at first only with the British Isles population, but was upgraded to include the world population and was published in early 1993, by the studbook keeper, Dr Miranda Stevenson of Edinburgh Zoo.

The size of the total captive population was 188, on 31 December 1993. Animals were maintained in 77 collections in 22 countries, of which 24 collections were breeding the species (31% of the holders). Further analysis showed that the wild-caught founders were slowly dying out, and that therefore the future of the population depends on the breeding of captive bred stock. There have been no imports since the mid seventies.

Demographic Analysis

A demographic analysis has indicated that the age distribution across the population, for breeding purposes, is reasonably healthy; reproductive parameters of males and females are similar, with the best breeding occurring between 7–18 years old. The analysis has also shown that the carrying capacity of 200 animals would suffice to meet genetic and demographic requirements of the breeding programme. In other words, there are enough animals alive to achieve the goal of a self-sustaining captive population.

In the past, inbreeding and hybridization have created breeding problems. The studbook omitted all known hybrids from breeding programmes.

An evaluation of captive husbandry has been undertaken. The analysis indicates that the species does not appear to have any special requirements. Once the breeding group or pair have been established they seem to continue to breed; the problems occur in the initial formation of the group. Dietary information shows that the Diana monkey's requirements do not appear to be complex and are similar to those of other old-world primates.

Taxon Advisory Group

The studbook keeper makes recommendations on further breeding to the Taxon Advisory Group of the Federation of Zoos which oversees the coordinated management of the species.

(*Source*: M. *Stephenson, Edinburgh Zoo.*)

protection, by improving overseas captive breeding and management programmes in less-developed countries (see **Box 6.17**).

Animal gene banks are much more costly to set up and maintain at the correct scientific (and legal) level than are plant gene banks. In the UK the Rare Breeds Survival Trust (see Chapter Five Section 5.3.5.3) plays an important role in ensuring that strains of domesticated animals which, in the past, have been (or in the future may be) important in UK animal husbandry are maintained (e.g. Soay and Loghan sheep). It also manages a semen bank and a biodiversity database. The main animal breeding in the UK currently centres on Roslin Institute, Midlothian, where gene banking and research into pigs, cows and poultry take place (see Chapter Four).

Box 6.16 Examples of species genetically and demographically managed in zoos under the auspices of the Federation of Zoos of Great Britain and Ireland

(Data from Brouwer, K, Rietkerk, , & Smits, S (eds). (1994): EEP Yearbook 1993/1994 and from P.J. Olney)

Species	Status*(1)	JMSP*(2) No Collections	JMSP No Individs*(3)	EEP*(4) No Collections	EEP No Individs*(3)	International Studbook*(5)	Reintroduction Country*(6)
Pink Pigeon *Columba mayeri*	E	8	41/41(8) =90	11	49/45(8) =102	Yes	Mauritius
Hyacinthine macaw *Andorhynchus hyacinthinus*	V/R	19	30/27(3) =60	78	123/124(37) =284	No	
Bali Starling *Leucopsar rothschildi*	E	21	64/44(16) =124	65	180/166(68) =414	No	Bali
Golden lion tamarin *Leontopithecus rosalia*	E	13	32/40(3) =77	35	78/78(4) =160	Yes	Brazil
Gorilla *Gorilla gorilla*	V	12	33/54 =87	62	131/188 =319	Yes	
Orangutan (Bornean) *Pongo p. pygmaeus* (Sumatran) *Pongo p. abeli*	E	7 5	7/19 =26 6/8 =14	36 30	52/79 =131 45/73 =118	Yes Yes	Borneo Sumatra
Cheetah *Acinonyx jubatus*	V	14	40/44 =84	62	135/157 =292	Yes	
Snow leopard *Panthera uncia*	E	14	26/35 =61	65	102/117 = 219	Yes	
Sumatran tiger *Panthera tigris sumatrae*	E	7	5/6 =11	49	47/61=108	Yes	
Black rhinoceros *Diceros bicornis*	E	3	8/8 =16	12	21/33 =54	Yes	
Scimitar-horned oryx *Oryx dammah*	E	8	23/87 =110	34	102/189(8) =299	No	Tunisia
Arabian oryx *Oryx leucoryx*	E	6	8/9 =17	18	34/44 =78	Yes	Jordan Oman Saudi Arabia
Partula snails Partulidae	E/V	8				No	Moorea

1 Data from Groombridge, B (ed). (1993) 1994 IUCN *Red List of Threatened Animals*, IUCN, Gland, Switzerland and Cambridge, UK. E = Endangered; R = Rare; V = Vulnerable.
2 JMSP=Joint Management of Species Programmes of Federation of Zoological Gardens of Great Britain and Ireland. In this context Britain and Ireland are considered a sub-region of the European region and most JMSPs would be included in EEP Programmes.
3 Sexes are denoted as male/female(unknown).
4 EEPs=European Endangered Species Programmes of the European Association of Zoos and Aquaria; these programmes cover all of Europe, inclusive of Britain and Ireland.
5 International Studbooks are listed annually in the *International Zoo Yearbook* (ed. Peter J.S. Olney), Zool. Soc. London.
6 Reintroduction, Restocking, Rehabilitation – the country is given when a project is underway. Otherwise no action has yet been taken.

A few universities and other institutions maintain living collections of animals for research, e.g. the Natural History Museum holds living collections of water snails (vectors carrying *Schistosomiasis*), mainly to study the taxonomy of the snails.

Zoological Society of London

Arabian Oryx (*Oryx leucoryx*) in Shahaniyah, Qatar.

6.5 Collections of microorganisms

Since 1947, eleven national culture collections for microorganisms have been established in the UK, each collection being housed with parent institutions offering relevant expertise, and with a remit to provide cultures and related services to the scientific community. These institutions hold approximately 68,000 strains of algae, animal cells, bacteria, fungi, phages, plasmids, plant cells, and protozoa. (see **Box 6.18**). The Genetic Resources Collection at the International Institute of Mycology is an important one, with over 18,500 living isolates (including yeast and over 1,000 bacteria). Information on some major international collections of microorganisms is summarised by Hawksworth and Ritchie (1993).

Box 6.17 The Zoo Federation's Overseas Conservation Programme

In late 1993, the Federation of Zoological Gardens of Great Britain and Ireland received a grant from the Darwin Initiative to investigate the role that the British zoo community could best play in conserving biodiversity in the developing world. The report resulting from Dr Brian Bertram's feasibility study—'British Zoos and Conservation Overseas'—completed in April 1994, was accepted by the Department of the Environment and by the Council of the Federation. The latter resolved, in the autumn of 1994, that the Federation should set up its Overseas Conservation Programme, subject to funding and, again subject to funding, Dr Brian Bertram was asked to coordinate it,.

The background to the Overseas Conservation Programme is the scale and variety of conservation projects in which British zoos are already involved. In addition to their extensive captive conservation work, much of it is internationally coordinated, and in addition to their lively education programmes, zoos in Britain are undertaking or supporting some 70 *in-situ* projects and contributing well over a million pounds a year in doing so. The Federation's Overseas Conservation Programme is designed to complement and assist, not to substitute for, these important projects, and to concentrate on those areas of activity where zoos can achieve more collectively than individually.

The main focus of attention will be in Africa, primarily but not exclusively; this is for reasons of geography, need and alternative sources of help. The Federation's Overseas Conservation Programme will concentrate particularly on zoos and other wildlife organisations there, and will emphasise especially public awareness and all aspects of training. It is hoped to build institutional linkages between individual African and British zoos, for the benefits of both participants and their surrounding communities. Collaboration will of course be a two-way learning process.

In the developing world, small amounts of money can go a very long way. Funding is being sought from as wide a variety of sources as possible, because the objective is very much to involve large numbers of the 'not-yet-committed' in conservation work overseas. Among the sources are, of course, the 8 million visitors each year to the Federation's 56 member zoos.

The view is taken that we can help zoos in Africa to achieve better the shared education and conservation objectives that the World Zoo Conservation Strategy 1993 lays out for them. These include involvement with *in-situ* conservation. The conserving of wildlife in the wild habitats where it belongs depends critically on the support of the human populace of the countries of origin. As in Britain too, most ordinary people are disturbingly unaware of conservation issues, both local and planetary, and of problems, trends and solutions. We can help to ensure that zoos in Africa, which are (sadly but realistically) visited by vastly more people than can ever possibly visit their own National Parks, can and do provide focus for a growing conservation understanding among the peoples of Africa. It is upon their interest and votes the wildlife of the continent will ultimately and crucially depend.

(*Source*: P.J. *Olney*, FZGGBI.)

Cultures are required for identification purposes, particularly in the case of bacteria, yeasts and some filamentous fungi where the living 'types', that fix the application of names, must be maintained. Many isolates held would be very difficult to find again in their natural habitat, and *ex-situ* collections provide the only practical option for making them available for reference for identification, and for research and screening potential uses.

In order to improve the availability of these strains, and to ensure their long-term security, there is a need to consider a network of UK `national' collections. The Office of Science and Technology set up a committee to review the status and future organisation of the UK collections of microorganism Substantial recommendations were made concerning marketing and rationalisation of sites and services (Goodfellow, 1995). It is proposed that

Box 6.18 The holdings and location of UK culture collections of microorganisms

This table is based on that published by HMSO in *Review of UK microbial culture collections: an independent review commissioned by the Office of Science & Technology* (1994) pages 15–16.

Collection	Parent organisation	Public funding source	Location	Historical notes	Major culture holdings	Staff
International Mycological Institute (IMI)	CABI	International UK: CABI membership: ODA and DOE projects: (DTI, now ceased)	Egham	Formerly Commonwealth Mycological Institute, Kew	Fungi, yeasts, plant bacteria (18,060)	67 (12 managing collns)
National Collection of Industrial and Marine Bacteria Ltd (NCIMB)	Aberdeen University Research and Industrial Services, (AURIS) Ltd	MAFF (terminated 3/95: OST responsible for the collection)	Aberdeen	Formerly at Torry Research Institute, Aberdeen: commercialised 1983	General, industrial and marine bacteria (6,000)	12
European Collection of Animal Cell Cultures (ECACC)	Microbial Research Authority (MRA)	EC; UK: MRC/ NERC/BBSRC; DH via MRA; (DTI, now ceased)	Porton Down, Salibury	MRA (as CAMR) and ECACC were (till 1994) part of the PHLS	Animal cell cultures (1,400)	29
National Collection of Type Cultures (NCTC)	Public Health Laboratory Service (PHLS)	DH	Central Public Health Laboratory, Colindale, London		Bacteria pathogenic to humans and animals (5,000)	6
National Collection of Pathogenic Fungi (NCPF)	PHLS	DH	Since 1993, Mycology Reference Laboratory, PHL, Bristol	1948 School of Hygiene and Tropical Medicine 1985 CPHL, Colindale	Fungi pathogenic to humans and animals (1,630)	0.3
National Collection of Plant Pathogenic Bacteria (NCPPB)	Plant Health Group, Central Science Laboratory (CSL)	MAFF	CSL, York, formerly at Harpenden		Bacteria pathogenic to plants (3,800)	0.8
National Collection of Wood Rotting Fungi (NCWRF)	Building Research Establishment (BRE)	DOE	BRE, Watford	Formerly Forest Products Research Princess Risborough	Wood rotting fungi (550)	0.2
National Collection of Yeast Cultures (NCYC)	Institute of Food Research (IFR)	BBSRC	IFR, Norwich	Formerly the Brewing Industry Research Foundation Collection	Non-pathogenic yeasts, particularly for brewing (2,500)	2.1
National Collection of Food Bacteria (NCFB)	Institute of Food Research (IFR)	MAFF	IFR, Reading	Formerly the Institute for Research in Dairying Collection, Reading	General bacteria, including dairy organisms (1,600)	1.2
Culture Collection of Algea and Protozoa (CCAP), Oban	Dunstaffnage Marine Laboratory (DML)	NERC	Scottish Marine Biology Association, Dunstaffnage	Formerly at CCAP, Institute of Terrestrial Ecology, Cambridge	Marine algae (490)	2.7
CCAP (as above) Windermere	Institute of Freshwater Ecology (IFE)	NERC	Freshwater Biology Association, Windermere	As above	Protozoa (340), fresh-water algae and cyanobacteria (1,090)	4.6

Box 6.19 The Culture Collection of Algae and Protozoa (CCAP): marine algae at Dunstaffnage

The Culture Collection of Algae and Protozoa has its origins in Central Europe, in the collection of living cultures of freshwater algae started by Ernst Georg Pringsheim at the German University of Prague during the late 1920s. In 1970, the expanded collection formed the basis for the Culture Centre of Algae and Protozoa funded by the Natural Environment Research Council in a purpose built building in Cambridge. During 1986, the cultures and their associated activities were transferred to the Institute of Freshwater Ecology (IFE) at Windermere (freshwater algae and all protozoa) and the Dunstaffnage Marine Laboratory (DML) near Oban (marine algae).

Marine organisms were unrepresented during the early years. In the 1960s a few marine algae were acquired from the Marine Biological Association at Plymouth and many more were added from this source in the early 1970s. A major boost to the holdings of marine algae was provided, at this time, by the accession of cultures established by R.W. Butcher of the Fisheries Laboratory, Burnham-on-Crouch. Most of the cultures from these two sources were from British coastal waters and many were of great importance as representatives of type material.

Another major source of material for the CCAP was the collection of axenic strains, useful to physiologists and biochemists, held for many years as a research collection by M.R. Droop of the Scottish Marine Biological Association, first at Millport and from 1969 at Dunstaffnage. Other important additions to the CCAP's holdings include a collection of small red seaweeds, made by Vrije Universiteit, Amsterdam, diatoms, including Antarctic species collected by staff of th Alfred-Wegener Institute, Bremerhaven, and a large number of strains from various parts of the world, assembled by Ralph Lewin.

The table shows numbers of taxa currently held. Most of the strains have been collected from the neritic Eastern North Atlantic, but there are others from the Atlantic coast of North America, the Pacific, the Mediterranean and Black Seas and from several inland saline and soda lakes. Additional ecological niches from which the organisms have been collected range from mangrove swamps, salt marshes and supralittoral rock pools to the icy waters of the Southern Oceans.

Marine algae held by CCAP

Division/Class	**Number of strains**
Cyanophyta (Cyanobacteria)	32
Chlorophyta	226
Chlorarachniophyta	1
Euglenophyta	5
Phaeophyta	2
Chrysophyta	
Chrysophyceae	12
Prymnesiophyceae	43
Eustigmatophyceae	8
Raphidophyceae	3
Bacillariophyceae	70
Pyrrhophyta	20
Rhodophyta	59
Cryptophyta	37

Taken together with the freshwater algae and protozoa, CCAP's holdings are arguably the most genetically diverse of any single collection of organisms. The principal concern of the custodians of this collection is that its diversity should be utilised by as wide a range of users as possible. Some ideas of the CCAP's potential interface with biotechnology is given in Day, J.G. & Turner, M.F., (1992) Algal culture collections and biotechnology. *Proc. Symp.Culture Collection of Algae, Tsukuba* 1991 (M.M. Watanabe, ed.), pp. 11–27, National Institute for Environmental Studies, Tsukuba, Japan.

(*Source*: *Michael F. Turner, Curator, CCAP, Dunstaffnage Marine Laboratory*.)

the Biotechnology and Biological Sciences Research Council (see Chapter Four) should assume overall responsibility for the UK culture collections. **Box 6.18** gives a summary of the major national collections. There is an increasing number of biotechnological companies, as well as many established pharmaceutical and microbiological companies and university departments, now maintaining collections of microorganisms some of which have the status of national collections (e.g. National Collection of Industrial and Marine Bacteria for industrially significant bacteria). The Natural History Museum has living cultures of diatoms and other microscopic algae.

Many networks are being established at the European level and 32 UK member organisations of the European Culture Collections Organisation (ECCO), had total holdings of 253,765 strains in 1992, 120,646 of which were held in UK institutes (HMSO, 1994). The UK is playing a significant role in several international initiatives (e.g. Banque Européene de Glomales). See also Volume 2, Chapter Three, Box 3.1. The Secretariat of the international information and communications network – Microbial Strain Data Network (MSDN) is based in Cambridge (see Appendix Two). The databases, holding information on microorganisms and cultured cells, are distributed between different parts of the world. They are listed in Chapter Seven, Box 7.13.

6.6 References and bibliography

Anon. (1904). *The history of the collections in the Natural History Departments of the British Museum*, Vol. 1 *Libraries, Department of Botany, Department of Geology and Department of Minerals*; Vol. 2 *Department of Zoology*. Trustees of the British Museum, London.

Arnett, R.H. and Samuelson, G.A. (1986). *The insect and spider collections of the world*. E.J. Brill, Gainsville.

Arnold-Forster, K. (1993). *Held in Trust: Museums and collections of universities in northern England*. AMSEC, London.

Barbour, S. (1993). *Museums Yearbook 1993/1994, including a directory of museums and galleries of the British Isles*. Museums Association: Rheingold, London.

Bate, C. (1862). *Catalogue of the Amphipodous Crustacea in the British Museum (Natural History)*. Trustees of the British Museum, London.

Bateman, J., McKenna, G. and Timberlake, S. (1993). *Register of natural science collections in South-East England*, Cambridge Museum Documentation Association, Vols 1-3 (including. computer disc and microfiche issues), Cambridge.

Bell, T. (1855). *Catalogue of Crustacea in the British Museum (Natural History)*. Trustees of the British Museum, London.

Benson, E.E., Harding, K. and Smith, H. (1989). Variation in recovery of cryopreserved shoot-tips of *Solanum tuberosum* exposed to different pre- and post-free regimes. *Cryo-Letter*, 10: 323-344.

BGCS (Botanical Gardens Conservation Secretariat). (1990). *Botanic Gardens and Arboreta of Britain and Northern Ireland*. BGCS, Kew.

Boulenger, G.A. (1882a). *Catalogue of Batrachia Gradientia, s. Caudata and Batrachia Apoda in the collection of the British Museum*. Trustees of the British Museum, London.

Boulenger, G.A. (1882b). *Catalogue of Batrachia Salientia, s. Ecaudata in the collection of the British Museum*. Trustees of the British Museum, London.

Boulenger, G.A. (1889). *Catalogue of chelonians*. Trustees of the British Museum, London.

Boulenger, G.A. (1885-1889). *Catalogue of lizards*. 3 Vols. Trustees of the British Museum, London.

Boulenger, G.A. (1893-1896). *Catalogue of snakes*. 3 Vols. Trustees of the British Museum, London.

Clark, H.L. (1925). *A catalogue of recent sea urchins (Echinoidea) in the collection of British Museum (Natural History)*. BM(NH), London.

Clarke, J.F.G. (1955-70). *Catalogue of type specimens in the British Museum (Natural History), described by Edward Meyrick*. 8 vols. BM (NH), London.

Clokie, H.N. (1964). *An account of the herbaria of the Department of Botany in the University of Oxford*. Oxford University Press, Oxford.

Cook, F.E.M. (1995). *Economic Botany Data Collection Standard*. Royal Botanic Gardens, Kew.

Dance, S.P. (1986). *A history of shell collecting*. E.J.Brill, Leiden.

Davies, D.A.L. and Hull, G.M. (1976). *The zoological collections of the Oxford University Museum, a historical review and general account with a complete donor index to the year 1975*. Oxford University Press, Oxford.

Davies, D.A.L. and Tobin, P. (1984-85). *The dragonflies of the world – A systematic list of the extant Odonata*. Vols 1&2. Soc. Internat. Odonatologica Rapid Communications, Nos 3 & 5 (Suppl.).

DoE (Department of the Environment). (1994). *Biodiversity: The UK Action Plan*. HMSO, London.

Desmond, R. (1995). *Kew: The History of the Royal Botanic Gardens*. Harvill Press and The Royal Botanic Gardens, Kew.

Eastop, V.F. & Lambers, D.H.R. (1976). *Survey of the world's aphids*. Junk, The Hague.

Fabre, J. and Dereuddre, J. (1990). Encapsulation dehydration: a new approach to cryopreservation of *Solanum* shoot tips. *Cryo-Letter*, 11: 413-426.

Fletcher, H.J. & Brown, W.H. (1970). *The Royal Botanic Garden Edinburgh*, 1670-1970. HMSO, Edinburgh.

Gislen, T. (1928). Notes on some Crinoids in the British Natural History Museum. *Svenka Vet. Akad. Arch. Zool.*, 19: 24-30.

Goodfellow, M. (1995). Guest editorial: Review of UK microbial culture collections. UK *Federation for Culture Colls. Newsl.* 24:1-2.

Gray, J.E. (1849) *Catalogue of Mollusca (including fossil forms) in the collections of the British Museum, Pt 1: Cephalopoda Antepedia*. Trustees of the British Museum, London

Gray, J.E. (1888). *Catalogue of Marsupials*. Trustees of the British Museum, London.

Günther, A. (1912). *The history of the collections contained in the Natural History Departments of the British Museum*. Vol.2, *Appendix*. Trustees of the British Museum, London.

Hampson, G. (1898-1920). *Catalogue Lepidoptera Phalaenae*. Vols 1-16. British Museum, London

Hancock E.G. and Morgan, P.J. (1888). *A survey of zoological and botanical material in museums and other institutions of Great Britain*. Biology Curators Group, Cardiff.

Hawkes, J.G. (1990). *The potato: Evolution, biodiversity and genetic resources*. Smithsonian Institution Press, Washington, DC.

Hawksworth, D.L. and Ritchie, J.M. (1993). *Biodiversity and biosystematic priorities: microorganisms and invertebrates*. CAB International, Wallingford.

Heywood, C.A., Heywood, V.H. and Wyse Jackson, P.J. (1990). *International directory of botanic gardens*. Koeltz, Koenigstein.

Hopkins, G. H. E. and Rothchild, M. (1953-1971). *Illustrated catalogue of the Rothchild Collections of fleas, vols* 1-6. British Museum (Natural History), London.

Hopkins, G. H. E. and Clay, T. (1952). *A checklist of the genera and species of Mallophaga*. :1-302. British Museum (Natural History), London.

Holmgren, P.K., Holmgren, N.H. and Barnett, L.C. (1990). *Index Herbariorum. Part 1: The herbaria of the world*, (8 edn.). (Regnum Vegetabile No. 120.) New York Botanic Garden, New York.

Hudson K. and Nicholls, A. (1987). *The Cambridge guide to the museums of Britain and Ireland*. Cambridge University Press, Cambridge.

Johnston, G. (1865). *A catalogue of the British non-parasitical worms in the collections of the British Museum*. Trustees of the British Museum, London.

Keirans, J.E. (1985). *George Nuttall and the Nuttall tick collection*. 1-1786. USDA, Washington.

Kirby, W.F. (1904-1910). *A synomic catalogue of Orthoptera*. Vols 1-6. British Museum, London.

Linington, S. (1994). *List of seeds* (1994, biennial). Royal Botanic Gardens, Kew.

Linneaus, C. (1753). *Species Plantarum*. Stockholm. [Facsimile (1957). Ray Society, London.]

Linneaus, C. (1767-70). *Systema Natura*. Vol 1, (10 edn). [Facimile. British Museum (Natural History), London.]

Lyddekar, R. (1913-1916). *Catalogue of ungulates*. Trustees of the British Museum, London.

Mardon, D. (1981). *Illustrated catalogue of the Rothschild Collection of fleas*, Vol. 6. British Museum (Natural History), London.

Masser, I. and Blackmore, M. (eds). (1991). *Handling geographical information: methodology and potential applications*. Routledge, London.

Miller, K. (1977). *Simple surveying techniques for the small expedition*. Royal Geographical Society, London.

Mound L. and Halsey, J. (1978). *Whitefly of the world*. British Museum (Natural History), London.

Napier, P.H. and Jenkins, P.D. (1970-1990). *Catalogue of Primates*. British Museum (Natural History), London.

NCCPG (National Council for the Conservation of Plants and Gardens). (1996). The national plant collections. Directory 1996. NCCPG, Wisley.

Newport, G. (1854). *Specimens of Myriopoda in the British Museum*. Trustees of the British Museum, London.

OST (Office of Science and Technology). (1994). *Review of UK microbial culture collections*. HMSO, London.

Oldfield-Thomas, H.R. (1906). *The history of the collection contained in the Natural History Department of the British Museum*. Trustees of the British Museum, London.

Olney, P.J. and Ellis, P. (1993). Lists of zoos and aquaria. *International Zoo Yearbook* 32: pp. 261-360. Zoological Society, London.

Pearl, M.C. (1995). Wildlife Preservation Trust International: Programme for 1995-96. *Dodo, J. Wildl. Preserv. Trusts*, 31:10-13.

Ritchie, J.M. (1987). Insect biosystematic services in Africa: current status and future prospects. *Insect science and its application* 8: 425-432.

Smit, F.G.A. (1987). *An illustrated catalogue of the Rothschild collection of fleas (Siphonaptera) in the British Museum (Natural History): with keys and short descriptions for the identification of the families, genera, species and subspecies of the Order*. Oxford University Press/BM(NH), London.

Stace, H.E., Pettitt, C.W. and Waterston, C.D. (1987). *Natural Science collections in Scotland* (Botany, Geology & Zoology). National Museums of Scotland, Edinburgh.

Stearn, W.T. (1957). An introduction of the Species Plantarum and cognate botanical works of Carl Linnaeus. Prefixed to Ray Society facimile of Linnaeus' *Species Plantarum*. Ray Society, London.

Stearn, W.T. (1981). *The Natural History Museum at South Kensington*. Heinemann, London.

Theodor, O. (1967). *An illustrated catalogue of the Rothschild Collection of Nycteribiidae (Diptera) in British Museum (Natural History)*. British Museum (Natural History), London.

Tittley, I. and Sutton, D.A. (1984). *A geographical index to the collections of Phaeophyta in the British Museum (Natural History)*. *British Museum (Natural History)* microfiche.

Townsend, B.C. (1990). A catalogue of the Types of bloodsucking flies in the British Museum (Natural History). *Occ. Papers Syst. Entom.* 7:1-37.

Vitt, D.H., Gradstein S.R. and Iwatsuki, Z. (1985). *Compendium of Bryology. A world listing of herbaria, collectors, bryologists, and current research*. (Bryophytorum Bibliotheca, 30). Cramer, Braunsweig.

Walker, F. (1854-1866). *List of Lepidopterous insects in the British Museum*. 35 Pts. British Museum, Department of Zoology, London.

Walter, K.S., Chamberlain, D.F., Gardner, M.F., McBeath, R.J.D., Noltie, H.J. and Thomas, P. (1955). *Catalogue of plants growing at the Royal Botanic Garden, Edinburgh* 1995. Royal Botanic Garden, Edinburgh.

Whalley, G.P. (1993). *Register of natural science collections in the Midlands of England*. AMCRU, Nottingham.

White, A. (1847). *List of the specimens of Crustacea in the collection of the British Museum*. Newman, London.

Williams, B. (1987). *Biological collections in the UK*. The Museums Association, London.

Williams, D.M. (1988). An illustrated catalogue of type specimens in the Greville Diatom Herbarium. *Bull. British Museum (Natural History), Bot.*, 18: 1-148.

Williamson, P. (ed.) (1986). *UK marine culture collections and culture-related research*. NERC, Swindon.

Wyse Jackson, P. (1995). The international context. *Proc. Conf. on Plant Collns Network of Brit. & Irel.*:1-3. PlantNet, Oxford.

OTHER INFORMATION RESOURCES:
Libraries, maps, and databases

7.1 **Introduction**

7.2 **Libraries, bibliographic databases and archives**

7.3 **Cartographic materials: maps, atlases and gazetteers**

7.3.1 Sources of maps

7.3.1.1 The Royal Geographical Society Map Room

7.3.1.2 Ordnance Survey International library resources

7.3.1.3 The British Library Map Library

7.3.1.4 The Oriental and India Office Collection, British Library

7.3.1.5 The National Library of Scotland

7.4 **Air photographs**

7.5 **Satellite imagery**

7.6 **Geographical Information Systems (GIS) and digital geographical data**

7.7 **Information sharing: the challenges**

7.7.1 The Biodiversity Clearing House Mechanism

7.7.2 The World Conservation Monitoring Centre

7.8 **References and bibliography**

Chapter authorship

This chapter has been drafted and edited by Shane Winser *with additional material and comments by* R. Banks, S. Fitzgerald, M. Gillman, C. Gokce, K. Grose, D. Hawksworth, C. Jermy, D. Kirkup, N. McWilliam, L. Nunn, M. Sands, R. Smith, P. Surrey, A. Tatham *and* R. Teeuw.

7.1 Introduction

The information needed to prepare a biodiversity assessment is extremely diverse and will come from many different sources. Earlier chapters have already described the resources within UK's research institutes, universities, museums and herbaria. This Chapter looks at some of the other collections of data that have been amassed in the UK, in some cases over a considerable period of time. It deals with sources of information held in libraries such as books, journals and maps, and computerised databases generated here in the UK. Emphasis is given to the means of disseminating this information internationally using advances in information technology.

The Guidelines for Country Studies on Biological Diversity (UNEP, 1993) and *Global Biodiversity Assessment* (UNEP, 1995) provide detailed recommendations on what data needs to be gathered to support these studies, and suggests ways in which this data might be collected, analysed and managed. (See also Volume Two, Chapter One).These guidelines recommend that data gathering can be significantly advanced if the main out-of country sources are accessed early in the planning stage.

Access to documentary information about species and habitats is a vital adjunct to research on biodiversity. Prior to the 1970s, such a statement implied access to a well-resourced library in which librarians attempted to acquire as comprehensive a collection of printed matter as was required to service the needs of in-house and visiting users. Since that date the accelerating development of computerised databases and electronic interchange of data, text and, increasingly, pictures has begun to make access to documentary resources less reliant on the traditional library concept, with the emphasis moving towards direct delivery of information to a desk-top computer via electronic networks.

A short guide to the 'what', 'why' and 'how' of using these electronic networks is given in **Box 7.1**. The best published introduction to using the Internet is the *Whole Internet Users Guide* (Krol, 1994), and for advice on the computer technology required *Connecting to the Internet* (Estrada, 1993).

Through the Internet and other global networks, such as APC (See **Box 7.2**) researchers can at least communicate and share information cheaply and quickly. Using networks is becoming easier and more software is being written to help locate information, and present it in more easily interpreted format, some of it interactively. World Wide Web is fast becoming the most popular medium for displaying information. Some useful Internet addresses are given in **Box 7.3**.

7.2 Libraries, bibliographic databases and archives

Valued though these electronic developments are in solving some of the problems of access to materials and speeding up communication, there is a long way to go before the accumulated printed output of over two hundred years is digitised and available instantly at a key-stroke. Almost alone in the sciences, research on many aspects of biodiversity, for example such as the underlying taxonomy, is dependent on access to early literature as much as to modern works. The factor of obsolescence of published information, prevalent in physics, chemistry and some other disciplines, is remarkably less a factor in this field, and there continues to be an important role for libraries in providing access to comprehensive collections of printed and archival material.

The national copyright libraries receive every work published in the UK and also hold vast collections of foreign material. They are: the British Library in London, National Library of Scotland in Edinburgh, National Library of Wales in Aberystwyth and the Libraries of Oxford

British Library

The British Library, Bloomsbury, London.

Box 7.1 The What, Why, and How of Electronic Networks

"Electronic communication", "e-mail", "Internet", "APC", "ftp" are all terms which are starting to appear more and more. Magazine and newspaper articles speak of "information highways", the "Web", and "interconnectivity" while business cards are beginning to feature e- mail addresses. This article provides a brief explanation of what is meant by electronic communication, offers some definitions of the concepts and terminology and describes how it can be used.

What are electronic networks?

Electronic networks are, in fact, chains of interconnected computers. The first networks were created in the 1970s in the United States where governmental and research institutions wanted to maximise the use of powerful computers located at different sites around the country. By using telecommunications lines to link these machines, it became possible for a scientist in California to access a computer in New York. Nowadays, it is possible for you to take advantage of this form of communication. In the same way that you dial a telephone number to call someone on the telephone network, you can dial another computer in the same city or on the other side of the world.

While the telephone system creates a voice connection, the electronic network packages into **packets** the data from your computer. The packets are *stamped* with the *address* and are dropped into an electronic mailbox. Once in the mailbox, streams of packets are sent over any number of paths to reach the right computer. This process is called **packet switching**.

To keep all of this information flowing in the right direction, there are certain rules which must be followed. They are called **protocols.** These protocols are standards which govern the software and hardware and ensure that different machines communicate with each other.

A common telecommunications protocol is **TCP/IP** (Transmission Control Protocol/Internet Protocol).

Indeed, the **Internet** is the international network which results from the interconnection of those networks using the TCP/IP protocols. More than 20 million computers are now connected to this network of networks!

There are three basic Internet services: **Telnet, E-mail** and **FTP.**

Telnet allows you to connect to and use a remote computer. Using Telnet you can "log on" to a remote host computer as though it were in the next room. To help locate information, databases and other services, you can "telnet" to a gopher. A **gopher** is a menu system that makes the resources of the Internet available to you on your computer screen. Also available is **Veronica** which allows you to search menu options across all gopher space.

A recent addition is **World Wide Web (WWW).** *Hypermedia* is the foundation of the WWW. Here (media refers to any type of data such as text, audio, or graphic). Hypermedia connects the media or computer data so that a document is linked to another. You can move between hypermedia documents and explore related documents at your own pace, navigating in whatever direction you chose. IUCN's Library and its Environmental Law Centre are cooperating with **CIESIN – The Consortium for International Earth Science Information Network** to develop a WWW guide to IUCN's information (the address is **http://www.ciesin.org/** or access the catalogue service at **ciesin.info@ciesin.org**).

E-mail allows you to send messages from one computer to another. By entering the **e-mail address**, you can type in the message and send it via the network. The message will be deposited into the addressee's electronic "mail box" where it can be read, stored, copied or forwarded.

Related to this, there are *Mailing Lists* where any e-mail sent to the mailing list is automatically resent to anyone having an address on the list. Those receiving a message from the list can then reply to it and a copy will be sent to everyone else. These lists are sometimes called *mail reflectors or unmoderated lists*. Mailing lists may also be moderated by a *list administrator* who weeds out duplicates or inappropriate messages. A more sophisticated version of a mailing

Box 7.1 *Continued*

list is a *List server* which has an automated subscription feature and offers other services much like a bulletin board. The World Conservation Monitoring Centre (WCMC) maintains a "Listserv" dealing with trade in endangered species and other CITES- related issues. For more information contact, **info@wcmc.org.uk**.

FTP is a method of downloading (transferring) files over the Internet. Thousands of files containing text, graphical and sound information as well as data sets are available free at FTP sites. Using search tools such as **archie** you can identify the files that are of interest and then "ftp" them to your computer. This method of document delivery not only saves time but has the added advantage that the information is in digital form and can be used on any personal computer (PC).

While Internet is fast becoming the *de facto* standard, there are other important networks. One is the **APC Network**. (*Association for Progressive Communications*) and is an international non-profit organisation dedicated to facilitating progressive social change through cooperative local and global computer networking. Unlike the Internet which is a "northern" phenomenon, the APC network represents a partnership of NGOs from north and south. Like Internet, APC networks have e-mail and file transfer facilities and maintain hundreds of **electronic conferences**. These conferences, such as "IUCN.News" on GreenNet (for more information contact, **support@gn.apc.org**) are **Bulletin Board Systems** where you can access a conference, check on newly posted items, send a reply or download the text of a posted item to your PC. The APC network provides a **gateway** or bridge that translates data between the APC network and the Internet.

Why are networks useful?

Using e-mail you can communicate quickly with people all over the world. If you require expert advice to solve a problem, or want to be kept informed, e-mail can help. IUCN Headquarters now has regular e-mail exchange with the IUCN regional offices for Eastern Africa, Southern Africa, Meso-America, and South America; the IUCN country offices in the Hungary, Nepal and United States; the IUCN Publications Services, TRAFFIC and WCMC in the UK; and the Environmental Law Centre in Germany. The Species Survival Commission Chairman in Chicago, USA, and that of the Commission on Environmental Strategy and Planning (Sacramento, USA) are also connected.

E-mail combined with the other information tools, services and resources, will help people communicate better, and perhaps more importantly, offer more people than ever before the opportunity to contribute to world debate by exchanging their ideas and views.

And the cost? Surprisingly low; indeed, some IUCN offices have found that it is cheaper for them to communicate using e-mail than telephone or fax. Remember though, that e-mail does have its disadvantages.

It is digital information and subject to corruption or alteration. It does not bear a signature and will probably not be considered legal. Confidentiality can, likewise, not be guaranteed.

How can you connect to electronic networks?

The simplest access requires only a PC loaded with a telecommunications software such as "Kermit" or "CrossTalk", a modem (a device which links computers to a telephone line) and a telephone line. You then dial the nearest network **node,** a local communications centre via which messages pass. The cost of accessing the networks is normally the cost of a local call to the node plus the cost of a node account or subscription. Nearly 140 countries have international network connectivity. To find out how you can connect, ask someone in a university computer science department near you, or contact an NGO with links abroad or the local Post and Telecommunications Office. Once you have identified the local node administrator, they will be able to help you connect. If you cannot find a local contact, you can write to GreenNet, 4th floor, 393-395 City Rd., London EC1V 1NE, United Kingdom, Tel. 44 171 713 1941, Fax. 44 171 833 1169, email. **support@gn.apc.org**.

Kevin Grose, Head, IUCN Library.
Reprinted, with permission, from interACT, 1995.

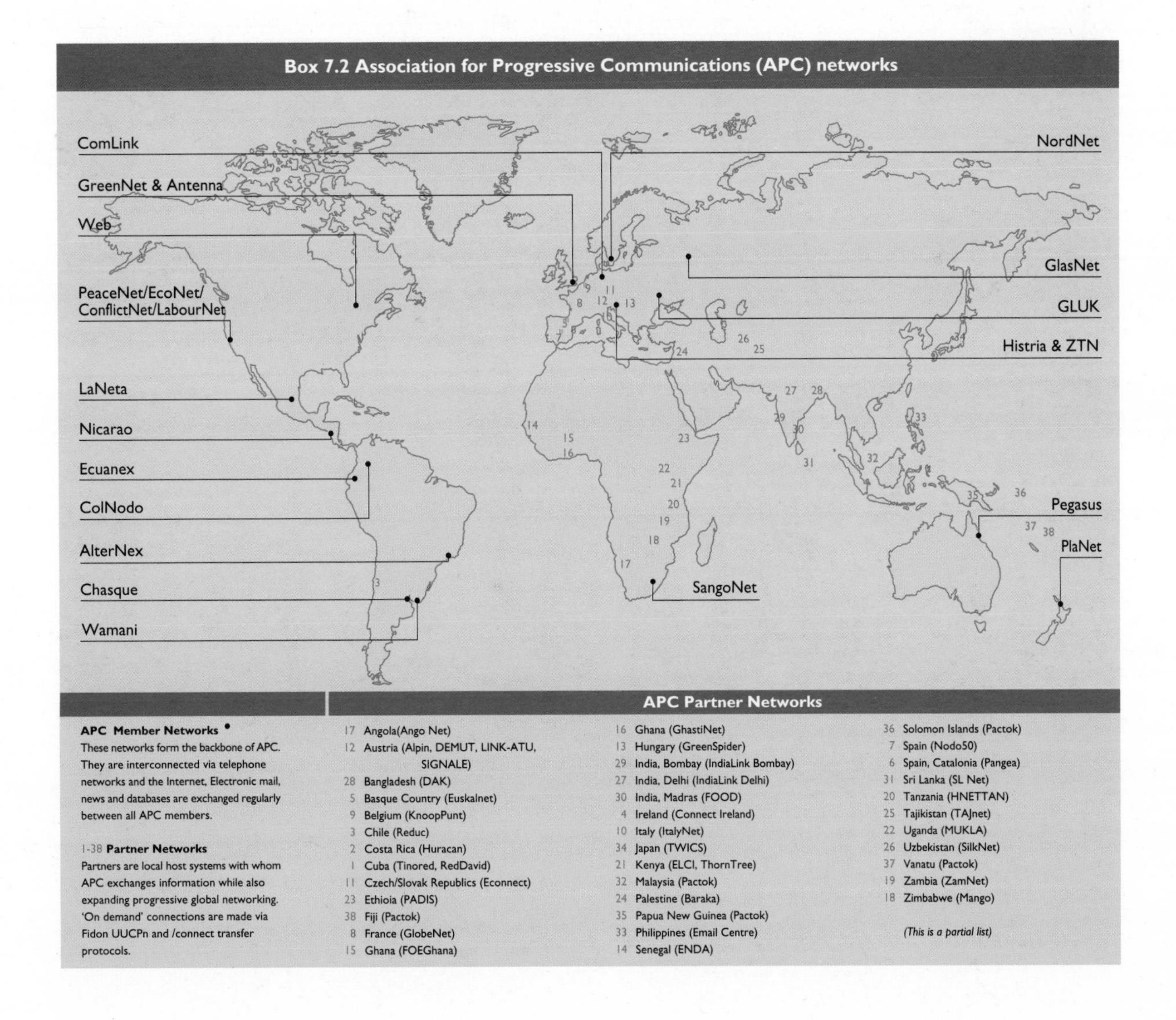

and Cambridge Universities. The Science Reference and Information Service (SRIS) is the UK's national reference library for modern science and technology which is part of the British Library and has the most comprehensive reference collection in western Europe. Its main purpose is to make scientific literature available to the public and no reader's ticket or prior arrangements are necessary to visit. The UK is also particularly well served by specialist libraries, many of them attached to learned societies and research centres (see **Box 7.4**). These are usually established to serve the needs of their members or staff, and as a result tend only to allow visitors by prior arrangement.

Within any given subject area there are a plethora of bibliographic aids to help researchers find their way through the literature including abstracting and indexing services, and current awareness journals. Fortunately many of these are now available in machine-readable form, and can be searched using a computer either on-line, or increasingly, CD-ROMS. International agriculture, for example, is particularly well served by CAB Abstracts (see **Box 7.5**) which provides a full range of these services. Biologists are served by BIOSIS, a bibliographic on-line publisher (see **Box 7.6**). There are many others, and there are a number of directories to help identify appropriate databases for individual needs. The majority are fee-paying services, and which do require a degree of familiarity with the database to get the best search results, and it may be more cost-effective if a librarian is asked to undertake the search. Most of the larger libraries offer this service including the British Library's Science Reference and Information Service. Furthermore, an increasing number of libraries are also making their library catalogues available for public use (e.g. the NHM Library Catalogue through Telnet). This can take the form of public-terminals in the library to enable readers to do their own electronic 'browsing', or over the Internet. ECO, the Environmental Information Trust, publishes the ECO *Directory of*

Box 7.3 Biodiversity-related resources on the Internet

Any reader with facilities and time to browse the Internet should refer to Biologists' guide to Internet resources by Una R. Smith mentioned below . Directories to specific disciplines may also be found on the Web itself, e.g. Entomology Index of Internet Resources. New 'bulletin boards' and 'Web pages' are appearing daily. Some may never be updated; others can be a useful source. The following shortlist of key sites provides useful starting points:

Australian Environmental Resources Information – http://www.erin.gov.au/

Biodiversity and Biological Collections http://muse.bio.cornell.edu/

Biologist's Guide to Internet Resources by Una R. Smith 1993 gopher://sunsite.unc.edu/1m/../.pub/academic/biology/ ecology+evolution/bioguide/bioguide.item

Ecology and Organismal Biology Resources, Harvard University http://www.digitas.org:80/harvard/

Entomology Index of Internet Resources http://www.public.iastate.edu/~entomology/ResourceList.html

Global Change Master Directory, NASA http://gcmd.gsfc.nasa.gov/

Institute of Terrestrial Ecology (ITE) http://www.nmw.ac.uk/ite/

Missouri Botanical Garden http://straylight.tamu.edu/MoBot/

National Biological Service (NBS) http://www.itsnbs.gov/nbs/

National Wetlands Inventory, US Fish & Wildlife Service Ecology Section http://www.nwi.fws.gov/Ecology.html

Natural History Museum (London) http://www.nhm.ac.uk/

Organismic and Evolutionary Biology, Harvard University (links to herbarium and arboretum gopher servers) http://oeb.harvard.edu/servers

Royal Botanic Gardens, Kew http://www.rbgkew.org.uk/

SMASCH Project http://www.calacademy.org/smasch.html

Smithsonian Natural History Home Page http://nmnhwww.si.edu/nmnhweb.html

U.S. Geological Survey – Data Available Online: http://www.usgs.gov/data/Index.html

World Conservation Monitoring Centre http://www.wcmc.org.uk/

Environmental Databases (Barlow *et al.*, 1995) in the UK which lists hundreds of information collections including those held in the voluntary sector and statutory agencies.

Archives contain much useful unpublished information which is relevant to biodiversity assessment. They are held by national institutions such as the Public Records Office and research organisations. Locations are recorded in the National Register of Archives (London and Edinburgh). See Appendix Two for addresses of these offices. There are also several published guides, e.g. *Natural History Manuscript Resources in the British Isles* (Bridson *et al.*, 1980). Guidance on the care and selection of archives is given in the Linnean Society's booklet, *Preserving the Archives of Nature* (1994).

7.3 Cartographic materials: maps, atlases and gazetteers

Maps are often overlooked as key sources of information, but are especially useful to biodiversity assessment studies, where much of the data collected needs to be spatially or geo-referenced.

Cartography has been defined as 'the science and technology of analysing, interpreting and communicating spatial relationships normally by means of maps'. As such maps and their associated products, air photographs and satellite imagery, are important sources of data

Box 7.4 UK libraries with significant holdings of relevance to biodiversity studies

LIBRARY	NO BOOKS/PERIODICALS	SUBJECT	ACCESSIBILITY
GENERAL REFERENCE			
British Library Science Reference and Information Service (SRIS) Aldwych Reading Room (for life sciences material) 9 Kean Street London WC2B 4AT Tel 0171 323 7288 Fax 0171 323 7217	243,000 books 67,000 periodicals 33,000,000 patents	Provides a comprehensive resource for world literature on the whole of science and technology. Increasingly its holdings of journals provide support for users from the libraries of specialist organisations no longer able to afford to keep up subscriptions for their own collections	Access is open to anyone on signing a visitors' book
British Library Document Supply Centre Boston Spa, Wetherby West Yorkshire LS23 7BQ Tel 01937 546060 Fax 01937 546333	3,000,000 books 242,000 periodicals	Offers the world's foremost national interlending and document supply service, covering all disciplines; 650 full-time staff, process around 3.7 million requests annually	Primarily though inter-library document supply. Reading room on site open to the public weekdays
National Library of Wales Aberystwyth Dyfed SY23 3BU Tel 01970 623816 Fax 01970 615709	350,000 books 2,600 periodicals	A legal depository library taking all British published literature, specialises in Welsh matters and other Celtic countries	Open to the public Monday to Saturday
National Sound Archive Wildlife Sounds Section 29 Exhibition Road London SW7 2AS Tel 0171 412 7440 Fax 0171 412 7441	>80,000 recordings of all classes of sound-producing animals from every zoogeographical region	The largest library of its kind in Europe and possibly the most comprehensive in the world. Includes tape recordings donated from private individuals and organisations such as the Royal Society for the Protection of Birds, the New Zealand Wildlife Service and the BBC Natural History radio broadcasts	There is a free Listening Service for visitors by appointment and copies of many recordings can be made available for research and other non-commercial uses
National Museum of Wales Library Cathays Park Cardiff CF1 3NP Tel 01222 397951 ext 202 Fax 01222 373219	115,000 books 1,200 current periodicals	Current collection embraces botany and zoology and there are a number of special collections relevant to biodiversity	Primarily serves the needs of the museum's staff, but is also open to non-members of staff by prior arrangement
Library of the Royal Museum of Scotland Chambers Street Edinburgh EH1 1JF Tel 0131 225 7534 ext 269 Fax 0131 220 4819	comprise c100,000 books 900 current periodicals	Particular strengths on Scottish natural history, with some other useful resources relevant to biological sciences	Open to non-members of museum staff by prior arrangement
Library of the Royal Society 6 Carlton House Terrace London SW1Y 5AG Tel 0171 839 5561 Fax 0171 930 2170	150,000 books 150 current periodicals	Strong on historical material and papers relating especially to the work of Fellows of the Society national academies of science, and science policy	It is available to non-Fellows on application
Scottish Science Library National Library of Scotland 33 Salisbury Place Edinburgh EH9 1SL Tel 0131 226 4531 Fax 0131 662 0644	19,700 books 2,500 current periodicals	A legal depository library taking all British published literature in science; particular collection on Scottish matters	Freely available for reference only but readers ticket required; on-line searches (fee-based), interlibrary loans
NATURAL HISTORY AND NATURE CONSERVATION			
Natural History Museum Cromwell Road London SW7 5BD Tel 0171 938 9191 Fax 0171 938 9290	>800,000 books 23,000 periodicals special collections of manuscripts and drawings 75,000 maps	The General Library is especially strong in holdings of the publications of academies, societies and institutions and the literature of travel and expeditions. The museum archives should also not be overlooked as a source of data on the museum's 67 million specimens. Specialist libraries in the museum are listed below with their subject fields	Available to external visitors by ticket and by appointment
CAB International Library Buckhurst Road Ascot, Berkshire SL5 7TA Tel 01344 872747 Fax 01344 872901	65,000 books 1,700 periodicals	Main CABI library dealing with information requests. Agriculture, horticulture, animal and plant breeding, forestry, veterinary science, parasitology, tourism, soils, land and water management	Not open to the public. Access only by prior arrangement with the librarian. CAB Abstracts available on CD-ROM (see Box 7.5)
English Nature Information and Library Services Northminster House Peterborough PE1 1UA Tel 01733 340345 Fax 01733 68834	40,000 books and pamphlets 1,500 current periodicals	Nature conservation, land use, planning, botany, zoology, ecology, geology and pollution. largely to do with UK and UK dependencies	Open to bona fide enquirers for reference only by appointment
Linnean Society of London Burlington House, Piccadilly London W1V 0LQ Tel 0171 434 4479 Fax 0171 287 9364	100,000 books 300 current periodicals	Founded on the library and collections of the Swedish naturalist Carl Linnaeus (1707–1778). Wealth of donated material from Fellows of the Society and from purchases. An historically important collection	Available to non-members of the Society by appointment

Box 7.4 *Continued*

LIBRARY	NO BOOKS/PERIODICALS	SUBJECT	ACCESSIBILITY
NATURAL HISTORY AND NATURE CONSERVATION *continued*			
Institute of Terrestrial Ecology Merlewood Res, Sta. Library Grange-over-Sands Cumbria LA11 6JU Tel 01539 532264 Fax 01539 534705	approx 10,000 books and pamphlets 800 current periodicals	Comprises a core resource for research on terrestrial ecological ecosystems, environmental protection and management	It is available to bona fide enquirers for reference
Natural Resources Institute Central Avenue Chatham Maritime Chatham, Kent ME4 4TB Tel 01634 883410/11 Fax 01634 880066/77	> 300,000 items of printed books, reports, pamphlets 2,500 current periodicals	On all aspects of tropical agriculture, including pest species	Enquiries are welcomed by appointment
BOTANY			
Botany Library Natural History Museum Cromwell Road London SW7 5BD Tel 0171 938 9421 Fax 0171 938 9290	>100,000 books 1,200 current periodicals	Plant and fungi taxonomy, floras of the world, the history of botany and botanical exploration, botanic gardens, medicinal plants and a major reference collection of botanical illustrations	Ticket and by appointment for reference only
CAB International, International Mycological Institute, Bakenham Lane, Egham Surrey TW20 9TY Tel 01784 470111 Fax 01784 470909	7,000 books 600 periodicals	Specialist library of free-living and pathogenic fungi, lichenised fungi and microorganisms	Open to bona fide research workers by arrangement
Cambridge University Botanic Garden (Cory) Library Cory Lodge Bateman Street Cambridge CB2 1JF Tel 01223 336265 Fax 01223 336278	5,600 books 62 periodicals	Plant identification, taxonomy of vascular plants, horticulture and the history of botany and horticulture	Advance appointment required
Oxford University Plant Sciences and Oxford Forestry Institute Library South Parks Road Oxford OX1 3RB Tel 01865 275082 Fax 01865 275074	10,000 books 1,800 current periodicals	Forestry, agriculture, plant taxonomy, biochemistry, genetics, ecology, physiology and molecular biology	Open to the public for reference only
Royal Botanic Garden Edinburgh Library Inverleigh Row Edinburgh EH3 5LR Tel 0131 5527171 Fax 0131 5520382	100,000 books 1,700 current periodicals	Plant and fungi taxonomy, amenity horticulture, history and current practice of gardening, conservation, botanical illustration, botanical travel and exploration and floras of the world	Open to all for reference only
Royal Botanic Gardens, Kew, Library and Archives Kew Richmond Surrey TW9 3AE Tel 0181 332 5414/5 Fax 0181 332 5278	>120,000 books 4,000 periodicals (1,650 of them current) 140,000 pamphlets 11,000 maps 10,000 microforms c. 185,000 MSS & drawings	Plant and fungi taxonomy, distribution and conservation, horticulture, economic botany, plant anatomy, propagation and seed storage, genetics and molecular biology, biochemistry and biological interactions	Open to bona fide research workers only, who should write to Head of Library & Archives for permission. It is for reference only
ENTOMOLOGY			
Entomology Library Natural History Museum Cromwell Road London SW7 5BD Tel 0171 938 9491 Fax 0171 938 9290	>90,000 books 1,000 current periodicals	Insect and arachnid taxonomy, insect faunas, medical and veterinary entomology, insect pests and the history of entomology including an extensive manuscript collection	Open to external visitors by ticket and by appointment for reference only
The Hope Library The University Museum Parks Road Oxford OX1 3PW Tel 01865 272982/50 Fax 01865 272970	21,000 books 150 current periodicals	Taxonomic zoology particularly entomology (as well as geology, palaeontology and mineralogy)	Require a letter of recommendation from a recognised academic institution or supervisor
Royal Entomological Society Library 41 Queen's Gate London SW7 5HR Tel 0171 584 8361 Fax 0171 581 8505	>10,000 books 250 current periodicals	All aspects of entomology with emphasis on taxonomic entomology especially in the Western Palaearctic Region	Open to non-fellows by appointment for reference only

Box 7.4 *Continued*			
LIBRARY	NO BOOKS/PERIODICALS	SUBJECT	ACCESSIBILITY
ZOOLOGY			
The Balfour and Newton Libraries Department of Zoology University of Cambridge Downing Street, Cambridge CB2 3EJ Tel 01223 336648 Fax 01223 336676	20,000 books 360 current periodicals	Zoology, entomology, ornithology, zoogeography, ecology, behaviour, physiology, cell biology, taxonomy and evolution	Open by written application to the librarian
Zoology Library Natural History Museum Cromwell Road London SW7 5BD Tel 0171 938 9191 Fax 0171 938 9290	>160,000 books 1,250 current periodicals	All branches of the animal kingdom except insects and arachnids (see Entomology Library) and birds (see Ornithology Library, Tring). The core of the collection is taxonomic zoology as well as faunas, anatomy, behaviour, ecology, parasitology, physiology, zoogeography and the history of zoology	Open to external visitors by ticket and by appointment for reference only
Zoological Society of London Library Regent's Park, London NW1 4RY Tel 0171 722 3333 Fax 0171 483 4436	approx 200,000 books 1,300 current periodicals	Zoology, zoological gardens, conservation, comparative medicine, physiology, nutrition, reproduction and education	Open on application to the librarian and payment of a fee for reference only
ORNITHOLOGY			
The Alexander Library Edward Grey Institute of Field Ornithology University of Oxford South Parks Road, Oxford OX1 3PS Tel 01865 271143 Fax 01865 310447	> 18,000 books 550 current periodicals	All aspects of global ornithology especially field ornithology including behaviour, ecology, migration and distribution. The collections include the British Ornithologists Union Library and the British Falconers' Club Library	Open to bona fide ornithologists by appointment for reference only
BirdLife International (formerly the International Council for Bird Preservation) Wellbrook Court Girton Road Cambridge CB3 0NA Tel 01223 277318 Fax 01223 277200	some 4,500 books 500 current periodicals 10,000 reports and pamphlets	The status of bird species world wide, data on all endangered bird species, conservation and international conventions. BirdLife is also a significant publisher in its own right producing regional checklists and guides of threatened birds, technical publications and field study reports. They also have extensive databases including the World Bird Database, European Bird Status Database, and Important Bird Database	Only available through direct enquiries to staff
The Ornithology Library & Rothschild Library The Natural History Museum Akeman Street Tring, Herts HP23 6AP Tel 01442 824181 Fax 01442 890693	> 70,000 books 450 current periodicals	Taxonomic ornithology, nomenclature, distribution, migration, anatomy, physiology, behaviour, zoology, nidology and the history of ornithology	Open to external visitors by ticket and by appointment for reference only
FRESHWATER AND MARINE BIOLOGY			
The Freshwater Biological Association Library The Ferry House Far Sawrey, Ambleside Cumbria LA22 0LP Tel 01539 442468 Fax 01539 446914	7,000 books 100,000 offprints and pamphlets 750 current periodicals	Limnology, freshwater biology (hydrobiology), microbiology, freshwater algae, invertebrates, fish, and the physics and chemistry of lakes and rivers	Open to members and bona fide workers in this and related fields by appointment.
National Oceanographic Library Southampton Oceanography Centre Southampton SO14 3ZH Tel 01703 596116 Fax 01703 596115	14,000 books 1,000 current periodicals	All aspects of marine science including marine biology, chemistry, engineering, physics and geology	Open to bona fide researchers by appointment
Plymouth Marine Laboratory and Marine Biological Association Library Citadel Hill, Plymouth PL1 2PB Tel 01752 222772 Fax 01752 226865	50,000 books 1,250 current periodicals	Marine biology, ecology, pollution, fisheries, oceanography and related subjects, also collects and indexes all relevant publications on marine and estuarine pollution	For visiting research workers but is also open by appointment for marine scientists
POLAR REGIONS			
British Antarctic Survey Library High Cross, Madingley Road Cambridge CB3 0ET Tel 01223 61188 ext 510 Fax 01223 62616	>6,000 books and pamphlets 300 current periodicals	Botany, zoology, ecology, geology, glaciology, marine biology, meteorology, climatology and upper atmosphere physics, all with an emphasis on Antarctic studies	Open to bona fide research workers for reference only by arrangement with the librarian
Scott Polar Research Institute Library Lensfield Road, Cambridge CB2 1ER Tel 01223 336557 Fax 01223 336549	25,000 books 40,000 pamphlets 23,000 maps and charts >1,000 current periodicals	The largest single collection in the world of published and unpublished literature on the Arctic and Antarctic and includes whaling, sealing and exploration	Open to all with a genuine interest on application to the librarian

Box 7.5 CAB Abstracts: a worldwide service to agriculture

The Centre for Agriculture and Biosciences International (CABI) is an intergovernmental organisation dedicated to the dissemination of information in support of agriculture and environmental management, especially relevant to the needs of developing countries. (See also Chapter Four).

CAB International maintains a computerised database of agricultural and biological information which contains all records from the 45 abstract journals published by CABI from 1973 to the present. These journals include the following subjects: animal breeding, animal disease, arid lands, entomology, plant breeding and pathology, rural recreation and tourism, soils, weeds and world agricultural economics. Over 8,500 journals in 37 different languages are scanned for inclusion, as well as books, reports, theses, conference proceedings, patents, annual reports and guides. The database contains over 2.8 million abstracts and is updated monthly. Outputs from these databases are distributed in printed form, and in electronic form on diskette, CD-ROM, and magnetic tape. Online access is provided by commercial database vendors. Copies of source documents are available through CABI's Document Delivery Service.

When the **CAB Abstracts** database was made available on CD-ROM in 1990, it established a new generation of electronic publishing. By the end of 1990 these disks were in use in 40 countries where, in the case of developing countries, researchers are often without access to reliable telecommunications and adequate library facilities. The CD-version has the advantage over the on-line Abstracts in that the CD version searches on any word in the record rather than just the author, title and keyword fields.

CABI also publishes books, primary journals, newsletters and training manuals.

Research and development is undertaken in innovative information systems and training in information management is offered, especially for developing countries. Where necessary, partnerships with development assistance agencies are sought to help developing countries acquire the products and services they need.

In CABI's Library four centuries of international agricultural publishing are spanned by the 200,000 volume library holdings. This forms the backbone of a document delivery service, which is supported by a network of collaborating major libraries throughout the world.

for field research, as tools for analysis, and for the presentation of results. Their usefulness largely depends on choosing the most appropriate products for the tasks to be performed. These tasks are likely to include:

- topographic and thematic information – as a source of both logistic and scientific information;
- sampling design requirements – to provide a spatial framework for sampling;
- base-map functions for plotting data collected;
- ground-truthing to correlate information collected in the field (on the ground) with remotely sensed imagery, e.g. aerial photographs and satellite imagery;
- navigation for basic route-finding in the research area; and
- graphic representation of results.

To help identify the maps needed for a particular research area, potential suppliers will need to have the location clearly and unambiguously identified by being given the latitude and longitude, and/or directions and distances to large and easily identified features or places. Places which do not appear on general maps, may be located by using national or international gazetteers.

Map scale should be equally carefully specified. The terms 'large scale' and 'small scale' are ambiguous, since they are interpreted differently in different countries: a representative fraction (e.g. 1:50,000 or 1:100,000) is by far the best way to describe scale. Detailed mapping at scales of 1:2,500 to 1:10,000, appropriate for field plotting, is rarely available in for remote areas. It is therefore highly likely that for base maps or for

Box 7.6 BIOSIS: a biological on-line bibliographic publisher

BIOSIS is an independent, not-for-profit publisher of printed and computer-readable products in the Life Sciences. Established in 1926, it is based in Philadelphia, USA and governed by a Board of Trustees drawn from both industrial and academic sectors of the community. Probably best known as the publisher of *Biological Abstracts*, since 1980 it has also had a wholly-owned subsidiary in York, England responsible for the compilation of *Zoological Record*.

These two major bibliographic databases – **Biological Abstracts** and **Zoological Record** – now cover some 675,000 references each year, drawn from over 12,000 serials together with books, conference proceedings and other original publications from virtually all the countries in the world. These databases are available in print form, as on-line files through commercial vendors, and on CD-ROM. There is a third smaller database, **BioBusiness**, covering the economic implications and business applications of biological and biomedical research, which is only available on-line. There is also a number of smaller specialist publications drawn from the main bibliographic database.

BIOSIS has two on-line systems of its own, the Life Science Network (LSN) which provides a specialised gateway to the commercial systems, and the TRF which is a microbiology electronic bulletin board and database.

The LSN is designed for individual researchers, not professional searchers, and offers access to 80 mainstream life science databases. It includes the BIOSIS Connection – a collection of smaller databases supported by BIOSIS and affiliated organisations, together with all the BIOSIS databases. The LSN uses specially written software to simplify searching. It has no minimum charges, can scan single or multiple databases, and has extensive help facilities (including an on-line human operator). It is available through BT Tymnet, Sprintnet and the CompuServe network, and will shortly be available through the Internet.

Other BIOSIS activities include close co-operation with the International Commission on Zoological Nomenclature in the development of a *List of Generic Names in Use in Zoology*, and links with similar activities which are taking place in botany.

BIOSIS *Garforth House*, 54 *Micklegate*,
York YOI ILF, UK.
Tel: 01904 642816 *Fax*: 01904 612793

digitising, enlarged copies of smaller scale maps will have to be made either photographically or using the scale-changing equipment available in some university or college geography, geology or surveying departments. In these cases, it is important to remember that the resolution is limited by the original copy.

The use of almost all mapping, in whatever format (paper, electronic, digital, etc.) is restricted by national copyright legislation. This varies from country to country, but map librarians will be able to provide information on the legislation relevant to their holdings. The penalties for copyright infringement are usually heavy fines.

Topographic maps out-number all other types of maps by about four to one, but soil and land capability mapping, geology, vegetation and land use maps will all have special relevance to biodiversity assessment surveys. Unlike topographic mapping, this is rarely considered a priority by national mapping authorities and identifying sources of these maps is likely to be more difficult. UNESCO has supported international efforts in vegetation mapping at a continental scale, including Africa and South America, and through the production of the handbook *International classification and mapping of vegetation* (UNESCO, 1973). The British Land Resources Development Centre (now part of the Natural Resources Institute) developed an integrated land systems mapping methodology. The difficulty of using traditional cartographic methods for this type of integrated mapping, has largely been superceded by the use of computerised data handling techniques, i.e.

Geographic Information Systems (GIS). *The International bibliography of vegetation maps* (Küchler, 1965–1970) is the only international bibliography on the subject and is rapidly becoming dated.

Maps are produced and sold by a bewildering array of organisations from national mapping agencies, government departments and research organisations to commercial publishers. Accordingly, tracking down and acquiring these maps is notoriously difficult. Up-to-date catalogues or reference services should be consulted to find what material is in print at a given scale, and how it can be purchased. Furthermore, such consultation may yield information about survey dates, sources and accuracy. Such indexes are held by map librarians, map agents or retailers, and can be used to specify the exact map series, scale, sheet number and order reference number or code.

In many parts of the world public availability of maps and images is either limited or prohibited. Tourist maps might be available for use in the country but export may be forbidden. In others, the photocopying of maps may be illegal. Do not be tempted to circumvent such regulations: the penalties for breaching them may be severe (including imprisonment). Where necessary, request written authorisation to carry detailed maps and, indeed, for any activity which might be construed as 'surveying'.

National atlases are often overlooked as significant sources of spatial information. In many countries where larger-scale map series are restricted or otherwise unobtainable, the national atlas may contain useful topographic and a range of thematic information at scales as large as 1:250,000. As many as 65 national atlases are probably in print at the present time, and several are in production. An increasing number are in digital form.

7.3.1 Sources of maps

World Mapping Today (Parry & Perkins, 1987) gives listings and addresses of current topographic and resource mapping arranged alphabetically on a country by country basis, with additional sections on the oceans and polar areas. Graphical indexes of major map series enable sheet numbers within series to be identified. These listings are backed by several chapters giving background information on national mapping activities and trends in world mapping. A new edition of this most important reference is due out in 1996/97.

GEOBASE, an on-line database produced by the publishers of *Geographical Abstracts*, the leading abstracting journal for geography, is accessible via DIALOG and includes cartography amongst its 25 subject areas. Their other on-line service GEOREF is largely concerned with geology and geological mapping.

Parry and Perkins (1987) describe a major decline in world map production and availability. This is confirmed by the limited number of international map retailers who hold stocks of detailed mapping, or even issue catalogues. The exception is Geo Center in Germany which produces the two-volume *GeoKatalog*. Volume 1 focuses on tourist maps, and Volume 2 on more detailed mapping of 'scientific' interest. However, even Geo Center has experienced great difficulties in supplying maps listed. The British map retailers Stanfords specialise in worldwide touring and survey maps and travel guides, which are useful for initial planning, but they have found it almost impossible to provide worldwide coverage at 1:250,000, due to the many constraints imposed on the commercial sale of maps outside the country of origin. However, they do supply worldwide coverage of Operational Navigational Charts, ONC (1:500,000) and Tactical Pilotage Charts, TPC (1:1 million). In many cases the only way to obtain the necessary mapping is to make a personal visit to the national survey authority of the country concerned (listed in Parry and Perkins (1987) and Böhme (1989-93)) or the national hydrographic office (listed in *Catalogue of Admiralty Charts and other hydrographic publications*, Anon, 1995) of the country concerned.

This limited availability of even published maps emphasises the importance of the major international map libraries (for addresses of these and map suppliers, see **Box 7.7**). Of the major British map collections only those of the British Library, the National Library of Scotland and the Royal Geographical Society are considered comprehensive in terms of world-wide coverage of printed maps and are truly open to the public.

7.3.1.1 The Royal Geographical Society Map Room

The Royal Geographical Society maintains a collection of 850,000 sheets of maps and charts, 2,500 atlases and 700 gazetteers, forming part of what is probably the world's largest private map collection. Since its foundation in 1839, the RGS has concentrated on acquiring scientifically surveyed maps produced by official agencies in the UK and abroad, through donations and bequests. Donations are regularly received from the Ministry of Defence, the Directorate of Military Survey and the Hydrographic Office in the UK and from a number of overseas map and chart agencies. The Map Room also has carto-bibliographies, catalogues of both modern and antiquarian maps, as well as of map dealers. Gazetteers are held in various languages throughout the world, dating from the eighteenth century to the present day . These materials are publicly available for consultation and out-of-copyright material may be photocopied. The public service from given by the Map Room is provided with the assistance of an annual grant from the UK Government. Specialist staff are available to assist enquirers and personal visits are encouraged.

7.3.1.2 Ordnance Survey International library resources

OS International holds extensive collections of air photography, maps, survey data and books dealing with the developing world, especially the tropical Commonwealth countries. These include library holdings of the former Directorate of Overseas Surveys accrued between 1946-84. In Africa, the collections embrace all the Commonwealth countries, plus Ethiopia, Liberia and Sudan; in the Americas, Belize and Guyana; in Asia, Hong Kong, Nepal and Yemen, as well as the islands of the Caribbean and Indian Ocean. Some other areas are also covered, including the former French West Africa and East European countries. OS International holds about 100,000 maps, from single-sheet world maps, through basic 1:50,000 scale topographic mapping, to large scale rural and urban mapping. Although most of the maps are topographic, in some areas there are significant holdings of geological, land use, cadastral and international boundary maps. For a more detailed account of the collection see *The library resources of* OS *International* (Porter, 1993).

7.3.1.3 The British Library Map Library

The British Library holds the national map library of Great Britain, receiving the complete range of British map production through copyright deposit, and acquiring foreign topographical mapping, together with thematic and general atlases and maps from a range of sources. The Map Library is one of the British Library's Special Collections (as is the Oriental and India Office Library Collection, see below). The BLML holds some two million map sheets, the majority of which are concerned with the British Isles. There is some overlap between the collections of the BLML and the RGS, but in general the RGS is considered to have more up-to-date and detailed overseas material. An on-line database of all maps acquired by BLML since 1974, BLAISE-line, is accessible via Internet.

7.3.1.4 The Oriental and India Office Collections

The British Library also houses the archives of the former East India Company and its successors, with some 80,000 maps covering South/South-east Asia, and many other parts of the world in which the East India Company traded. It also holds sea charts from Deptford to Vladivostock and land mapping of areas such as the Gulf States, South Africa, Malaysia, China and Japan. Its chief strength lies in its large scale topographic mapping of South Asia including India, Pakistan, Myanamar (Burma), Bangladesh, Afghanistan, Iran and Nepal. New mapping of these countries is being obtained when possible.

7.3.1.5 The National Library of Scotland, Edinburgh

The National Library of Scotland is one of five copyright libraries and the largest library in Scotland. Its forerunner, the Advocates' Library, collected maps and atlases, both British and foreign, from the middle of the 18th century. The NLS aims to be a comprehensive world-wide reference collection, and includes a reference collection of overseas maps, particularly of anywhere with Scottish interest. Regular deposits are received from mapping agencies in Canada, New Zealand and Australia, as are the various map series produced by the US Geological Survey. All visitors are welcome. Photocopy facilities are available.

7.4 Air photographs

Air photographs can provide invaluable support to a scientific programme. However, they should be regarded with caution by inexperienced users because they are almost always unrectified (i.e. they incorporate marked warping related to the imaging system, camera tilt and ground relief). If used alone they are inaccurate representations of position, distance and direction. These errors generally increase in relative importance with increasing ranges of ground relief and photo scale (i.e. they are worst in mountainous areas and at scales of 1:10,000 or 1:20,000, which are often taken from low flying aircraft).

Interpretation is greatly improved by the use of stereoscopic pairs of overlapping photographs, and simple portable lens stereoscopes, suitable for field use, can be purchased relatively cheaply. However, correct training in their use and a degree of experience is needed before they can be used reliably. Air photographs may give a very season-specific picture of the landscape and vegetation. It is therefore important to check the date carefully in areas of marked seasonality.

Outside the major Western countries air photo coverage is usually sparse and very restricted in its public availability. A major collection for the tropical zone of the Commonwealth (parts of Africa, Asia, Caribbean and Latin America) is held by OS International at the Ordnance Survey (Southampton) (see Section 7.3.1.2), while the Institute Geographique in Paris has a good collection of material from francophone countries,. However, though on-site reference is possible, there may be prohibitions on the supply of copies of these photographs.

Box 7.7 Addresses of UK map libraries and suppliers

To consult maps

BRITISH LIBRARY MAP LIBRARY
The British Library, Great Russell Street, London WC1B 3DG (0171-636 1544).

THE ORIENTAL AND INDIA OFFICE COLLECTIONS, BRITISH LIBRARY
197 Blackfriars Road, London SE1 8NG (tel: 0171 412 7873; fax: 0171-412-7870).

NATIONAL LIBRARY OF SCOTLAND MAP ROOM
137 Causewayside, Edinburgh EH9 1PH (0131-226 4531).

ORDNANCE SURVEY INTERNATIONAL LIBRARY RESOURCES
Romsey Road, Maybush, Southampton SO9 4DH (tel: 01703-792659, fax: 01703-792230).

ROYAL GEOGRAPHICAL SOCIETY MAP ROOM
1 Kensington Gore, London SW7 2AR (tel: 0171-589 5466, fax: 0171-584 4447).

To purchase maps

GEO CENTER
Internationales Landkartenhaus, Postfach 800 830, D-7000 Stuttgart 80, Germany (49 711-735031).

LONDON MAP CENTRE
Cook, Hammond & Kell, 22-24 Caxton Street, London SW1H OQU (0171-222 2466).

STANFORDS INTERNATIONAL MAP CENTRE
12-14 Long Acre, London WC2E 9LP (tel: 0171-836 1321; fax: 0171-836 0189).

THE MAP SHOP
15 High Street, Upton-on-Severn, Worcs. WR8 OHJ (0168-46-3146).

For many parts of the world restrictions on public access to air photographs are more stringent than those pertaining to maps. Application can be made to the appropriate National Survey and Mapping Department with precise details of the area of interest and the exact purposes for which the photography would be used. Border areas may be particularly sensitive. As in the case of map enquiries to official bodies, researchers should allow for a considerable time to elapse before an answer is received – six months is suggested.

7.5 Satellite imagery

Space provides a unique vantage point from which to view the planet. Satellite remote sensing has revolutionised surveys of the Earth: huge amounts of data, conveying a great quantity of environmental information, and covering vast tracts, can be processed to produce valuable maps in minutes rather than years. Remotely sensed data is increasingly being used as a substitute for conventional mapping. As with conventional mapping, identifying the most appropriate satellite data will depend on its final use.

Resolution, as quoted in **Box 7.8**, gives a crude ranking of the detail shown on a particular image but has severe limitations in this respect. It would be quite normal for an object to need to be two or three pixels (picture elements) in size before it could be relied upon to appear on an image, whilst surfaces that are particularly well differentiated from their surroundings may appear strongly even though they are less than one pixel in size. Linear features are especially easy to distinguish (roads, railways or rivers), but only as long as they contrast with their surroundings. A dirt road across the bush or a dry river bed in a semi-arid area may be difficult to detect.

Like air photographs, satellite images incorporate geometrical distortions which prevent them acting as precise maps of the ground surface. This warping results partly from the rotation of the Earth during the production of the image and partly from the nature of the imaging system. Recent advances in operational processing allow digital data to be supplied in a standard geocoded format for a chosen map projection. Care must be taken to establish whether the data or image has been corrected before they are purchased. Such correction seems very useful, but be aware that it inevitably means that the raw data have been recalculated (often changing the pixel size) so that the way in which the ground surface is depicted will be subtly altered.

However, for other than superficial purposes a considerable specialist knowledge (from courses or texts) is required to make the most of satellite data. Ideally a project team should include a remote sensing specialist who has access to image processing facilities. With such support, satellite data can make a significant contribution to a survey. In most cases such work will grow out of existing links between the team and remote sensing scientists and facilities. Many universities run courses on remote sensing, and there is an active Remote Sensing Society (details from the RSS Secretary, Geography Department, University of Nottingham; tel. 01602-587611).

Despite their limitations, even single images can provide an invaluable synoptic view of large areas for navigational or interpretative purposes. They may be more recent, more informative and certainly more easily acquired than the equivalent maps. A number of agencies are now regularly producing maps from satellite data. Image search and browse

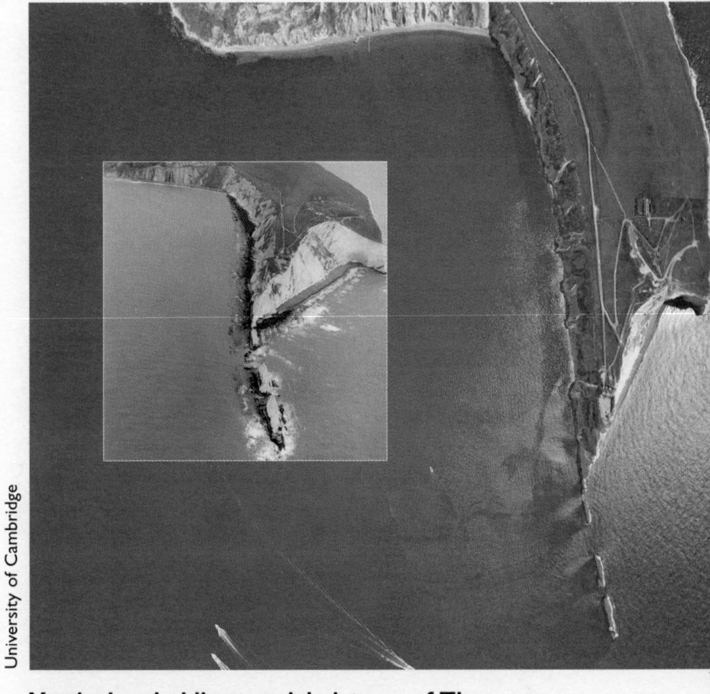
University of Cambridge

Vertical and oblique aerial pictures of The Needles, Isle of Wight.

Box 7.8 Approximate scales for various types of satellite imagery

Typical scales for:	Possible satellite data from:
Continental coverage, 1:1 M to 1:250,000	Weather satellites: AVHRR, CZCSRegional
coverage, 1:250,000 to 1:100,000	Landsat: MSS, TM; SPOT: HRV
Local coverage, 1:100,000 to 1:25,000 ?	Landsat TM; SPOT: HRV, Pan;
Detailed coverage at more than 1:25,000 ?	SPOT Pan; Landsat 6 Pan, Russian imagery

facilities in the UK are available at the British Library Map Library (see **Box 7.7**) and at the National Remote Sensing Centre (see **Box 7.9**).

Remote sensing products are expensive. However, once access is gained to the data and suitable processing facilities, colour ink-jet or colour dye sublimation printers/plotters can be used to produce low-cost outputs for field annotation.

Enquiries about satellite image cover for a given region in the world can be made to the National Remote Sensing Centre Ltd, Delta House, Southwood Crescent, Southwood, Farnborough, Hampshire GU14 ONL (tel: 01252-541464, fax: 01252-375016) which is the principal UK supplier of remotely sensed imagery. NRSC is a licenced distributor for Landsat and SPOT data, and the UK point of contact for the distribution of images available through the European Space Agency (see **Box 7.9**).

7.6 Geographical Information Systems (GIS) and digital geographical data

The use of Geographical Information Systems (GIS), and computer systems for handling all types of spatial information, especially satellite imagery, has revolutionised conventional mapping techniques.

Most map users employ the paper map as a backdrop for plotting other kinds of information. As maps have come to be made by computers, a shift in the perception of the map has become apparent. Maps, when stored in the computer, have taken on a dynamic form. Their content, symbolisation and the area covered can now be changed rapidly and easily. It can be argued that maps are now better thought of as databases, or collections of information linked by their spatial relationships. In this concept a map can be seen simply as one expression of a spatial database. Maps showing different datasets for the same area can also be overlaid on top of each other to generate many new integrated maps. Hence, GIS offers much greater potential for spatial analysis, and much more efficient storage for spatial information. In the context of biodiversity assessments, GIS ought to be seen as the new standard for the archiving and checking of spatial information both before, during and after fieldwork. Raper and Green (1993) have produced an interactive computer learning package GIS *Tutor* published by Longman, which gives an excellent introduction to GIS, whilst the book by Burroughs (1986) is a useful introductory text.

More and more geographical data is becoming available in computer-readable form. Digital data is easy and cheap to store, access and disseminate, and has the primary advantage of being readily integrated with other data in a biodiversity study using GIS. Indeed some data sets are being prepared specifically for easy import into particular GIS software packages.

However, care should be taken to obtain information about the original data source and its accuracy (metadata), for example, as being developed by the Federal Geographic Data Committee (FGDC) (Web page (1995): http://fgdc.er.usgs.gov/fgdc.html) (see also Section 7.7).

Digital geographical data has two basic elements. First, an indication of location, or *geo-reference*, typically based on a two-dimensional co-ordinate system. Second, that which is at the particular location, i.e. *attribute* data. In the context of biodiversity assessment, the attributes may relate directly to biodiversity. Species ranges, for example, may be represented by a presence/absence on a grid, by a line marking the limits of the range, or by records of individual sightings. Alternatively, the digital data may relate to other topics in understanding and managing biodiversity. For example:

- elevation data (is a species restricted in altitude?);
- protected areas (is a species adequately protected?);
- political units (which authorities to liaise with?);
- climate (what effect would climate change have?);
- vegetation (how is a species' range correlated with vegetation and can its range be modelled?); and/or
- human features (what effect does/would a road/oil refinery have here?).

A comprehensive directory of digital geographical data sources in the UK (mostly commercial) is given in *The AGI Source book for geographic information systems* 1995 (Green & Rix, 1994). International coverage is in the 1994 *International GIS Sourcebook* (Rodcay, 1993). Some relevant sources are given in **Box 7.10**. Many digital maps have been put into the public domain through UNEP GRID. Other UN agencies are following suit, publishing their own maps in digital form, for example, the FAO soil maps of the world.

GIS is a powerful tool which can easily combine individually known locations of a species (specimen data) with its distribution range (taxon data). Further by combining the specimen data with other attributes such as soil, vegetation, and meteorological data for the same locations, analysis can identify common features correlating with the species distribution. By searching for other areas which also share the common features, GIS makes the identification of further areas of likely occurrence for the species comparatively easy and relatively accurate (see **Box 7.11**). GIS systems also offer considerable advantages in integrating data

Box 7.9 The UK National Remote Sensing Centre

The National Remote Sensing Centre Limited (NRSC) is one of Europe's largest suppliers of products and services based on information derived from remote sensing satellites and aircraft. Formed in 1989, the company employs some 125 people based at Farnborough in Hampshire, with another office in Brussels.

In April 1992, NRSC received a contract from the British National Space Centre with the remit to establish a commercially viable remote sensing industry within the UK. The Company has corporate shareholders drawn from the aerospace and remote sensing industries.

NRSC produces information products such as satellite images, maps, statistics and digital elevation models. Typical customers include gas and oil companies which use imagery to highlight areas for exploration on the ground, cartographers who use imagery to up-date and produce maps, governments and companies which use imagery and derived information for a wide range of applications including land use mapping and environmental audits. NRSC is also involved in the design and specification of information networks and operation of satellite data centres.

The company's large team of specialist applications consultants are expert in extracting and applying information covering a variety of environmental disciplines. Equipped with some of the most advanced image processing equipment available including Geographic Information Systems (GIS), they handle work ranging from advice on interpreting a single image to undertaking major projects. Examples of these are: their recent environmental audit of the coastal region of the United Arab Emirates, monitoring environmental pollution associated with oil extraction in Siberia; and mapping land-use in water catchment areas.

National Remote Sensing Centre Limited
Delta House, Southwood Crescent,
Farnborough, Hampshire GU14 ONL
tel: 01252 362033
fax: 01252 375016

Box 7.10 Some sources of digital geographical data

Bartholomew Data Sales (12 Duncan Street, Edinburgh EH9 1TA, tel. 0131 667 9431, fax 0131 662 4282). Products include Bartholomew UK 1:250,000 database with contours, drainage, national parks, roads, urban areas etc.

Global Resources Information Database (GRID). (PO Box 30552, Nairobi, Kenya, tel +254 2 230 800, fax +254 2 226 491). Division of United Nations Environment Programme (UNEP) which maintains world-wide geo-referenced environmental data sets with global, continental and national coverage. Global coverage includes soils, climate, land use and vegetation. These are distributed at nominal cost for research purposes.

National Geophysical Data Center (325 Broadway, Boulder, CO 80303-3328, tel. +1 303 497 6338, fax +1 303 497 6513). In the US (in contrast to UK) most government scientific data is considered 'public domain' and is available at nominal cost. NGDC has a wide range of such data products including CD-ROMs of global relief, ecosystems and NDVI.

NPA Group Ltd., Edenbridge, Kent TN8 6HS, tel. 01732 865 023, fax 01732 866 521). Consultants and providers of a wide range of digital data products, including DCW (the digital chart of the world at scale 1: 100,000), bathymetric data, topographic data etc.

Ordnance Survey, Digital Sales (Romsey Road, Maybush, Southampton SO9 4DH, tel. 01703 792 000, fax 01703 792 404). Comprehensive UK thematic coverage at 1:50,000 and smaller scales. Limited availability of 1:1,250-1:10,000 data.

gathered in the inventory phase of any national action plan.

Journals useful for following developments in data processing techniques and applications, and advertisements for digital data and software include:

Cartographic Journal; GIS *World*; GIS *Europe*; *Mapping Awareness & GIS in Europe*; *Earth Observation Magazine*; *Journal of Photogrammetry and Remote Sensing*; *International Journal of Geographical Information Systems*; and *International Journal of Remote Sensing*.

Professional organisations concerned with cartography, remote sensing and geographical information systems are given in **Box 7.12**.

7.7 Information sharing: the challenges

The challenge for those charged with implementing the Biodiversity Convention is to ensure that the many and disparate sources of information are made available to those who need them. Article 18(3) of the Convention on Biological Diversity clearly states the need to establish 'a clearing house mechanism' for the exchange of information.

7.7.1 The Biodiversity Clearing House Mechanism

In the discussion document The Biodiversity Information Clearing House – Concepts and Challenges (WCMC, 1994), it is suggested that access and effective use of this information is hampered by:

- lack of shared knowledge of who has what information, technology or expertise, because it is held in a broad range of separate sectoral and regional institutions;
- lack of consistency in observational methodologies, classification standards, quality assurance methods, analytical models, so that inter-sectorial and cross-sectoral integration and comparison are very difficult; and
- lack of equity of access due to varying levels of information technology infrastructure, and 'information buying power'.

There is general agreement that, to be effective, the Biodiversity Clearing House Mechanism, (CHM) as far as possible, should build on existing institutions, start modestly and

Box 7.11 Using GIS and remote sensing techniques to assess forest canopy and butterfly diversity

Advances in remote sensing, geographic information systems (GIS) and ecological monitoring are now providing the opportunity to relate elements of biodiversity to tropical forest vegetation structure and extrapolate this information over wide areas.

An ongoing project in Guyana and Trinidad is on ground-truthing satellite data from tropical lowland forests with different levels of disturbance and relating tree canopy structure to the species richness and abundance of selected butterfly species.

Two major objectives are to estimate the total suitable habitat for selected butterfly species in the north-east Neotropics and to model changes in butterfly species richness and abundance with given scenarios of forest management. The technology and analysis in this project can be applied to other tropical forest regions in the New and Old World.

Radar data from the European Space Agency ERS-1 satellite are image-processed on IDRISI GIS, a low-cost pc-based GIS suitable for developing country institutes. Radar data have the advantage of being unaffected by cloud cover and are not dependent on sunlight. Each image covers 10 000 km^2 at a resolution of up to 12m x 12m. These data are ground-truthed with tree data which are in turn related to butterfly abundance. Tree and butterfly data are collected in grids which are located using global positioning systems in lowland forests of south-east Trinidad and central Guyana. Typically a grid is a minimum of 300m x 300m and divided into 100 30 x 30m plots. Grids are chosen to give representative coverage of all types of terrain and vegetation cover. Each plot is systematically surveyed, measuring and tagging all tree species with a dbh>10cm and the canopy is photographed. Only the dominant species are identified, as the emphasis is on vegetation structure, estimated as tree basal area per unit area of ground and canopy openness (taken from the photographs).

In Trinidad, we can relate disturbance to known levels of timber extraction in the periodic blocks of the Victoria-Mayaro reserve. It should be possible to apply these techniques to other timber management systems in the world.

Butterflies are sampled in a selection of the plots using fruit trapping and walk-and-count techniques. The increase in species richness with sampling effort is recorded so that predictions of the maximum number of species in a sampling area can be made. Butterflies were chosen because of the relative ease of identification, the relative speed of sampling, the fact that faunal lists are available for many areas in the world and because many of them are known or believed to exhibit predictable responses to even modest levels of disturbance. This has been confirmed by our preliminary work in Trinidad. Modelling work is being undertaken to predict the effects of changes in area and intensity of disturbance on butterfly diversity.

M.P. Gillman/Open University

***Caligo teucer*, a frequent canopy butterfly in the forests of the low altitude Trinity Hills, Trinidad.**

Practical and policy relevance

Sustainable utilisation of tropical forests requires conservation of both resource and associated biodiversity. In Guyana and Trinidad timber is an important natural resource which supplies the domestic market, with limited export. Both countries are also rich in biodiversity, e.g., there are more than 600 butterfly species in Trinidad and more than 1100 in Guyana. Many of

Box 7.11 Continued

these are associated with lowland forest. The benefit of the research is the linkage of forest disturbance due to selective felling or other activities with a quantitative assessment of an important component of biodiversity (butterflies). This can then be extrapolated over different spatial scales to an accuracy dependent on the correlation of radar data to canopy disturbance and canopy structure to butterflies. Combined with the modelling of butterfly dynamics, decisions on the area and intensity of timber extraction or other forest use can be made in the light of implications for biodiversity.

(*Source*: M. *Gillman*, *The Open University*; R. *Teeuw*, *Hertfordshire University*.)

Box 7.12 Professional organisations concerned with cartography, remote sensing and geographical information systems (GIS).

ASSOCIATION FOR GEOGRAPHICAL INFORMATION, c/o Royal Institute of Chartered Surveyors, 12 Great George Street, London SW1P 3AD (0171-222 7000 x226).

ASSOCIATION OF CONSULTING ENGINEERS, Alliance House, 12 Caxton Street, London SW1H 0QL (0171-222-6557).

BRITISH AIR SURVEY ASSOCIATION, c/o J.A. Storey and Partners, Chartered Land Surveyors, 92-94 Church Road, Mitcham, Surrey CR4 3TD (0181-640-1971).

BRITISH CARTOGRAPHIC SOCIETY, Sales and Distribution Manager: Dr T Adams, LaserScan Laboratories Ltd., Cambridge Science Park, Milton Road, Cambridge CB4 4FY (01223 420414).

MILITARY SURVEY DEFENCE AGENCY, Elmwood Avenue, Feltham, Middlesex TW13 7AH (0181-890 3622).

INSTITUTION OF CIVIL ENGINEERS, 1-7 Great George Street, London SW1P 3AA (0171-222 7722).

INTERNATIONAL CARTOGRAPHIC ASSOCIATION (ICA)
Secretary General, Jean Philippe Grelot, Institute Geographique National, 136 bis, rue de Grenelle, F-75700 Paris, France.

NATIONAL ASSOCIATION. OF AERIAL PHOTOGRAPHIC LIBRARIES, The Mond Building, Free School Lane, Cambridge CB2 3RF.

NATIONAL REMOTE SENSING CENTRE, Delta House, Southwood Crescent, Southwood, Farnborough, Hants. GU14 ONL (01252-541464).

NERC THEMATIC INFORMATION SYSTEMS (NUTIS), Department of Geography, University of Reading, Whiteknights, PO Box 227, Reading RG6 2AD (01734 318741).

ORDNANCE SURVEY, Romsey Road, Maybush, Southampton SO9 4DH (01703-775555).

REMOTE SENSING SOCIETY c/o Department of Geography, University of Nottingham, University Park, Nottingham NG7 2RD (01602-587611).

ROYAL INSTITUTE OF CHARTERED SURVEYORS, 12 Great George Street, Parliament Square, London SW1 3AD (0171-222 7000).

SCHOOL OF MILITARY SURVEY, Hermitage, Thatcham, Nr. Newbury, Berkshire RG18 9TP (01635-204215) (Main library, direct dial).

Box 7.13 Microbial Strain Data Network (MSDN)

The MSDN is an international information and communications network for microbiologists and biotechnologists. It is sponsored by four sectors of the International Council of Scientific Unions. These are the International Union of Microbiological Societies (IUMS), the World Federation for Culture Collections (WFCC), the Committee on Data for Science and Technology (CODATA) and the Committee for Biotechnology (COBIOTECH). It is a not-for-profit Company Limited by Guarantee.

The MSDN has a staff of three, but many of the functions of the network are distributed between different parts of the world. The result is an international effort, sharing the workload and the costs. Thus, many of the databases are hosted on the Tropical Database computer in Brazil, whilst specialised software for microbial database management is maintained by colleagues in the USA. Regional networks have been established in Russia, India, Australia, Latin America. Specialist scientific support (e.g. for viruses, genetic strains) is provided by collaborating laboratories worldwide.

The MSDN is supported by grants from a number of agencies and also recovers about 30% of its core costs from a surcharge placed on usage. The data is in the main free. A link to the academic network, INTERNET, has recently been established, and some 400 mail boxes have been issued to scientists from 32 countries.

The databases hold information on microorganisms and cultured cells (bacteria, filamentous fungi, yeasts, algae, protozoa, viruses, plasmids, genetically marked strains, animal cells, hybridomas) as well as general biotechnology and culture collection information. The databases are directories, strain databases, catalogues, bibliographic and general biotechnology and culture collection databases. Some are maintained on the Brazilian computer and use a common search programme 'INFO'; others are maintained on remote computers and are linked to the MSDN via electronic gateways. Links have been established to the DataStar bibliographic databases, the INTERNET system, and the IRRO network (on the release of organisms into the environment).

Databases accessible through the microbial strain data network

'+' denotes databases using INFO software

MICROBIAL STRAIN DATA NETWORK (MSDN) DIRECTORY+
A directory for locating laboratories and culture collections having information on microbial cultures with specific, scientifically defined properties. Produced and maintained by the MSDN secretariat, Cambridge, UK.

AMERICAN TYPE CULTURE COLLECTION (ATCC) DATABASES+
Databases produced and maintained by the ATCC, Rockville MD, USA and include the ATCC Cell Lines Catalogue, ATCC, Recombinant Materials Database, and ATCC Bacteria, Phages and Media Catalogue.

EUROPEAN COLLECTION OF ANIMAL CELL CULTURES DATABASE (ECACC)+
Produced from Porton Down, Salisbury, UK.

UK NATIONAL COLLECTIONS OF YEASTS & FOOD BACTERIA (NCYC/NCFB)
Produced by NCYC, Norwich, UK and maintained on the Institute Vax computer.

UK CULTURE COLLECTIONS DATABASES (MICIS)
Maintained on the DSM/GBF computer in Braunschweig, Germany. The MICIS database contains information about the test results (strain data) for cultures held in the UK service culture collections. Also available are the UK Culture Collection of Algae and Protozoa (CCAP) database and a database of contact information for European collections.

DEUTSCHE SAMMLUNG VON MIKROORGANISMEN UND ZELLKULTUREN (DSM) DATABASES
Maintained on the DSM/GBF computer in Braunschweig, Germany. Catalogues and strain data for yeasts, bacteria and filamentous fungi, as well as the Approved List of Bacterial Names.

NETHERLANDS CULTURE COLLECTION DATABASES (CBS/NCC)
Maintained on the computer of the Centraalbureau voor Schimmelcultures in the Netherlands. Catalogue and strain data for yeasts, bacteria and filamentous fungi are available.

Box 7.13 *Continued*

FRANCE (MINE) CULTURE COLLECTION DATABASES
Available through SUNIST in France which form a part of the MINE Network. Catalogue and strain data for bacterial and fungi are available, and can be searched either in French or in English.

HYBRIDOMA DATA BANK DIRECTORY (HDB) THROUGH CAN/SND
The CODATA/IUIS HDB is maintained on the Canadian Scientific Numeric Database System (CAN/SND). Information on over 20,000 records describing individual hybridomas and/or their monoclonal antibody products is available.

WORLD DATA CENTER (WDC) DATABASES
Maintained at RIKEN, Japan. A directory to culture collections worldwide and the species maintained, the HDB database (see above), and databases on algal collections worldwide, hybridomas and bibliographic information on plant tissue and cell cultures are available.

TROPICAL DATABASES BRAZIL (BDT)+
Maintained at the Base de Dados Tropical (BDT) system, Campinas, Brazil. Includes the national Brazilian catalogue of strains, information about research activities and contacts for Brazilian collections. Information concerning bacteria, yeasts, fungi, algae, cell lines and viruses is available. Searches can be carried out in either Portuguese or English.

DATASTAR DATABASES
Access is provided to the bibliographic databases available through the commercial database host, Datastar, located in Berne, Switzerland. A wide range of scientific (including biomedical, chemical, and biotechnology) databases are available as well as business and reference databases.

CYCLOPEAN GATEWAY SERVICE (CGS) DATABASES
Easy access is provided to the databases (over 850) distributed by 13 worldwide commercial bibliographic database hosts. Broad subject areas covered include science and technology, medicine and health care, business, patents, law, social sciences, education, arts, people, literature and religion. The CGS system is provided by BT North America.

BIOINDUSTRY ASSOCIATION (BIA) DATABASES+
Produced by the UK Bioindustry Association. These databases cover the field of UK and EC regulatory issues concerning biotechnology and exports assistance for the UK. Information about forthcoming conferences, trade missions, etc. concerning biotechnology is available; the latest BIA online news bulletin; and contact information for National Biotechnology Associations.

BIOTECHNOLOGY COURSES (BEMET) DATABASES+
Produced by BEMET 'Biotechnology in Europe: Manpower, Education and Training'. The database currently describes biotechnology courses from academic institutions in the UK, but is being extended to include Europe-wide courses.

BIOTECH KNOWLEDGE SOURCES (BKS) DATABASE+
A listing of new publications and forthcoming conferences in the field of biotechnology. Produced by BioCommerce Data Ltd and updated monthly. Covers books and other media, including software, videos, databases etc. Conference coverage is worldwide, up to 12 months from the current date.

CZECHOSLOVAK CATALOGUE OF FILAMENTOUS FUNGI (CCF)
Produced by the Culture Collection of Fungi, Charles University, Prague. Contains records of about 1600 strains of Zygomycetes and Ascomycetes in both teleomorphic and anamorphic states.

CZECHOSLOVAK CATALOGUE OF ALGAE AND CYANOBACTERIA (CCALA)
Produced by the Culture Collection of Autotrophic Organisms at the Institute of Botany, Czechoslovak Academy of Sciences, Trebon. Incorporates the Uhlir and Pringsheim collection established at Charles University in 1913. Includes cyanophytes, algae, mosses, liverworts, ferns and duckweeds.

Box 7.13 *Continued*

Under development
Animal Virus Database, Czech Collection of Microorganisms catalogue, International Mycological Institute Catalogue. For IRRO network: BIOTRACK OECD database on releases, BIOCAT database on the interaction between insects and insects, UNIDO Guidelines for the release of organisms.

MSDN Bulletin Boards
MSDN BULLETIN BOARD
– edited by MSDN. This has categories on general information relating to databases, culture collection news, microbial and monoclonal antibody exchange, user notices, jobs and MICROIS software information. Also the MSDN newsletter is available on this Bulletin Board.

EUROPEAN BIOTECHNOLOGY INFORMATION SERVICE (EBIS)
– edited by the Concertation Unit for Biotechnology in Europe (CUBE, DGX I 1, SDM-2/66, 200 Rue de la Loi, 10 14 Brussels, Belgium). This has categories on editorials, community activities, member states, international developments, feature articles, association news, reports and books, press reviews, meeting agendas.

EUROPEAN BIOTECHNOLOGY STANDARDS
– edited by the CEN TECHNICAL COMMITTEE 233 WORKING GROUP 4. It has a closed category for committee members and an open category for discussions on standards in biotechnology and related areas. This Bulletin Board is new and additional categories will be added as the initiative develops.

INFORMATION RESOURCE FOR THE RELEASE OF ORGANISMS into the environment (IRRO) – edited by the IRRO Steering Committee:

(*Source*: *Microbial Strain Data Network* (MSDN))

concentrate in the first instance on reliable information, including metadata (i.e. information about existing data sources) (see also under Section 7.6).

The intention is to develop a decentralised electronic network, linking existing national and regional databases, information services and networks, perhaps with central coordination (for example in a United Nations agency). Focussed on facilitating access, e.g. through the development of indices, directories and other sign posting tools, networks of distributed databases are beginning to appear in a number of disciplines. The Microbial Strain Data Network, servicing microbiologists and biotechnologists in more than 30 countries around the world (see **Box 7.13**) is an example.

7.7.2 The World Conservation Monitoring Centre

WCMC in Cambridge has been managing information in the field of biodiversity conservation for more than 15 years, working in close collaboration with both national and international organisations in the development of information products and services. **Box 7.14** lists the WCMC Biodiversity Databases.

In the process of developing these databases, WCMC has gained considerable experience in the application of modern technology to the provision of biodiversity information services. For example, it has taken a strong lead in the application of Geographic Information Systems (GIS) to the management of information on biodiversity conservation, with the development of its Biodiversity Map Library.

A great deal of the information on species, protected areas and habitats occurs in text form, and considerable expertise has also been accumulated in using text searching and indexing tools, and standard data dictionaries to integrate conventional databases with narrative information. An integrated metadatabase has been found to be a powerful and convenient tool for searching and locating information across the disparate spatial, textual and conventional databases. WCMC has recently opened a World Wide Web server providing hypertext, text indexing and database

Box 7.14 World Conservation Monitoring Centre Biodiversity Map Library and Databases

Tropical forest database

A geographic information system (GIS) of tropical forest maps, at scale of 1:1 million now forms a tropical forest database. Some of the information has already been published as the *Conservation Atlas of Tropical Forests*. Coverage includes rain, monsoon, montane, swamp and mangrove forests. The flexibility of the GIS enables overlay of the forest cover data with other relevant information, such as national parks coverage or population density, as a means of analysis and assessment.

Protected areas database

WCMC has recently completed a digital database containing maps and statistical data for 37,000 of the world's protected areas. This complements the database which has been built up over a number of years in collaboration with IUCN. Publications based on this data include the United Nations List of National Parks and Protected Areas, and the four-volume *Directory of the National Parks and Protected Areas of the World*. In addition, the Centre has used the GIS to map all those Forest Reserves that are managed for environmental or conservation purposes.

World Conservation Monitoring Centre

The WCMC Biodiversity Map Library and Database on screen.

Biodiversity Map Library

The digital coverages of protected areas and forests are augmented by a wide range of other thematic GIS data including a number of biogeographical classifications, global potential vegetation maps, maps of globally important wetlands and certain other ecosystems (particularly coastal habitats such as coral reefs), maps of critical conservation sites noted for their biodiversity but lacking legal protection, and many others. A catalogue of these data is available.

Plants database

Taxonomic, distribution, and conservation information is held on more than 81,000 kinds of plants. Data is recorded on single-country endemic, tropical timber species, wild relatives of crop plants, plants of ethnobotanical or pharmaceutical value, and plants present in protected areas. All data is linked, where possible, to a computerised bibliography of 17,000 references dealing with plant conservation, and will be used in the publication of the IUCN *Red List of Threatened Plants*.

Animals database

WCMC has compiled the IUCN *Red List of Threatened Animals* in collaboration with the IUCN Species Survival Commission. All the data on the 6,000 animal species currently regarded as threatened globally are held on computer, including their nomenclature, common names, distribution and conservation status. Additional databases list country endemic species. The Centre is active in preparing digital distribution maps for threatened animals and has developed maps for marine turtles and endemic freshwater fishes.

Trade database

The CITES *Trade Database* holds some two million records on trade in wildlife species and their derivative products. It is managed for the Secretariat of the Convention on International Trade in Endangered Species, of which there are currently 126 member states. The information spans from 1975 to the present and information is updated from annual reports submitted by CITES parties.

(*Source: World Conservation Montoring Centre, Cambridge.*)

Box 7.15 World Conservation Monitoring Centre
Assisting others in information management

The quantity, complexity and variability of biodiversity information can make it difficult to handle, even with sophisticated computer equipment. The information must be collated, synthesised and presented in a way that facilitates rational and objective decision-making based on factual information.

World Conservation Monitoring Centre

Participants from nine countries at the completion of the Biodiversity Data Management course, Cambridge, UK, November 1995.

WCMC has significant experience in the management of biodiversity information, and is well placed to assist others through information management advice, establishment of standards such as documentation and transfer formats, systems analysis, software development, development of electronic communications and technical training. The following examples demonstrate how WCMC has supported individual organisations in this area:

- **Advice to the Indira Ghandi Conservation Monitoring Centre, India**

  At the request of WWF India and the UK Overseas Development Administration, WCMC staff have undertaken various missions to Delhi to assist WWF India in the development of this Centre. In particular, WCMC support has covered development of plans and strategies for implementation of the Centre, advice on resource needs (and in particular, computer requirements), and identification of key datasets.

- **Development of an integrated biodiversity database at the National Museums of Kenya**

  The National Museums of Kenya has been internationally recognised as a developing centre of excellence in the fields of natural history, taxonomy and biodiversity, and has established a Centre for Biodiversity to provide an integrated focus on biodiversity issues. WCMC has undertaken detailed review of biodiversity information within the museum, and has developed models showing the inter-relationships between biodiversity data and activities of a range of departments and projects. The Centre has used this review to identify procedures to be followed by the Museum, which will lead to appropriate biodiversity information management.

- **Developing GIS capacity at the Haribon Foundation**

  During May 1994, WCMC organised a GIS training and acquisition programme for the Haribon Foundation, the leading environmental NGO in the Philippines. Funded by the British Council, the programme comprised training and work experience in the UK, and the acquisition of a complete GIS system. This is now being used by the Haribon Foundation to strengthen its work on integrating local people and protected areas and other conservation programmes in the Philippines.

queries of WCMC data. This provides WWW users with an attractive graphic interface within which they can interactively view text, maps and graphics on biodiversity and provides the Centre with global access to other centres of excellence, such as the Environmental Resource Information Network (ERIN) in Australia.

WCMC's initial experiences with the new advanced telecommunications technology have demonstrated the great potential for the cooperative exchange of information, and for sharing of views, experiences, strategies and technology in a manner which is easily accessible to all.

Box 7.15 *Continued*

- **Information management for the Goeldi Museum, Brazil**

  WCMC was requested by the UK Overseas Development Agency to provide technical consultancy on GIS and database systems to the Projeto Mamiraua based at the Goeldi Museum in Belem. Project aims were to establish a reserve management plan in which computers facilitate management decisions. The purpose of the consultancy was to design an appropriate hardware/software configuration to deliver tools for decision making.

- **Russian Arctic Programme**

  WCMC has negotiated funding from the UK Environmental Knowhow Fund to assist the Faculty of Geography at Moscow State University to collect environmental data on the Russian Arctic and format it into a GIS database.

(*Source*: *World Conservation Monitoring Centre, Cambridge*.)

WCMC has been very active in supporting development of in-country information management, and is the hub of a network preparing guidelines and materials for capacity building. These activities build on earlier collaboration between WCMC and UNEP on the development of *Guidelines for Country Studies on Biological Diversity* (UNEP, 1993). WCMC is providing the support necessary for developing and implementing the national biodiversity strategies and action plans called for by the Convention on Biological Diversity (see **Box 7.15**).

There are many other models of information provision. Progress in recent years has been rapid, and many UK institutes have initiated and are participating in innovative international initiatives that will make the possibility of a Biodiversity Clearing House Mechanism a reality.

The United Nations Environment Programme and the World Conservation Monitoring Centre have published *The Resource Inventory*, an ambitious attempt to present a wide range of background and reference material on biodiversity data management in electronic form. This includes extensive information sources and reference materials. The emphasis is on CD-ROM, on-line databases, and the Internet, plus metadatabases, key addresses and a series of organisation's 'Profiles' for key agencies involved in biodiversity data management. Distributed on computer discs this information is fully searchable and can be edited for adaptation to local needs.

7.8 References and bibliography

Anon. (quarterly). *Remote sensing of natural resources: a quarterly literature review*. Technology Application Centre, University of New Mexico, Albuquerque. [Abstracts and citations on satellite cartography].

Anon. (annual). *Catalogue of Admiralty Charts and other hydrographic publications* (N.P. 131), Hydrographer of the Navy, M.O.D., Taunton.

Avery, T.E. and Berlin, G.L. (1985). *Interpretation of aerial photographs*. (edn. 4). Burgess Publishing Company, Minneapolis.

Barlow, M., Fleming, P. and Button, J. (1995). ECO: *Directory of databases in the United Kingdom* 1995/6. Eco Environmental Education Trust, Bristol.

Böhme, R. (compiler) (Eng. transl. by R. Anson). *Inventory of world topographic mapping*. International Cartographic Association/Elsevier Applied Science, London.

– (1989) Vol. 1 *Western Europe, North America and Australasia*.

– (1991) Vol. 2 *South America, Central America and Africa*.

– (1993) Vol. 3 *Eastern Europe, Asia, Oceania and Antarctica*.

Bridson, G.D.R., Phillips, V.C. and Harvey, A.P. (1980). *Natural history manuscript resources in the British Isles*. Mansell, London.

British Library (annually). *Guide to libraries and information units in Government departments and other organisations*. The British Library, London.

Burroughs, P.A. (1986). *Principles of geographic information systems for land resources evaluation*. Oxford University Press, Oxford.

Canhos, V., Lange, D., Kirsop, B. E., Nandi and Ross, E. (eds.) (1992). *Needs and specifications for a Biodiversity Information Network*. Proceedings of an International Workshop held at the Tropical Database, Campinas, Brazil 26-31 July, 1992. UNEP, Nairobi.

Chibnall, J. (1995). *A directory of UK map collections*. (edn. 3). Map Curators Group Publication 4. British Cartographical Society, Cambridge.

Cracknell, A.P. and Hayes, L. (1991). *Introduction to remote sensing*. Taylor & Francis, London.

Cracknell, A.P. (Annual). *Remote sensing Yearbook*. Taylor & Francis, London.

Cross, P.A. (1991). GPS for GIS's. *Mapping Awareness and GIS*, 5(10): 30-34.

Curran, P.J. (1985). *Principles of remote sensing*. Longman, Harlow.

David, F.W., Stoms, D.M., Estes, J.E., Scepan, J. and Scott, M. (1990). An information systems approach to the preservation of biological diversity. *Int. J. Geogr. Inform. Syst.*, 4 (1): 55-78.

Drury, S.A. (1990). *A guide to remote sensing: Interpreting images of the earth*. Oxford University Press, Oxford.

Estes, J.E., Ehlers, M., Malingreau, J.P., Noble, I.R., Raper J., Sellman, A., Star, J.L. and Weber, J. (1992). *Advanced data acquisition and analysis technologies for sustainable development*. MAB Digest 12. UNESCO, Paris.

Estrada, S. (1993). *Connecting to the Internet*. O' Reilly, Sebastopol, California.

Geelan, P.J.M. (1973). The collection of place names by small expeditions. *Geogr.* J. 139(1): 104-106.

Geo Catalog. 3 Vols, looseleaf (intermittent updates). Geo Centre Internationales Landkartenhaus, Stuttgart. [By subscription only].

Green, D.R., Rix, D. and Corbin, C. (eds.) (1995). *The AGI source book for geographic information systems* 1996. Association for Geographic Information.

Groombridge, B.(ed.). (1994). *Biodiversity data sourcebook*. World Conservation Monitoring Centre, Cambridge.

Hollis, S. and Brummitt, R.K. (1992). *World geographical scheme for recording plant distributions* (*Plant taxonomic database standards no*. 2). Hunt Institute for Botanical Documentation, Pittsburg.

International Cartographic Association (1973). *Multilingual dictionary of technical terms in cartography*. Steiner, Wiesbaden.

International Geographical Union (1992). *Orbis Geographicus* 1992/93. *World Directory of Geography*. Franz Steiner, Stuttgart. [Lists national cartographic agencies and major map collections worldwide].

Krol, E. (1994). *Whole Internet users guide*. O' Reilly, Sebastopol, California.

Küchler, A.W. (ed.) (1965–1970). *International bibliography of vegetation maps*. 4 Vols. University of Kansas, Lawrence.

Library Association (Annual). *Libraries in the United Kingdom and the Republic of Ireland*. (Occasional publication). Library Association, London.

Linnean Society. (1994). *Preserving the archives of nature: A guide for the owners of papers on nature conservation*. Linnean Society, London.

Maguire, D.J., Goodchild, M.F. and Rhind, D.W. (1991). *Geographical information systems: principles and applications*. 2 Vols. Longman Scientific and Technical, London.

Olivieri, S., Harrison, J. and Busby, J.R. (1995). Data and information management and communication. In UNEP, *Global biodiversity assessment*. Cambridge University Press, Cambridge.

Parry, R.B. and Perkins, C.R. (1987). *World mapping today*. Butterworths, London.

Perkins, C.R. and Parry R.B. (eds.) (1990). *Information sources in cartography*. Bowker-Saur, London.

Porter, R. (1993). *The library resources of* OS *International*. Ordnance Survey, Southampton.

Raper, J. and Green, N. (1993). GIS *Tutor 2 for Windows* 6.1. Longman, London.

Rittner, D. (1992). *EcoLinking: Everyone's guide to Online Environmental Information*. Peachpitt Press, Berkeley, California.

Rodcay, G.K. (ed.) (1994). *International* GIS *sourcebook: Geographic Information System technology*. (edn 5). GIS World, Fort Collins.

Teeuw, R. (1992). *Data sources for expeditions using* GIS (GIS/GPS for Expeditions, Seminar notes). Dept. of Environmental Sciences, University of Hertfordshire, Hatfield.

WCMC (1994). *The Biodiversity Information Clearing House – concepts and challenges*. WCMC Biodiversity Series No 2. World Conservation Press, Cambridge.

Wright, J. (1973). Air photographs for small expeditions. *Geogr.* J. 139(2): 311-322.

UNEP (1993). *Guidelines for country studies on biological diversity*. UNEP, Nairobi.
UNEP (1995). *Global biodiversity assessment*. Cambridge University Press/United Nations Environment Programme, Cambridge.
UNESCO (1973). *International classification and mapping of vegetation*. UNESCO, Paris.
WILSON, K. (ed.) (1994). *Global plant checklist project plan*. International Organisation for Plant Information, Canberra and Oxford.

APPENDICES

Appendix One

Universities and centres of further education in the UK

University of Aberdeen
Aberdeen
AB9 1FX

Abertay Dundee University
Bell Street
Dundee DD1 1HG

Anglia Polytechnic University
East Road
Cambridge CB2 1PT

Anglia Polytechnic University
Chelmsford Campus
Victoria Road South
Chelmsford CM1 1LL

Aston University
Aston Triangle
Birmingham B4 7ET

Bath College of Higher Education
Newton Park
Newton Loe
Bath BA2 9BN

University of Bath
Claverton Down
Bath BA2 7AY

Birkbeck College
(University of London)
Malet Street
London WC1E 7HX

University of Birmingham
Edgbaston
Birmingham B15 2TT

Bolton Institute of Higher Education
Deane Road
Bolton BL3 5AB

University of Bournemouth
Fern Barrow
Poole
Dorset BH12 5BB

University of Bradford
Bradford BD7 1DP

University of Brighton
Lewes Road
Brighton BN2 4AT

University of Bristol
Bristol BS8 1TH

Brunel University
Uxbridge UB8 3PH

University of Buckingham
Hunter Street
Buckingham MK18 1EG

University of Cambridge
Science Departments
Downing Street
Cambridge CB2 1TN

Cardiff Institute of Higher Education
Llandaff Centre
Western Avenue
Cardiff CF5 2SG

University of Central England in Birmingham
Perry Barr
Birmingham B42 2SU

University of Central Lancashire
Preston PR1 2HE

Cheltenham & Gloucester
College of Higher Education
The Park
Cheltenham GL50 2QF

City University
Northampton Square
London EC1V 0HB

Coventry University
Priory Street
Coventry CV1 5FB

De Montfort University
The Gateway
Leicester LE1 9BH

University of Derby
Kedleston Road
Derby DE22 1GB

University of Dundee
Dundee DD1 4HN

University of Durham
Old Shire Hall
Durham DH1 3HP

University of East Anglia
Earlham
Norwich NR4 7TJ

University of East London
Romford Road
Stratford
London E15 4LZ

University of Edinburgh
Old College
South Bridge
Edinburgh EH8 9YL

University of Edinburgh
The King's Buildings
Mayfield Road
Edinburgh EH9 3JH

University of Essex
Wivenhoe Park
Colchester CO4 3SQ

University of Exeter
Exeter EX4 4QJ

University of Glamorgan
Treforest
Pontypridd CF37 1DL

University of Glasgow
Glasgow G12 8QQ

Glasgow Caledonian University
Cowcaddens Road
Glasgow G4 0BA

Goldsmiths' College
(University of London)
Newcross
London SE14 6NW

University of Greenwich
Wellington Street
Woolwich
London SE18 6PF

University of Hertfordshire
College Lane
Hatfield AL10 9AB

University of Huddersfield
Queensgate
Huddersfield HD1 3DH

University of Hull
Cottingham Road
Hull HU6 7RX

University of Humberside
Cottingham Road
Hull HU6 7TR

Imperial College of Science, Technology & Medicine (University of London)
South Kensington
London SW7 2AZ

University of Keele
Keele ST5 5BG

University of Kent at Canterbury
Canterbury CT2 7NZ

King's College London
(University of London)
Strand
London WC2R 2LS

University of Kingston upon Thames
Penrhyn Road
Kingston upon Thames KT1 2EE

University of Lancaster
Lancaster LA1 4YW

University of Leeds
Leeds LS2 9JT

Leeds Metropolitan University
Calverley Street
Leeds LS1 3H3

University of Leicester
University Road
Leicester LE1 7RH

University of Liverpool
PO Box 147
Liverpool L69 3BX

Liverpool John Moores University
Rodney House
70 Mount Pleasant
Liverpool L3 5UX

University of London
Senate House
Malet Street
London WC1E 7HU

London Guildhall University
117/119 Houndsditch
London EC3A 7BU

London School of Economics & Political Science
(University of London)
Houghton Street
London WC2A 2AE

London School of Hygiene and Tropical Medicine
(University of London)
Keppel Street
London WC1E 7HT

Loughborough University of Technology
Loughborough LE11 3TU

University of Luton
Park Square
Luton LU1 3JU

The Manchester Metropolitan University
Manchester M15 6BH

University of Manchester
Oxford Road
Manchester M13 9PL

University of Manchester Institute of Science and Technology
PO Box 88
Manchester M60 1QD

Middlesex University
All Saints
White Hart Lane
London N17 8HR

Millport Marine Station
(University Biological Station)
Millport
Isle of Cumbrae KA28 0EG

Moray House College
Heriot-Watt University
Edinburgh EH14 4AS

Napier University
Craiglockart
219 Colinton Road
Edinburgh EH14 1DJ

Nene College
Northampton NN2 7AL

University of Newcastle upon Tyne
Newcastle upon Tyne NE1 7RU

University of North London
166-220 Holloway Road
London N7 8DB

University of Northumbria at Newcastle
Ellison Place
Newcastle upon Tyne NE1 8ST

University of Nottingham
University Park
Nottingham NG7 2RD

The Nottingham Trent University
Burton Street
Nottingham NG1 4BU

Open University
Walton Hall
Milton Keynes MK7 6AA

University of Oxford
University Offices
Wellington Square
Oxford OX1 2JD

University of Oxford
Forestry Institute and Plant Biology
South Parks Road
Oxford OX1 3RB

University of Oxford
Department of Zoology
South Parks Road
Oxford OX1 3PS

Oxford Brookes University
Gipsy Lane
Headington
Oxford OX3 0BP

University of Paisley
High Street
Paisley PA1 2BE

University of Plymouth
Drake Circus
Plymouth PL4 8AA

University of Portsmouth
Winston Churchill Avenue
Portsmouth PO1 2UP

Queen Margaret College
Clerwood Terrace
Edinburgh EH12 8TS

Queen Mary & Westfield College
(University of London)
Mile End Road
London E1 4NS

The Queen's University Belfast
University Road
Belfast BT7 1NN

University of Reading
PO Box 217
Reading RG6 2AH

Robert Gordon University
Schoolhill
Aberdeen AB9 1FR

Royal Holloway & Bedford New College
(University of London)
Egham TW20 0EX

University of St. Andrews
College Gate
St. Andrews KY16 9AJ

University of Salford
Salford M5 4WT

School of Oriental and African Studies
(University of London)
Thornhaugh Street
London WC1H 0XG

University of Sheffield
Sheffield S10 2TN

Sheffield Hallam University
Pond Street
Sheffield S1 1WB

South Bank University
103 Borough Road
London SE1 0AA

University of Southampton
Highfield
Southampton SO9 5NH

Staffordshire University
College Road
Stoke on Trent ST4 2DE

University of Stirling
Stirling FK9 4LA

University of Strathclyde
16 Richmond Street
Glasgow G1 1XQ

University of Sunderland
Langham Tower
Ryhope Road
Sunderland SR2 7EE

University of Surrey
Guildford GU2 5XH

University of Sussex
Falmer
Brighton BN1 9RH

University of Teeside
Middlesbrough TS1 3BA

University of Ulster
Coleraine
Co. Londonderry BT52 1SA

Thames Valley University
St Mary's Road
London W5 5RF

University College London
(University of London)
Gower Street
London WC1E 6BT

University College of Wales
Aberystwyth SY23 2AX

University College of North Wales
Bangor LL57 2DG

University of Wales, Cardiff
PO BOX 68
Cardiff CF1 3XA

University of Wales, Swansea
Singleton Park
Swansea SA2 8PP

University of Warwick
Coventry CV4 7AL

University of the West of England in Bristol
Coldharbour Lane
Bristol BS16 1QY

University of Westminster
309 Regent Street
London W1R 8AL

University of Wolverhampton
Wulfruna Street
Wolverhampton WV1 1SB

Wye College
(University of London)
Wye, Ashford
Kent TN25 5AH

University of York
Heslington
York YO1 5DD

Appendix Two

Research institutes, museums, botanic gardens, zoos and other useful addresses

NB. Addresses of Societies with an interest in biodiversity given in Volume 1, Box 5.9; those of the Museums Area Councils in Box 6.8; and for addresses of libraries see Boxes 7.4 and 7.7.

Research Institutes

Babraham Institute
Babraham Hall
Babraham
Cambridge CB2 4AT

British Antarctic Survey
High Cross
Madingley Road
Cambridge CB3 0ET

Building Research Establishment
Garston
Watford WD2 7JR

Centre for Arid Zone Studies
School of Agriculture, Forestry and Allied Sciences
University College of North Wales
Bangor LL57 2UW

Centre for Coastal and Marine Sciences
Prospect Place
West Hoe
Plymouth PL1 3DH

Centre for Ecology and Hydrology
Maclean Building
Crowmarsh Gifford
Wallingford OX10 8BB

Centre for Population Biology
Imperial College
Silwood Park
Ascot SL5 7PY

Directorate of Fisheries Research
Fisheries Laboratory
Lowestoft NR33 OHT

Dunstaffnage Marine Laboratory
PO Box 3
Oban
Argyll PA34 4AD

Environmental Information Centre
(Institute of Terrestrial Ecology)
Monks Wood
Abbots Ripton
Huntingdon PE17 2LS

Fish Genetics Programme (ODA)
School of Biological Sciences
University of Wales
Singleton Park
Swansea SA2 8PP

Forestry Commission
Forest Research Station
Alice Holt Lodge
Wrecclesham
Farnham GU10 4LH

Forestry Commission
Northern Research Station
Roslin
Midlothian EH25 9SY

Hannah Research Institute
Ayr KA6 5HL

Horticulture Research International
Wellesbourne
Warwick CV35 9EF

HRI East Malling
East Malling
West Malling ME19 6BJ

HRI Efford
Efford
Lymington SO14 0LZ

HRI Hop Research Unit
Wye College
Ashford TN25 5AH

HRI Kirton
Willington Road
Kirton
Boston PE20 1NN

HRI Littlehampton
Worthing Road
Littlehampton BN17 6LP

HRI Stockbridge House
Cawood
Selby YO8 0TZ

Institute for Animal Health
Compton
Newbury RG20 7NN

Institute of Arable Crop Research
Rothamsted
Harpenden AL5 2JQ

Institute of Food Research
Earley Gate
Whiteknights Road
Reading RG6 6BZ

Institute of Freshwater Ecology
Windermere Laboratory
The Ferry House
Far Sawrey
Ambleside LA22 0LP

Institute of Grassland and Environmental Research
Plas Gogerddan
Aberystwyth SY23 3EB

Institute of Hydrology
Maclean Building
Crowmarsh Gifford
Wallingford OX10 8BB

Institute of Oceanographic Science
Southampton Oceanography Centre
Empress Dock
Southampton SO14 3ZH

Institute of Terrestrial Ecology
Monks Wood
Abbots Ripton
Huntingdon PE17 2LS

Institute of Virology and Environmental Microbiology
Mansfield Road
Oxford OX1 3SR

Institute of Zoology
Regents Park
London NW1 4RY

International Institute of Entomology
56 Queen's Gate
London SW7 5JR

International Institute of
Biological Control
Silwood Park
Buckhurst Road
Ascot SL5 7TA

International Institute of
Parasitology
395 Hatfield Road
St Albans Al4 0XU

International Mycological
Institute
Bakeham Lane
Egham TW20 9TY

John Innes Centre
Norwich Research Park
Colney Lane
Norwich NR4 7UH

Long Ashton Research Station
Long Ashton
Bristol BS18 9AF

Macaulay Land Use Research
Institute
Craigiebuckler
Aberdeen AB29 2QL

Moredun Research Institute
408 Gilmerton Road
Edinburgh EH17 7JH

National Institute of Agricultural
Botany
Huntingdon Road
Cambridge CB3 0LE

Natural Resources Institute
Centre Avenue
Chatham Maritime ME4 4TB

Oxford Forestry Institute
Oxford University Department of
Plant Sciences
South Parks Road
Oxford OX1 3RB

Plant Breeding Institute
Maris Lane
Trumpington
Cambridge CB2 2LQ

Plymouth Marine Laboratory
Prospect Place
West Hoe
Plymouth PL1 3DH

Proudman Oceanographic
Laboratory
Bidston Observatory
Birkenhead
Merseyside L43 7RA

Public Health Laboratory
61 Colindale Avenue
London NW9 5DF

Rowett Research Institute
Greenburn Road
Bucksburn
Aberdeen AB2 9SB

Scottish Agricultural Science
Agency
East craigs
Edinburgh EH12 8NJ

Scottish Crop Research Institute
Invergowrie
Dundee DD2 5DA

Scottish Office Agriculture and
Fisheries Departmeent
Pentland House
47 Robb's Loan
Edinburgh EH14 1TW

Sea Mammal Research Unit
C/o British Antarctic Survey
Madingley Road
Cambridge CB3 0ET

Silsoe Research Institute
Wrest Park
Silsoe
Bedford MK45 4HS

Museums

Ashmolean Museum
Wellington Square
Oxford OX1 2JD

Biology Curators Group
C/o Towneley Hall Museum
Towneley Park
Burnley BB11 3RQ

Birmingham Museum and Art
Gallery
Chamberlain Square
Birmingham B3 3DH

Cambridge University Museum
of Zoology
Downing Street
Cambridge
CB2 3EJ

Glasgow Art Gallery & Museum
Kelvingrove
Glasgow G3 8AG

Hancock Museum
(University of Newcastle upon
Tyne)
Barras Bridge
Newcastle upon Tyne NE2 4PT

Kirkcaldy Museums & Art Gallery
War Memorial Gardens
Kirkcaldy KY1 1YG

Manchester Museum
The University
Oxford Road
Manchester M13 9PL

Museums Association
34 Bloombury Way
London WC1A 2SF

Museum Documentation
Association
347 Cherry Hinton Road
Cambridge CB1 4DH

National Museum of Wales
Cathays Park
Cardiff CF1 3NP

The Natural History Museum
Cromwell Road
South Kensington
London SW7 5BD
(See also Walter Rothschild
Museum)

Nottingham Natural History
Museum
Wollaton Hall
Nottingham NG8 2AE

Pitt Rivers Museum
Linacre College
Wellington Square
Oxford OX1 2JD

Royal Museum of Scotland
Chambers Street
Edinburgh EH1 1JF

Walter Rothschild Zoological
Museum
(The Natural History Museum)
Akeman Street
Tring HP23 6AP

Botanic Gardens

Bath Botanical Gardens
Bath City Council
Department of Leisure
Pump Room
Bath BA1 1LZ

Birmingham Botanical Gardens
Westbourne Road
Edgbaston
Birmingham B15 3TR

Botanical Gardens Conservation International
Descanso House
199 Kew Road
Richmond TW9 3AH

Chelsea Physic Garden
66 Royal Hospital Road
London SW3 4HS

Cruickshank Botanic Garden
St Machar Drive
Old Aberdeen AB9 2UD

Dawyck Botanic Garden
(see RBG Edinburgh)

Dyffryn Gardens
St Nicholas
Cardiff CF5 6SU

Glasgow Botanic Gardens
Glasgow G12 0UE

Harlow Car Gardens
The Northern Horticultural Society
Harrogate
North Yorkshire HG3 1QB

Sir Harold Hillier Gardens and Arboretum
Jermyns Lane, Ampfield
Romsey SO51 0AQ

City of Liverpool Botanic Gardens
The Mansion House
Calderstones Park
Liverpool L18 3JD

Logan Botanic Garden
(see RBG Edinburgh)

National Council for the Conservation of Plants and Gardens
The Pines
Wisley Gardens
Woking GU23 6QB

University of Oxford Botanic Garden
Rose Lane
Oxford OX1 4AX

Plant Net
(Consortium of UK botanic gardens)
C/o University of Oxford Botanic Garden

The National Pinetum
Bedgebury
Nr Goudhurst
Cranbrook TN17 2SL

Paignton Botanical Gardens
Totnes Road
Paignton TQ4 7EU

Royal Botanic Garden Edinburgh
Inverleith Row
Edinburgh EH3 5LR

Royal Botanic Garden Edinburgh
Dawyck Botanic Garden
Stobo
Pebbles EH45 9JU

Royal Botanic Garden Edinburgh
Logan Botanic Garden
Port Logan
Stranraer DG9 9ND

Royal Botanic Garden Edinburgh
Younger Botanic Garden
Benmore
Dunoon PA23 8QU

Royal Botanic Gardens Kew
Kew, Richmond
Surrey TW9 3AB

Royal Botanic Gardens Kew
Wakehurst Place
Ardingley
Haywoods Heath RH17 6TN

Royal Horticultural Society's Garden
Wisley
Woking GU23 6QB

Sheffield Botanic Gardens
Clarke House Road
Sheffield S10 2LN

South London Botanical Institute
323 Norwood Road
London SE24 9AQ

Thorp Perrow Arboretum
Bedale
N. Yorks DL8 2PR

Treborth Botanic Garden
University College North Wales
Bangor LLS7 2RQ

Tresco Abbey Gardens
Tresco
Isles of Scilly TR24 8QQ

University of St Andrews Botanic Gardens
St Andrews KY16 8RT

University of Birmingham Botanic Garden
The University
Birmingham B15 2TT

University of Bristol Botanic Garden
Woodland Road
Bristol BS8 1UG

University of Cambridge Botanic Garden
Bateman Street
Cambridge CB2 1JF

University of Dundee Botanic Garden
516 Perth Road
Dundee DD2 1LW

University of Durham Botanic Garden
Hollow Drift
Green Lane
Durham DH1 3LA

University of Hull Botanic & Experimental Garden
57 Thwaite Street
Cottingham HU16 4QX

Univerity of Leicester Botanic Garden
Stoughton Drive South
Leicester LE2 2NA

University of Liverpool Botanic Gardens
Ness
South Wirral L64 4AY

The University of Keele Botanical Garden
Keele ST5 5BG

University of London Botanic Garden
Elm Lodge
Egham TW20 OBN

University of Nottingham Arboretum
Sutton Bonington
Loughborough LE12 5RD

University of Reading Plant Science Botanic Garden
Whiteknights
Reading RG6 2AS

University of Sheffield Botanic Gardens
26 Taptonville Road
Sheffield S10 5BR

University of Southampton Botanic Garden
The University
Southampton SO9 5NH

University College, Swansea Botanic Garden
Singleton Park
Swansea SA2 8PP

Ventnor Botanic Garden
Steephill Road
Ventnor
Isle of Wight PO38 1UL

Wakehurst Place
(see RBG Kew)

Winkworth Arboretum
Hascombe Road
Godalming GU8 4AD

Westonbirt Arboretum
(Forestry Commission - Research)
Tetbury GL8 8QS

Younger Botanic Garden
Benmore
(see RBG Edinburgh)

Zoological Gardens

Banham Zoo
The Grove
Banham
Norwich NR16 2HB

Belfast Zoo (City of Belfast Zoo)
Hazelwood
Antrim Road
Belfast BT36 7PN

Birdworld
Holt Pound
Farnham GU10 4LD

Blackpool Zoo Park
East Drive
Blackpool FY3 8PP

Bristol, Clifton & West of England Zoological Society
Bristol BS8 3HA

Chester Zoo (The North of England Zoological Society)
Upton-by-Chester CH2 1LH

Colchester Zoo
Maldon Road
Stanway CO3 5SL

Dudley & West Midlands Zoological Society
2 The Broadway
Dudley DY1 4QB

Edinburgh Zoo (Royal Zoological Society of Scotland)
Scottish National Zoological Park
Murrayfield
Edinburgh EH12 6TS

Federation of Zoological Gardens of Great Britain and Ireland
Zoological Gardens
Regent's Park
London NW1 4RY

Glasgow Zoo (Zoological Society of Glasgow & West of Scotland)
Calderpark Zoological Gardens
Uddington
Glasgow G71 7RZ

Harewood Bird Garden
Harewood House
Leeds LS17 9LF

Hawk Conservancy
Weyhill
Andover SP11 8DY

Jersey Wildlife Preservation Trust
Les Augres Manor
Jersey
Channel Islands JE3 5BF

Knowsley Safari Park
Prescot
Merseyside L34 4AN

Liverpool Museum Aquarium and Vivarium
William Brown Street
Liverpool L3 8EN

Lotherton Hall Bird Garden
Towton Road
Nr. Aberford
Leeds LS25 3EB

Marwell Zoological Park
Colden Common
Winchester SO21 1JH

National Birds of Prey Centre
Newent GL18 1JJ

Owl Centre
Muncaster Castle
Ravenglass CA18 1RQ

Paignton Zoological and Botanical Gardens
Paignton TQ4 7ED

Regent's Park Zoo (Zoological Society of London)
Regent's Park
London NW1 4RY

Southport Zoo and Conservation Centre
Princes Park
Southport
Merseyside PR8 1RX

Thrigby Hall Wildlife Gardens
Thrigby Hall
Gt. Yarmouth NR29 3DS

Twycross Zoo
Norton-juxta-Twycross
Atherstone CV9 3PX

Welsh Mountain Zoo (Zoological Society of Wales)
Colwyn Bay LL28 5UY

West Midlands Safari Park
Spring Grove
Bewdeley DY12 1LF

Whipsnade Wild Animal Park (Zoological Society of London)
Nr. Dunstable LU6 2LF

Other useful addresses

Africa Educational Trust
Africa Centre
38 King Street
London WC2E 8JS
(for students from African countries)

Association of Commonwealth Universities
36 Gordon Square
London WC1H 0PF

BIOSIS
Garforth House
Mickelgate
York YO1 1LF

Centre for Social and Economic Research on the Global Environment
University of East Anglia
Norwich NR4 7TJ

Conservation International
1015 18th St., NW, Suite 10000,
Washington DC 20036, USA.

Department of Education
Further and Higher Education,
Branch 3
Sanctuary Buildings
Great Smith Street
London SW1P 3BT

Department for Education
Further and Higher Education
(for postgraduate grants)
Branch 3 (Room 214)
Mowden Hall
Staindrop Road
Darlington DL3 7PJ

Department of the Environment
2 Marsham Street
London SW1P 3PY

The Foundation of Ethnobiology
North Parade Chambers
75 Banbury Road
Oxford OX2 6PE

Foreign and Commonwealth Office
Environment, Science and
Engergy Dept.
London SW1H 9NF

International Centre for Conservation Education
Greenfield House
Guiting Power
Cheltenham GL54 5TZ

IUCN UK Secretariat
Nature Conservation Bureau
38 Knigfisher Court
Hambridge RG14 5SJ

Joint Nature Conservation Committee
Monkstone House
City Road
Peterborough PE1 7JY

Microbial Strain Data Network
C/o Biotechnology Centre
University of Cambridge
307 Huntingdon Road
Cambridge CB3 0JX

National Register of Archives
Quality Court
London WC2A 1MP

National Remote Sensing Centre
Southwood Crescent
Southwood
Farnborough
Hants GU14 0NL

Overseas Development Agency
Natural Resources Research Department
94 Victoria Street
London SW1E 5JL

Public Record Office
Ruskin Avenue
Kew, Richmond
Surrey TW9 4DU

The Remote Sensing Society
C/o Geography Department
University of Nottingham
Nottingham NG7 2RD

UK Tropical Forest Forum
c/o Natural Resources Institute
Central Avenue
Chatham Maritime ME4 4TB

World Conservation Monitoring Centre
219c Huntingdon Road
Cambridge CB3 0DL

World University Service
20/21 Compton Terrace
London N1 2UN
(for information and advice on the education rights and needs of refugees)

World Wildlife Fund for Nature
Cattleshall Park
Godalming
Surrey GU7 1XR

Appendix Three

Offices of the British Council throughout the world

Mail for those marked * should be sent via UK H.Q., London (below).
Those marked FCO Bag Room should be marked for the B.C. in the appropriate country and sent to FCO, King Charles St., London SW1A 2AH.

Headquarters

10 Spring Gardens
London SW1A 2BN
Tel +44 171 9308466
Telex 8952201 bricon g
Fax 8396347

11 Portland Place
London W1N 4EJ
Tel +44 171 9308466
Fax 3893199

Information Section
The British Council
Medlock Street
Manchester M15 4AA
Tel +44 161 9577000
Fax 9577111

UK offices

Northern Ireland
1 Chlorine Gardens
Belfast BT9 5DJ
Tel +44 1232 666706/
666770/683880
Fax 665242

Scotland
3 Bruntsfield Crescent
Edinburgh EH10 4HD
Tel +44 1314474716
Fax 4528487

Wales
28 Park Place
Cardiff CF1 3QE
Tel +44 1222 397346/7/8/9
Fax 237494

Overseas offices

Albania
The British Council Eastern Adriatic is responsible for work in Albania.

Liason Officer: British Council
Resource Centre
University of Tirana
Tirana
Tel +873 144 5577 (Embassy)
Fax 5601 (Embassy)

Algeria
c/o The British Embassy
7 Chemin des Glycines
BP 43, Alger Gare 16000
Algiers
Tel +213 2 692601/
692831/692411/692038
Fax 692410

Argentina
Marcelo T. de Alvear 590
(4th Floor)
1058 Buenos Aires
Tel +54 1 3119814/3117519
Fax 3117747

Armenia
The British Council Russia is responsible for work in Armenia

Australia
Edgecliff Centre
401/203 New South Head Road
PO Box 88, Edgecliff
Sydney, NSW 2027
Tel +61 2 3262022
Fax 3274868
e-mail bcsydney@
bc-sydney.sprint.com

Austria
Schenkenstra(e 4, A-1010
Vienna
Tel +43 1 5332616/7/8
Fax 533261685

Azerbaijan
The British Council Turkey is responsible for work in Azerbaijan.

ELT Consultant
Bakü Institute of
Social Management and
Politology
74 Lermntov Street
Bakü
Tel +994 12989236
Fax 989236

Bahrain
AMA Centre
PO Box 452
146 Shaikh Salman Highway
Manama 356
Tel +973 261555
Fax 241272

Baltic states
(Latvia, Lithuania and Estonia)

*British Council (Director Baltic states)
Lazaretes iela 3
Riga 226010, Latvia
Tel +371 2320468
Fax 8830031
e-mail bc.riga@
british-council.sprint.com

Lithuania Liason Officer:
British Council Resource Centre
Vilniaus 39
Vilnius, Lithuania
Tel +370 2 616607
Fax 221602
e-mail bc.vilnius@
british-council.sprint.com

Estonia Liason Officer:
British Council Resource Centre
Vana Posti 7, Tallinn
Estonia
Tel +372 2 441550
Fax 313111
e-mail bc.tallinn@
british-council.sprint.com

Bangladesh
5 Fuller Road
PO Box 161
Dhaka 1000
Tel +880 2 868905/6/7/868867/8
Fax 863375
e-mail bcdhaka!barlow@
pradeshta.net

Barbados
see Caribbean

Belarus
The British Council Russia is responsible for work in Belarus

British Council Resource Centre
Institute of Foreign Languages
Ul Zakharova 21
220662 Minsk
Tel +7 0172 367953
Fax 367953
[UK mail via FCO Bag Room]

Belgium and Luxembourg
Rue Joseph II/
Joseph 11 straat 30
1040 Brussels
Tel +32 2 2193600
Fax 2175811
e-mail bc.brussels@
british-council.sprint.com

c/o United Kingdom Research
and Higher Education
European Office
rue de la Loi 83
BP10, 1040 Brussels
Tel +32 2 2305275
Fax 2304803
e-mail ukeo@bbsrc.ac.uk

Bosnia-Herzegovina
c/o The British Embassy
8 Tina Ujevica
Sarajevo
Tel +387 71 444429/663922
Fax 444429/663922

Botswana
c/o British High Commission
Queen's Road
The Mall, PO Box 439
Gaborone
Tel +267 3 53602
Telex 2368 brico bd
Fax 56643

Brazil
SCRN 708/9
Bloco F Nos 1/3
Caixa Postal 6104
70740-780 Brasilia DF
Tel +55 61 2723060
Fax 2723455
e-mail bc_brasilia@
mcimail.com

Av. Domingos Ferreira 4150
Boa Viagem
Caixa Postal 4079
51021-040 Recife PE
Tel +55 81 3266640
Fax 3264880
e-mail bc_recife@mcimail.com

Rua Elmano Cardim 10 Urca
Caixa Postal 2237
22291 Rio de Janeiro RJ
Tel +55 21 2957782
Fax 5413693
e-mail bc_rio@mcimail.com

Rua Maranhão 416
Higienópolis
Caixa Postal 1604
01240-902 São Paulo SP
Tel +55 11 8264455
Fax 663765
e-mail bc_sao_paulo@
mcimail.com

Brunei
The British Council Singapore is responsible for work in Brunei.

Liason Officer:
Room 505, 5th floor
Hong Kong Bank Chambers
Jalan Pemancha
Bandar Seri Begawan 2085
PO Box 3049, Bandar Seri
Begawan 1930
Negara Brunei Darussalam
Tel +673 2 227480/227531
Fax 241769

Bulgaria
7 Tulovo Street
1504 Sofia
Tel +359 2 467133/476215/
443148/467233/420098
Fax 4920102

Burma
c/o The British Embassy
80 Strand Road
PO Box 638, Rangoon
Tel +95 1 81700/2/3/95300/9
Fax 89566/50292/83895

Cameroon
Avenue Charles de Gaulle
BP 818, Yaoundé
Tel +237 211696/203172
Telex 8408 bricon kn
Fax 215691

Canada
c/o British High Commission
80 Elgin Street, Ottawa
Ontario K1P 5K7
Tel +1 613 2371530
Fax 5691478
e-mail af572@
freenet.carleton.ca

1000 ouest rue de la Gauchetière
Bureau 4200, Montréal
Quebec H3B 4W5
Tel +1 514 8665863
Fax 8660202

Caribbean
PCMB Building
64 Knutsford Boulevard
Kingston 5, Jamaica
Tel +1 809 9296915/9297049
Fax 9297090

c/o British High Commission
19 St Clair Avenue, St Clair
PO Box 778, Port of Spain
Trinidad and Tobago
Tel +1 809 6281234/6222748
Fax 6224555

Chile
Eliodoro Yáñez 832
Casilla 115 Correo 55
Santiago
Tel +56 2 2361199/
2360193/2356660
Fax 2357375

China
c/o The British Embassy
4th floor, Landmark Building
8 North Dongsanhuan Road
Chaoyang District
Beijing 100006
Tel +86 10 5011903
Fax 5011977
e-mail russel@mimi.cnc.ac.cn

c/o British Consulate-General
Qi Hua Tower 5B
1375 Huai Hai Zhong Lu
Shanghai 200031
Tel +86 21 4714849
Fax 4333115

for South China, use the
Hong Kong office:
Room 1202
Easey Commercial Building
255 Hennessy Road
Wanchai, Hong Kong
Tel +852 28795136
Fax 25075563

Colombia
Calle 87 No. 12-79
Apartado Aéreo 089231
Santafé, de Bogotá
Tel +571 6180175/
2187518/2576188
Fax 2187754
e-mail firstname.lastname@
sprintcol.sprint.com

Croatia
Ilica 12/1
PO Box 55
41000 Zagreb
Tel +385 41 273491/2/424888
Fax 421725

Cyprus
3 Museum Street
1097 Nicosia
PO Box 5654
1387 Nicosia
Tel +357 2 442152
Fax 477257

Czech Republic
Narodni 10, 125 01
Prague 1
Tel +42 2 2491 2179/83
Fax 2491 3839
e-mail bc.prague@
britcoun.anet.cz

Denmark
see Nordic countries

East Jerusalem
(West Bank and Gaza)
Al-Nuzha Building
2 Abu Obeida Street
PO Box 19136, Jerusalem
Tel +972 2 282545/
271131/894392
Fax 283021

Ecuador
Avda. Amazonas
1646 y Orellana
Casilla 17-07-8829, Quito
Tel +593 2 540225/
225421/508282/508284
Fax 508283/223396
e-mail erey@britcoun.org.ec

Egypt
192 Sharia el Nil
Agouza, Cairo
Tel +20 2 3453281/4
Telex 21534 brico un
Fax 3443076

Eritrea
Lorenzo Ta'zaz Street
No. 23, PO Box 997
Asmara
Tel +291 1 123415
Fax 116620

Estonia
see Baltic states

Ethiopia
Artistic Building
Adwa Avenue
PO Box 1043
Addis Ababa
Tel +251 1 550022
Telex 21561 bc et
Fax 552544

European Commission
European Commission
Relations Office
Britannia House
rue Joseph II/Jozef 11
straat 30
1040 Brussels
Tel +32 2 2193600
Fax 2199391
e-mail bc.brussels@
british-council.sprint.com

Finland
see Nordic countries

France
9/11 rue de Constantine
75007 Paris
Tel +33 1 49557300
Fax 47057702
e-mail bc.paris@
bc-paris.sprint.com

Georgia
The British Council Russia is
responsible for work in Georgia.

*British Council Resource
Centre
Ministerstvo obrazovaniya
Ul Uznadze 52
380002 Tbilisi
Tel +095 915 3511
Fax 975 2561

Germany
Hahnenstra(e 6, 50667
Cologne
Tel +49 221 206440
Fax 2064455
e-mail bc.cologne@
british-council.sprint.com

The British Council Eastern
Länder
Hardenbergstra(e 20
10623 Berlin
Tel +49 30 3110990
Fax 31109920

The British Council Leipzig
Lumumbastra(e 11-13
04105 Leipzig
Tel +49 341 5647153
Fax 5647152
e-mail bc.leipzig@
british-council.sprint.com

Ghana
Liberia Road
PO Box 771, Accra
Tel +233 21 663414/
663979/233415
Telex 2369 brico gh
Fax 663337
e-mail director@
britcoun.aau.org

Greece
17 Plateia Philikis Etairias
PO Box 3488 Kolonaki Square
102 10, Athens
Tel +30 1 3633211/2/3/4/5/
3606011/2/3/4
Fax 3634769/3609164
e-mail british.council@
bc-athens.sprint.com

Hong Kong
Easey Commercial Building
255 Hennessy Road
Wanchai, Hong Kong
Tel +852 28795138
Telex 74141 bcoun hx
Fax 25075731
e-mail bc.hongkong@
britcoun.org.hk

Hungary
Budapest VI
Benczur Utca 26
H-1068, Budapest VI
Tel +36 1 3228246/
3214039/3420127
Fax 3425728/2696594
e-mail bc.budapest@
british-council.sprint.com

India
c/o British High Commission
17 Kasturba Gandhi Marg
New Delhi
Tel +91 11 3711401/
3710111/3710555
Telex 3165460 bcnd in
Fax 3710717/3782016
e-mail delhi@bcdd.ernet.in

West India
c/o British Deputy High
Commission
Mittal Tower, C Wing
Nariman Point
Bombay 400021
Tel +91 22 2823560/
2823530/2823484/2823445
Telex 1186991 bcby in
Fax 2852024
e-mail bombay@bcdb.ernet.in

East India
c/o British Deputy High
Commission
5 Shakespeare Sarani
Calcutta 700 071
Tel +91 33 2425370/8/9/
2429144/2429108
Telex 215984 bcca in
Fax 2424804
e-mail calcutta@bcdc.ernet.in

South India
c/o British Deputy High
Commission
737 Anna Salai
Madras 600 002
Tel +91 44 8525002/8525412/
8525422/8525432/8522593
Telex 417775 bcms in
Fax 8523234
e-mail postmast@
bcdm.iitm.ernet.in

Indonesia
S. Widjojo Centre
Jalan Jenderal
Sudirman 71
Jakarta 12190
Tel +62 21 2524115/22
Fax 2524129
e-mail bc.indonesia@
bc-jakarta.sprint.com

Ireland
Newmount House
22/24 Lower Mount Street
Dublin 2
Tel +3531 6764088/6766943
Fax 6766945

Israel
140 Hayarkon Street
PO Box 3302
Tel Aviv 61032
Tel +972 3 5222194/
5/6/7/5242558/5241350/1
Fax 5221229

Italy
Via Quattro Fontane 20
00184 Rome
Tel +39 6 478141
Fax 4814296
e-mail bc.rome@
british-council.sprint.com

Jamaica
see Caribbean

Japan
2, Kagurazaka 1-chome
Shinjuku-ku, Tokyo 162
Tel +81 3 32358031
Fax 32358040

Western Japan
77 Kitashirakawa
Nishimachi
Sakyo-ku, Kyoto 606
Tel +81 75 7917151
Fax 7917154

Jordan
Rainbow Street (off First Circle)
PO Box 634
Amman 11118
Tel +962 6 636147/8/
638194/624686
Fax 656413

Kazakhstan
The British Council Russia is
responsible for work in
Kazakhstan.

Liason Officer:
Ul Pamfilova, 158/17
480064 Almaty
Tel +7 3272 633543
Fax 506260

Kenya
ICEA Building
Kenyatta Avenue
PO Box 40751, Nairobi
Tel +254 2 334855/6/7/
334811/81/85
Telex 23212 britco ke
Fax 339854
e-mail bc.nairobi@
british-council.sprint.com

Korea
1st floor, Anglican Church
Foundation Building, 3-7
Chung-dong, Choong-ku
100-120 Seoul
Tel +82 2 7377157
Fax 7379911
e-mail bc.seoul@
british-council.sprint.com
[UK mail via FCO Bag Room]

6th Floor, Kyobo Building
536-6 Boojeon-dong
Pusanjin-ku
Pusan 614-030
Tel +82 51 8074612/3
Fax 8074611

Kuwait
2 Al Arabi Street, Block 2
PO Box 345, 13004 Safat
Mansouriya
Tel +965 2533204/
2515512/2533227/2520067/8
Fax 2520069/2551376
e-mail bc.kuwait@
british-council.sprint.com

Kyrgyzstan
The British Council Russia is
responsible for work in Kyrgyz
Republic.

Latvia
see Baltic states

Lebanon
Sidani Street, Azar Building
Beirut
Tel +961 1 864534
Fax 864534

Lesotho
Hobson's Square
PO Box 429, Maseru 100
Tel +266 312609
Fax 310363

Lithuania
see Baltic states

Macedonia FYR of
see Yugoslavia

Madagascar
see Mauritius

Malawi
Plot No. 13/20, City Centre
PO Box 30222
Lilongwe 3
Tel +265 783244/783419
Telex 44476 bricoun mi
Fax 782945

Librarian James Hamisi
Angoni Street
PO Box 456
Blantyre

Malaysia
Jalan Bukit Aman
PO Box 10539
50916 Kuala Lumpur
Tel +60 3 2987555
Fax 2937214/2930807
e-mail brcokl@britkl.po.my

Sarawak
Bangunan WSK
PO Box 615 93712 Kuching
Jalan Abell
93100 Kuching
Sarawak
Tel +60 82 256271/256044/
242632/237704
Fax 425199

Sabah
Wing On Life Building
1st Floor, 1 Lorong Sagunting
PO Box 10746
88808 Kota Kinabalu
Sabah
Tel +60 88 248055/
248298/222059
Fax 238059

Penang
43 Green Hall
PO Box 595, 10770 Penang
10200 Penang
Tel +60 4 2630330
Fax 2633589

Johor Bahru
Unit 14.01, Wisma LKN
49 Jalan Wong AH Fook
PO Box 8, 80700
Johor Bahru
Tel +60 7 2233340
Fax 2233343

Maldive Islands
see Sri Lanka

Malta
c/o British High Commission
7 St Anne Street
Floriana, VLT 15
Tel +356 226227
Fax 226207

Mauritania
see Morocco

Mauritius
The British Council Mauritius is responsible for work in Madagascar and Seychelles.

Royal Road, PO Box 111
Rose Hill
Tel +230 4549550/1/2
Telex 4872 britcoun iw
Fax 4549553

Mexico
Maestro Antonio Caso 127
Col. San Rafael
Apdo Postal 30-588
Mexico City 06470 DF
Tel +52 5 5666144/5666191/
5666743/5666384
Fax 5355984

Moldova
The British Council Russia is responsible for work in Moldova.

Morocco
36 Rue de Tanger
BP 427, Rabat
Tel +212 7 760836
Fax 760850

Mozambique
Travessa da Catembe 21
PO Box 4178, Maputo
Tel +258 1 421571/2/3/4
Fax 421577
e-mail root@bcmaputo.uem.mz

Namibia
74 Bülow Strasse
Windhoek
PO Box 24224
(no street deliveries)
Tel +264 61 226878/226776
Fax 227530

Nepal
Kantipath
PO Box 640
Kathmandu or BFPO 4
(for mail from Britain and countries with access to BFPO)
Tel +977 1 221305/223796/
222698
Telex 2382 bricon np
Fax 224076

The Netherlands
Keizersgracht 343
1016 EH Amsterdam
Tel +31 20 6223644
Fax 6207389

New Zealand
c/o British High Commission
44 Hill Street
PO Box 1812, Wellington
Tel +64 4 4726049
Telex 3325 ukrer nt
Fax 4736261

Nigeria
11 Kingsway Road, Ikoyi
PO Box 3702, Lagos
Tel +234 1 2692188/
89/90/91/92
Telex 22071 brico ng
Fax 2692193/2690646/615047

John Udeh Teachers' House
Ogui Road
PO Box 330, Enugu
Tel +234 42 255577/
255677/258456
Telex 51144 brico ng
Fax 330158

British Council-Leventis
Foundation Library
Magazine Road
Jericho, Ibadan
Tel +234 22 400870/1/2

Yakubu Gowon Way
PO Box 81, Kaduna
Tel +234 62 201080/1/
236033/5
Telex 71159 britco ng
Fax 216330

10 Emir's Palace Road
PMB 3003 Kano
Tel +234 64 646652
Fax 632500

Nordic countries
(*Denmark, Norway, Finland, Sweden*)
The British Council Nordic Countries
Gammel Mont 12.3
1117 Copenhagen K
Tel +45 33112044
Fax 33321501

Finland
Hakaniemenkatu 2
00530 Helsinki
Tel +358 07018731
Fax 07018725
e-mail tuija.talvitie@oph.fi

Norway
Fridtjof Nansens Plass 5
0160 Oslo
Tel +47 22 426848
Fax 424039

Sweden
Strandvägen 57A
4tr S-115 23 Stockholm
Tel +46 8 6719190
Fax 6637172

Norway
see Nordic countries

Oman
Road One
Medinat Qaboos West
PO Box 73, Postal Code 115
Muscat
Tel +968 600548
Fax 699163

Al Fahya Street
PO Box 18249, Salalah
Tel +968 292080
Fax 294854

Al Hadiqah Street
PO Box 854
Postal Code 311, Sohar
Tel +968 843396
Fax 843398

Pakistan
Block 14, Civic Centre, G6
PO Box 1135, Islamabad
Tel +92 51 222105/9/825265/815760/822205
Telex 4644 brico pk
Fax 221211

20 Bleak House Road
PO Box 10410
Karachi 75530
Tel +92 21 520391/2/3/4/5/6/7
Telex 29570 brico pk
Fax 5683694

65 Mozang Road
PO Box 88, Lahore
Tel +92 42 6362497/8
Fax 6368674

North-West Frontier Province
35 Shahrah-e-Quaid-e-Azam
PO Box 49, Peshawar
Tel +92 521 275056
Telex 44337 brico pk
Fax 273687

Peru
Calle Alberto Lynch 110
San Isidro
Apartado 14-0114
Santa Beatriz, Lima 14
Tel +51 1 4704350/4704360/6
Fax 4215215
e-mail postmaster@britco.org.pe

Philippines
No. 7, 3rd Street
New Manila
PO Box AC 168, Cubao
Quezon City, Metro Manila
Tel +63 2 7211981/2/3/4
Telex 67789 britcoun pn
Fax 7211336
e-mail bc!root@uucp.admu.edu.ph

Poland
Al Jerozolimskie 59
00-697 Warsaw
Tel +482 6287401/2/3
6283663/6287188/6254272/3
Telex 812555 brin pl
Fax 6219955

Portugal
Rua de São Marçal 174
1294 Lisbon Codex
Tel +351 1 3476141/2/3/4/5/6/7
Fax 3476152

Qatar
Ras Abu Aboud Road
PO Box 2992, Doha
Tel +974 426185/426159/426193/4
Fax 320065
e-mail bcdoha@bc-doha.sprint.com

Romania
*Calea Dorobantilor 14
Bucharest
Tel +40 1 2105347/2100314
Telex 11295 prodm r
Fax 2100310

Russia
Russia is responsible for work in Armenia, Belarus, Georgia, Kazakhstan, Kyrgiz Republic, Moldova and Turkmenistan.

*Ul Nikoloyamskaya, 1
109189 Moscow
Tel +7 095 9153511
Fax 9752561

*Biblioteka im Mayakovskovo
Fontanka 46
191025 St Petersburg
Tel +7 812 1196073
Fax 1196074

Saudi Arabia
Tower B, 3rd floor
Al Mousa Centre
Olaya Street
PO Box 58012
Riyadh 11594
Tel +966 1 4621818
Fax 4620663

West Saudi Arabia
4th floor, Middle East Centre
Falasteen Street
PO Box 3424
Jeddah 21471
Tel +966 2 6723336/6701420
Fax 6726341

East Saudi Arabia
David Baldwin
Al-Moajil Building
5th Floor, Dhahran Street/Mohamed Street
PO Box 8387
Dammam 31482
Tel +966 3 8343484/ 8344381
Fax 8346895

Al-Huwaylat Shopping Centre,
1st Floor, Al-Huwaylat
PO Box 11363
Jubail Industrial City 31961
Tel +966 3 3419122/3
Fax 3419124

Senegal
34/36 Boulevard
de la Republique
Immeuble Sonatel
BP 6232, Dakar
Tel +221 222015/222048
Telex 21709 brico sg
Fax 218136
e-mail bc.dakar@
endadak.gn.apc.org

Seychelles
see Mauritius

Sierra Leone
Tower Hill
PO Box 124, Freetown
Tel +232 22 222223/7/
224683/4
Fax 224123

Singapore
30 Napier Road
Singapore 1025
Tel +65 4731111
Fax 4721010
e-mail britcoun@
britcoun.org.sg

Slovakia
Panská 17
PO Box 68
814 99 Bratislava
Tel +42 7 331793/
331185/331261
Fax 334705

Slovenia
Stefanova 1/III
61000 Ljubljana
Tel +386 61 1259032/
1259292
Fax 1259139

South Africa
76 Juta Street
PO Box 30637
Braamfontein 2017
Johannesburg
Tel +27 11 4033316
Telex 426428 brisa
Fax 3397806
e-mail bc.johannesburg@
british-council.sprint.com

1 Prieska Road
Sybrand Park
PO Box 493, Athlone
Cape Town 7760
Tel +27 21 6962914
Telex 523288 bcct
Fax 6964180

Spain
Paseo del General
Martínez Campos 31
28010 Madrid
Tel +34 1 3373500
Fax 3373573
e-mail bc.madrid@
british-council.sprint.com

Canary Is.
Bravo Murillo 25
35003 Las Palmas de
Gran Canaria
Tel +34 28 368300/
368323
Fax 382378

Balearic Is.
Edifici Sa Riera
Calle Miguel
dels Sants Oliver 2
07012 Palma de Mallorca
Tel +34 71 172550

Sri Lanka
The British Council Sri Lanka is
responsible for work in Maldive
Islands

49 Alfred House Gardens
PO Box 753
Colombo 3
Tel +94 1 581171/2/587078/
580301/502449
Fax 587079
e-mail ellwood@
dsl.britcoun.is.lk

178 DS Senanayake Veediya
Kandy
Tel +94 8 34284/22410
Fax 22410

Sudan
14 Abu Sin Street
PO Box 1253, Khartoum
Tel +249 11 80817
Telex 23114 bckht sd
Fax 74935

Swaziland
British High Commission
Building
Alister Miller Street
private bag, Mbabane
Tel +268 43101/43103/42918
Fax 42641

Sweden
see Nordic countries

Switzerland
c/o The British Embassy
Thunstra(e 50
PO Box 265
CH 3000 Berne 15
Tel +41 31 3527025
Fax 352 7029

Syria
Abd Malek Bin Marwan Street
Tasheen Tabba' Building
Al Malki, PO Box 33105
Damascus
Tel +963 11 3333109/ 3310631/2
Telex 413455 brico sy
Fax 3310630

Taiwan
The British Council Hong Kong
is responsible for Taiwan.

ATEC
7th floor, Fu Key Building
99 Jen Al Road, Section 2
Taipei 10625
Tel +886 2 3962238
Fax 3415749

Tanzania
Samora Avenue/ Ohio Street
PO Box 9100, Dar es Salaam
Tel +255 51 46486/7/8/9/46490
Telex 41719 brico tz
Fax 46034

Thailand
428 Rama I Road
Siam Square Soi 2
Pathumwan, Bangkok 10330
Tel +66 22526136/7/8/2535311
Telex 72058 bricoun th
Fax 2535312/2526111
e-mail oipmoss@
chulkn.chula.ac.th

198 Bumrungraj Road
Chiang Mai 50000
Tel +66 53 242103
Fax 244781

Trinidad and Tobago
see Caribbean

Tunisia
c/o The British Embassy
BP 229, 5 Place de la Victoire
Tunis 1015 RP
Tel +216 1 259053/351754
Fax 353411

Turkey
The British Council Turkey is responsible for work in Azerbaijan and Uzbekistan.

c/o The British Embassy
Kirklangiç Sokak No. 9
Gazi Osman Pasa
06700 Ankara
Tel +90 312 4686192/9
Fax 4276182
e-mail firstname.lastname@bc-ankara.sprint.com

c/o British Consulate
Örs Turistik Ís Merkezi
Ístiklal Caddesi 251/253
(Kat 2-6) Beyoglu
80060 Istanbul
Tel +90 212 2527474/8
Fax 2528682
e-mail firstname.lastname@bc-istanbul.sprint.com

Turkmenistan
The British Council Russia is responsible for work in Turkmenistan.

Uganda
IPS Building
Parliament Avenue
PO Box 7070, Kampala
Tel +256 41 257301/3
Telex 61202 ukrep kla (High Commission)
Fax 254853

Ukraine
*9/1 Bessarabska Ploshcha
Flat 9, Kiev 252004
Tel +380 44 2945528/2945578/29455188
Fax 2945507

United Arab Emirates

Abu Dhabi
Al-Sadaqa Tower
near Emirates Plaza Hotel
PO Box 46523, Abu Dhabi
Tel +971 2 788400
Fax 789516

Dubai
Tariq bin Zaid Street
near Rashid Hospital
PO Box 1636, Dubai
Tel +971 4 370109
Fax 370703

United States of America
c/o The British Embassy
3100 Massachusetts Ave. NW
Washington DC 20008
Tel +1 202 8984275/4621340/8984407
Telex 211427 ukemby (Embassy)
Fax 8984612
e-mail bcwashington@bc-washingtondc.sprint.com

Uzbekistan
The British Council Turkey is responsible for work in Uzbekistan.

Liason Officer:
Dept of International Relations
University of World Economy and Diplomacy
54 Buyuk Ipak Yuli Street
Tashkent
Tel +7 3712 688456
Fax 670900

Venezuela
Torre La Noria, Piso 6
Paseo Enrique Eraso
Las Mercedes/Sector San Román, Apartado 65131
Caracas 1065
Tel +58 2 915222/915343/915543/915443/915143
Telex 24473 brico vc
Fax 915943

Vietnam
1 Pho Ba Trieu Street
Hanoi
c/o The British Embassy
16 Pho Ly Thuong Kiet Hanoi
Tel +84 4 245681/2
Fax 245683
e-mail bchanoi@netnam.org.vn

Yemen

As-Sabain Street No. 7
PO Box 2157, Sana'a
Tel +967 1 244121/2/244153/4/5/6
Fax 244120

Yugoslavia
(Serbia and Montenegro)
The British Council Eastern Adriatic is also responsible for work in Albania and the former Yugoslav Republic of Macedonia.

The British Council Eastern Adriatic
Generala Zdanova
34-Mazanin
PO Box 248
11001 Belgrade
Tel +318 11 332441/2/3227910
Telex 11032 bribel yu
Fax 3249013/631664

Zambia
Heroes Place, Cairo Road
PO Box 34571, Lusaka
Tel +260 1 228332/3/4/223602
Telex 40750 brico za
Fax 224122/226756
e-mail cchatam@zamnet.zm

Buteko Avenue
PO Box 70415, Ndola

Zimbabwe
23 Jason Moyo Avenue
PO Box 664, Harare
Tel +263 4 790627/8/9/0 793792/3
Telex 24568 brctco zw
Fax 737877
e-mail bcharare@britcoun.stellar.org.zw

75 George Silundika Street
PO Box 557, Bulawayo
Tel +263 9 75815/6
Fax 75815

Appendix Four

Glossary of acronyms

(*see also Chapter Four, Box* 4.15)

ABTI	All-Biota Taxonomic Inventory
AES	Amateur Entomologists Society
AMCRU	Area Museums Collections Research Unit
ANN	Artificial Neural Network
APC	Association for Progress Communications
ASEANET	South East Asia BioNET
ATBI	All-Taxa Biodiversity Inventory
ATI	Appropriate Technology International
BAS	British Antarctic Survey
BAS	British Arachnological Society
BBS	British Bryological Society
BBSRC	Biotechnology and Biological Sciences Research Council
BCIS	Biodiversity Conservation Information System
BCG	British Chelonian Group
BCG	Biology Curators Group
BCS	British Cartographic Society
BDS	British Dragonfly Society
BES	British Ecological Society
BGCI	Botanical Gardens Conservation International
BGCS	Botanical Gardens Conservation Secretariat (now BGCI)
BHS	British Herpetological Society
BIOSIS	BIOSciences Information Services
BLOWS	British Library of Wildlife Sounds
BLS	British Lichen Society
BMS	British Mycological Society
BOU	British Ornithologists' Union
BPS	British Phycological Society
BPS	British Pteridological Society
BSBI	Botanical Society of the British Isles
BSS	Botanical Society of Scotland
BTO	British Trust for Ornithology
CAD	Computer Aided Design
CAFOD	Catholic Fund for Overseas Development
CARINET	Caribbean BioNET
CATIE	Centro Agronomico Tropical de Investigacion y Enseñanza
CCAMLR	Commission for the Conservation of Antarctic Marine Living Resources
CCAP	Culture Collection of Algae and Protozoa
CCC	Coral Cay Conservation
CCMS	Centre for Coastal and Marine Science
CGIAR	Consultative Group for International Agricultural Research
CI	Conservation International
CIAT	Centro Internacional de Agricultura Tropical
CIESIN	Consortium for International Earth Science Information Network
CIFOR	Centre for International Forestry Research
CIMMYT	Centro Internacional de Mejoramiento de Maiz y Trigo
CIP	Centro Internacional de la Papa
CITES	Convention on International Trade of Endangered Species
CODATA	Committee on Data for Science and Technology
CORINE	CoORdination of INformation on the Environment
CSGBI	Concological Society of Great Britain and Ireland
DH	Department of Health
DICE	Durrell Institute of Conservation and Ecology
DoE	Department of the Environment
DTI	Department of Trade and Industry
EAC	Expedition Advisory Centre
EAFRINET	East Africa BioNET
ECN	Environmental Change Network
ECTF	Edinburgh Centre for Tropical Forests
EFI	European Forest Institute
ESF	European Science Foundation
FAO	Food and Agriculture Organisation of the United Nations
FBA	Freshwater Biological Association
FCO	Foreign and Commonwealth Office
FFI	Fauna and Flora International
FORW	The Forest Conservation, Education and Research Service
FRA	Forest Resources Assessment
FSC	Field Studies Council
FSC	Forest Stewardship Council
FTP	File Transfer Protocol
GATT	General Agreement on Tariffs and Trade
GCOS	Global Climate Observing System
GEF	Global Environment Facility
GIS	Geographical Information System
GOOS	Global Ocean Observing System
GOP	Geographical Observatories Programme
GTOS	Global Terrestrial Observing System
HDRA	Henry Doubleday Research Association
IACR	Institute of Arable Crops Research
IAM	International Association of Meiobenthologists
IBPGR	International Board of Plant Genetic Resources (now IPGRI)

ICARDA	International Center for Agricultural Research in the Dry Areas
ICBP	International Council for Bird Preservation (now BirdLife)
ICIMOD	International Centre for Intergrated Mountain Development
ICLARM	International Center for Living Aquatic Resources Management
ICRAF	International Centre for Research in Agroforestry
ICRISAT	International Crops Research Institute for the Semi-Arid Tropics
ICSU	International Council of Scientific Unions
IDB	Inter-American Development Bank
IFE	Institute of Freshwater Ecology
IFPRI	International Food Policy Research Institute
IH	Institute of Hydrology
IIED	International Institute for Environment and Development
IIMI	International Irrigation Management Institute
IITA	International Institute of Tropical Agriculture
INBio	Instituto Nacional de Biodiversidad
IOPI	International Organisation for Plant Information
IOS	Initial Operational System
IOS	Institute of Oceanographic Sciences
IPGRI	International Plant Genetic Resources Institute
IRRI	International Rice Research Institute
ISNAR	International Service for National Agricultural Research
ITE	Institute of Terrestrial Ecology
ITTA	International Tropical Timber Agreement
ITTC	International Tropical Timber Council
ITTO	International Tropical Timber Organisation
IUBS	International Union of Biological Sciences
IUCN	International Union for the Conservation of Nature (now The World Conservation Union)
IUDZG	International Union of Directors of Zoological Gardens
IUFRO	International Union of Forestry Research Organisation
IUMS	International Union of Microbial Societies
IVEM	Institute of Virology and Environmental Microbiology
IWRB	International Waterfowl and Wetlands Research Bureau (now WI)
JIC	John Innes Centre
JWPT	Jersey Wildlife Preservation Trust
JNCC	Joint Nature Conservation Committee
LME	Large Marine Ecosystems
LOOP	Locally Organised and Operated Partnerships (BioNET)
MaB	Man and the Biosphere
MAFF	Ministry of Agriculture, Fisheries and Food
MBA	Marine Biological Association (of the United Kingdom)
MCS	Marine Conservation Society
MGC	Museums and Galleries Commission
MLURI	Macauley Land Use Research Institute
MRA	Microbial Research Authority
MRC	Medical Research Council
MSDN	Microbial Strain Data Network
MSL	Malacological Society of London
NGO	Non-governmental organisation
NRI	Natural Resources Institute
NCCPG	National Council for the Conservation of Plants and Gardens
NERC	Natural Environment Research Council
NFZGGBI	National Federation of Zoological Gardens of Great Britain and Ireland
NHM	Natural History Museum
NRSC	National Remote Sensing Centre
ODA	Overseas Development Administration
OECD	Organisation for Economic Cooperation and Development
OFI	Oxford Forestry Institute
ORSTOM	Organisation Recherche Scientifique et Technologique d'Outre-Mer
PACINET	South Pacific BioNET
PBR	Plant Breeding Rights
PCR	Polymerase Chain Reaction
PHLS	Public Health Laboratory Service
PML	Plymouth Marine Laboratory
PSGB	Primate Society of Great Britain
PTES	People's Trust for Endangered Species
RAP	Rapid Assessment Programme
RBGE	Royal Botanic Garden Edinburgh
RBGK	Royal Botanic Gardens Kew
RBST	Rare Breeds Survival Trust
RDFN	Rural Development Forestry Network
RES	Royal Entomological Society
RFLP	Restriction Fragment Length Polymorphism
RGS	Royal Geographical Society
RHS	Royal Horticultural Society
RIVPACS	River Invertebrates Prediction and Classification Scheme
RSGS	Royal Scottish Geographical Society

RSPB	Royal Society for the Protection of Birds
RSE	Royal Society of Scotland
RSS	The Remote Sensing Society
RZSS	Royal Zoological Society of Scotland
SA	Systematics Agenda (2000)
SAFRINET	Southern Africa BioNET
SCAR	Scientific Committee on Antarctic Research
SCRI	Scottish Crop Research Institute
SEPASAL	Survey of Economic Plants for Arid and Semi-Arid Lands
SMRU	Sea Mammal Research Unit
SPRI	Scott Polar Research Institute
SPRU	Science Policy Research Unit
SRIS	Science References and Information Service
SSC	Species Survival Commission
TBA	Tropical Biological Association
TCP	Transmission Control Protocol
TFAP	Tropical Forest Action Programme (formerly Tropical Forestry Action Plan)
TRIP	Trade-Related aspects of Intellectual Property
UNCED	United Nations Conference on Environment and Development
UNDP	United Nations Development Programme
UNECE	United Nations Economic Commission for Europe
UNEP	United Nations Environmental Programme
UNIDO	United Nations Industrial Development Organisation
UNU	United Nations University
VES	Visual Encounter Survey
WARDA	West Africa Rice Development Association
WCED	World Commission on Environment & Development
WCMC	World Conservation Monitoring Centre
WDCS	Whale and Dolphin Conservation Society
WFCC	World Federation for Culture Collections
WHO	World Health Organisation
WI	Wetlands International
WIPO	World Intellectual Property Organisation
WMO	World Meteorological Organisation
WPA	World Pheasant Association
WRI	World Resources Institute
WWF	World Wild Fund for Nature
WWT	Wildfowl and Wetlands Trust
WWLCT	World Wide Land Conservation Trust
WWW	World Wide Web
WZO	World Zoo Organisation

Printed in the United Kingdom for HMSO
Dd302723 C20 7/96